Measuring and Modeling by Examples

How Mathematical Functions
Can Be Used (and Misused) to
Describe the World

Measuring and Modeling by Examples

How Mathematical Functions
Can Be Used (and Misused) to
Describe the World

Koen Van de moortel

World Scientific

NEW JERSEY · LONDON · SINGAPORE · BEIJING · SHANGHAI · HONG KONG · TAIPEI · CHENNAI · TOKYO

Published by

World Scientific Publishing Co. Pte. Ltd.

5 Toh Tuck Link, Singapore 596224

USA office: 27 Warren Street, Suite 401-402, Hackensack, NJ 07601

UK office: 57 Shelton Street, Covent Garden, London WC2H 9HE

Library of Congress Control Number: 2024043046

British Library Cataloguing-in-Publication Data
A catalogue record for this book is available from the British Library.

MEASURING AND MODELING BY EXAMPLES
How Mathematical Functions Can Be Used (and Misused) to Describe the World

info@lerenisplezant.be
www.researchgate.net/profile/Koen-Van-De-Moortel

Software used:
- The regression graphs and the analyses were made with FittingKVdm, written by the author, see: www.lerenisplezant.be/fitting.htm.
- Other graphs were made with GeoGebra Classic 5, see: www.geogebra.org.
- The electrical circuits were drawn with www.circuitlab.com.

ISBN 978-981-12-9676-5 (hardcover)
ISBN 978-981-12-9677-2 (ebook for institutions)
ISBN 978-981-12-9678-9 (ebook for individuals)

For any available supplementary material, please visit
https://www.worldscientific.com/worldscibooks/10.1142/13948#t=suppl

Desk Editors: Nambirajan Karuppiah/Rok Ting Tan

Typeset by Stallion Press
Email: enquiries@stallionpress.com

About the Author

Koen Van de moortel (°Antwerpen, Belgium, 1962) is a master in experimental physics who worked in research, math and physics education (schools and private tutoring), software development and photography.

Preface

I am a difficult person. I don't believe people just because they are "an authority", especially not if they have commercial interests in their preachings, or if their thinking is troubled by dogmatic ideologies, or if they hide behind excuses like "that's the way we've always done things here". That has not made my life easier since many people "in charge" have oversized egos and don't like to be questioned, but so be it...

I like modest and creative people who are brave enough to follow their intuition to seek the truth and bring quality rather than superficial fame, like Isaac Newton, Albert Einstein, Alfred Wegener (the continental drift guy), John Harrison (inventor of the first stable clocks that could work on a ship)[1], Shuji Nakamura (Noble Prize 2014 for inventing blue LEDs)[2], my uncle Juul Waumans (not a professor but a man of practice, who used his common sense to develop one of the best sports floors in the world), etc.

Intuition is often misunderstood as "just a feeling", but feeling has nothing to do with it. Intuition, as I understand it, is "composted experience", the result of many years of observing, tinkering, and reflecting, processed in the mysterious depths of our subconscious mind. It makes us see connections and patterns in a blink, as seeds to be nourished by the logical part of our brain.

From the age of six, I knew for sure I wanted to be a scientist, to explore how and why things work. The "why" is a question I have abandoned as it is probably impossible to answer, but the "how" can be described by mathematical models, the formulas connected with patterns we can observe, and those have been a common thread in the biggest part of my 62 years on this planet, so I think I built up some intuition on this subject. I see many people struggling with this process of analyzing their data, while they give their trust more and more to "artificial intelligence" and sometimes forget to use their own intelligence, so that's why I decided to write this book, to share my experience.

Although I can appreciate the beauty of a theory absolutely, I studied experimental physics because I'm more an observer than a theorist, and I'll try to bring my message as intuitively as possible, hoping it will help you in a practical way! And of course I don't want you to believe me just because I say so! Just test it out!

I want to dedicate this book to Roger De Weerdt, Rik Verhulst, and Rudi Luyten, the high school teachers from the Pius X institute in Antwerpen, who fanned my mathematical and scientific fire and learned me to observe carefully and find order in chaos. And of course I thank my partner-in-life for more than 25 years by now, Dragana, for her patience with me (well... at least most of the time).

Koen Van de moortel
Ghent, Belgium
22 February 2024

---

[1]   See the movie "Longitude", the story of Harrison:
       www.youtube.com/watch?v=LHvt48S9l4w
[2]   Veritasium documentary about Nakamura:
       www.youtube.com/watch?v=AF8d72mA41M

Contents

Introduction

"Why would we do it?" is always a good question to ask. **Why would we want to see mathematical patterns in our observations?**

First of all, detecting patterns is a skill that is crucial to survive. Animals that don't recognize a predator from a visual, auditive, or smell pattern are doomed to get extinct. Not being able to distinguish food from poison will do the same.

As a human, you can go further. Detecting mathematical patterns will help you in a more and more complex environment. It helps you to predict the tides and to navigate. It allows you to foresee the best times to seed and to harvest or to hunt. You have an **evolutionary advantage** if you can read the signs that predict a coming storm or a dangerous epidemic.

A few centuries ago, it might have seemed a waste of time to figure out whether the attraction between two masses or two electrical charges was inversely proportional to the distance or the square or the third power of that distance, as Newton and Coulomb did, but now we realize we would not have had any of the technology we are so addicted to, without knowing those patterns (or "natural laws" if you will).

Finding the mathematical formula that describes your observation is a step in the **exploration** of the world. It might lead you to a **hypothesis about how things work**. Finding that the pressure in a gas was proportional to the temperature brought us to the idea that there might be invisible particles in the air that bump to the walls and store energy by vibrating, even if nobody could see those particles! Finding the formula describing the magnetic field caused by a change in electrical current led to a lot of technical inventions that made long-distance communications possible.

Once you know the mathematical pattern, the "model", you can start doing **predictions**. If they turn out to be correct, over and over, they confirm your hypothesis. If not, they force you to rethink the explanation you had in mind. Newton's formulas to describe gravitation were very precise until we started experimenting with very high speeds and masses. If Einstein and his colleagues hadn't improved the model, you wouldn't have this fantastic navigation tool in your smartphone now.

You can also start doing **optimizations**: once you know *how* the composition of a substance influences its qualities, or if you know how things influence the yield of a crop, you can make it better.

Finding relationships that connect different quantities can also just **make practical measurements easier or less expensive**: measuring temperature, humidity, salinity, wind speed, light intensity, concentration, distance, and so much more, in the 21st century, can all be reduced to measuring an electrical current. But first you have to develop the sensor and the mathematical formula to **calibrate** it!

Can models prove causal relationships? Yes and no... If you can control variable x and you see, an instant or some time later, a change in y as predicted by the model, you can be quite sure that x is the cause and y the consequence. In physics, chemistry, biology, etc. such experiments can often be done, but in other sciences, one has to use the available data and be happy if you find just correlations. As you probably know, a correlation by itself does not prove causality: the hands of the clock are very well correlated with the changes of day and night, but they don't cause them. Among women, higher education is significantly correlated with a higher risk for breast cancer, although the first doesn't cause the second (I hope), but this observation confirms a known causal mechanism, namely, higher educated women tend to have fewer children and at a later age, and that makes them more vulnerable. So, studying correlations is useful to make the pieces of the puzzle of causality fit.

The book has four parts:

(1) If you are not very familiar with the wide spectrum of those formulas, no problem. I take you on a journey in **the world of mathematical functions** first, well... at least some relatively simple and interesting kinds, and I'll discuss their useful features.

(2) Then I tell you some stories about **the art of measuring**. How to handle the uncertainties that are present in most data collections? And how to avoid or minimize them as much as possible? How to stick a number to observations that are not so easy to quantify, like "happiness", or "wine quality"? Some tricks and plain common sense can help you here!

(3) Next, I try to clarify **the process of finding the "best" possible**

function to use as a model for your observations. I also focus on some issues that are often neglected in this process, even by seasoned scholars, and introduce my improvements to existing techniques which I implemented in a software program "FittingKVdm". Questions like "How trustworthy is my model?" need to be asked certainly.

(4) The last part of this book consists of **many examples from the real world**, from physics to psychology, from biology to linguistics, from sports to economy. In some of them, I analyze easy-to-repeat experiments that can be done at home or in a classroom, while others use data from public sources.

I hope you enjoy the ride, and I'm certainly looking forward to hearing your comments and suggestions for improvement.

1

Some Useful Functions to Describe the World

What is a function?

Parameters influence the shape of a function.

Linear transformations: shift and stretch.

Learn to recognize the characteristics of different classes of functions that are interesting for modeling.

1.1. Functions and Their Linear Transformations

If you finished a few years of high school, you'll probably know what a mathematical "function" is. But, as I've seen with students, the concept hasn't always been stored in their minds as clearly as it should be, so let's straighten a few things out.

Suppose we have two sets A and B, containing some kinds of elements, and we construct a new set (R), consisting of a number of couples (ordered pairs) with the first object an element from A, and the second object an element from B. We now have a **"mathematical relationship"** R. Graphically, this can be depicted as two Venn diagrams and arrows between some of their elements; each arrow represents an ordered pair.
If there are no duplicates among the first elements, or in other words, if it never happens that more than one arrow starts from an element of A, we call R a **"function"**. The "John Doe" name for a function is usually "f" or one of the next letters of the alphabet. The first set is called the "domain" $(A = \text{Dom}(f))$ and the set of destination (B) is also called the "codomain" or the "image" of f.

Example: A *dictionary* can be considered as a function, see Fig. 1.1. Set A might be the set of all Dutch words and B the set of all English words, and f could be the set of all the translations, i.e., the couples of words that correspond in the two languages: $f = \{(huis, house), (papier, paper), (jongen, boy),...\}$.
We write: $f(huis) = house$, and in general, $f(x) = $ the English translation of x, where "x" stands for "some element of A".
The so-called **"independent variable"** ("IV") here is the Dutch word for a certain concept, and the **"dependent variable"** ("DV") is the English word for the same concept.

We can also invert the arrows of f, and then we get a new set of couples: $f^{-1} = \{(house, huis), (paper, papier),...\}$. This is called **the inverse relationship**, and if there are no couples with the same first element, it is also a function.

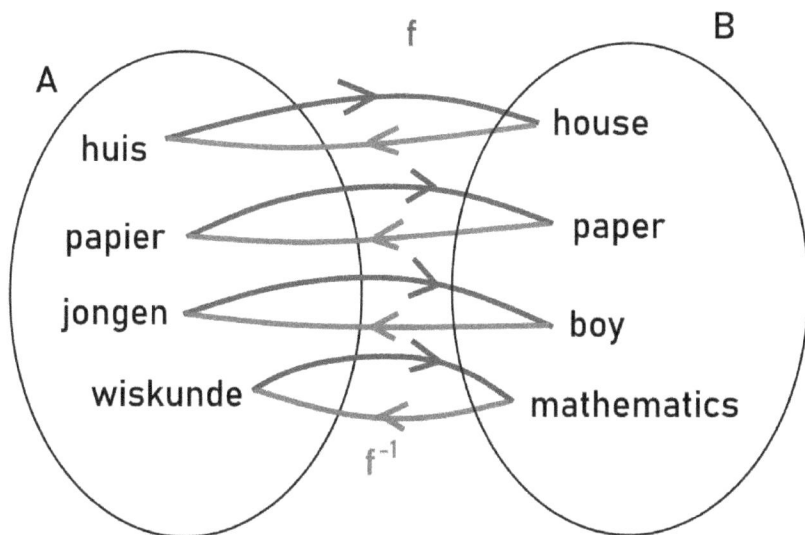

Fig. 1.1. A mathematical function and its inverse: a sample of a Dutch-English dictionary.

Another example of a function: the *menu* in a restaurant: $g =$ {(hamburger, 11€), (lasagne, 17€), (spaghetti, 17€),...}
Here: $g(x) =$ the price of menu item x. The "independent variable" is the name of an item, and the "dependent variable": the price of the same item. The inverse relationship g^{-1} is *not* a function here, since it's not clear what $g^{-1}(17€)$ would be.

When there are no arrows arriving in the same point, we have a so-called "**bijection**", a one-to-one relationship. Of course, bijections are always invertible.

Of course, there are functions with more than one independent variable, e.g., the value of a car depends on its type, age, and condition. Value $=$ f(type, age, condition).

In this book, we focus on "real" functions with one IV, i.e., functions that have (parts of) $\mathbb{R}$ (the set of real numbers) as their domain and co-domain. Usually, these functions are defined by a formula that says how you can calculate the value corresponding with a number x, e.g., "$f(x) =$ $5/x - 7$" or "$f(x) = 4x^3 + 2^x + 1$".

Real functions can be represented in a *graph*. A graph is nothing but the "locus" of all the points in the Cartesian plane whose location satisfies the condition $y = f(x)$. In other words, it's a "club" of points (x, y) and the membership condition is that y should be equal to $f(x)$. If $f(x) = x^3$, $(2,8)$ is a member of that club, and $(1,10)$ isn't.

We have a look at several interesting classes of functions that can be used to depict and approximate phenomena that occur in the real world. These classes usually have some basic "seed" function $f_0(x)$ that has to be adapted to coincide with reality by doing some shifting and scaling.

For example, the distance (d) an object has traveled after falling for a time t, is basically $d = t^2$, but if we use meters and seconds, and do this experiment near the earth's surface, we will measure approximately $d = 4.95t^2$.

This "shifting and scaling" is what we call a **"linear transformation"**. If the basic graph is defined by $y = f_0(x)$, and we shift it horizontally over a distance p, vertically over a distance q, and we apply a stretch factor r horizontally and s vertically $(r \neq 0 \neq s)$, we get a graph (Fig. 1.2) with locus:

$$\frac{y - q}{s} = f_0\left(\frac{x - p}{r}\right)$$

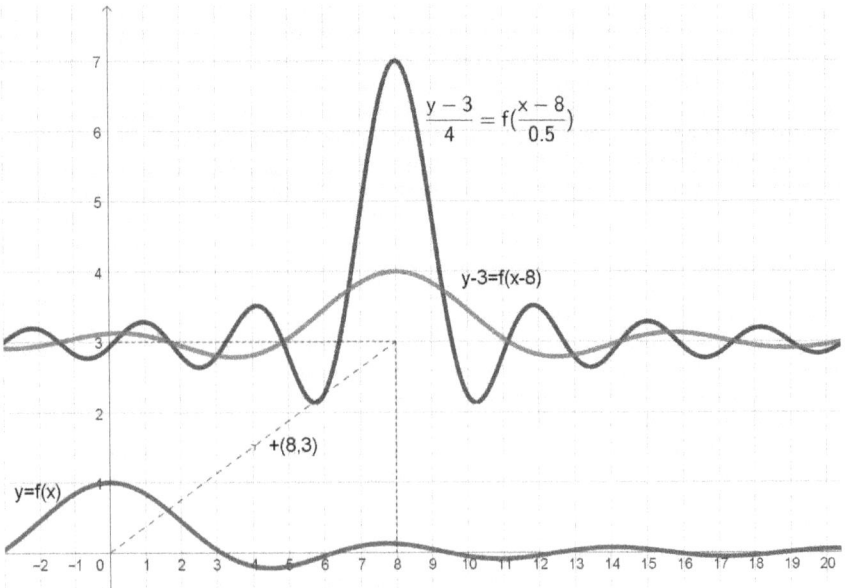

$$\frac{y - 3}{4} = f\left(\frac{x - 8}{0.5}\right)$$

y-3=f(x-8)

+(8,3)

y=f(x)

Fig. 1.2. Example of a function graph and a possible linear transformation.

The numbers p, q, r and s are called **parameters** of the function. If we change them, the graph will look a bit different, but still recognizable.

So, the generalized function looks like:

$$f(x) = s \cdot f_0\left(\frac{x-p}{r}\right) + q$$

The **inverse function** can easily be calculated, using the "**rule of the socks and the shoes**". (This is not an official name, but that's how my high school teacher Roger De Weerdt used to call it. First, you put on your socks, then your shoes; for the inverse operation, you do the opposite actions, but in reversed order.)

$$x \xrightarrow{-p} x - p \xrightarrow{/r} \frac{x-p}{r} \xrightarrow{f_0(\,)} f_0\left(\frac{x-p}{r}\right) \xrightarrow{\cdot s} s \cdot f_0\left(\frac{x-p}{r}\right) \xrightarrow{+q} s \cdot f_0\left(\frac{x-p}{r}\right) + q = f(x)$$

$$f^{-1}(x) = r \cdot f_0^{-1}\left(\frac{x-q}{s}\right) + p \xleftarrow{+p} r \cdot f_0^{-1}\left(\frac{x-q}{s}\right) \xleftarrow{\cdot r} f_0^{-1}\left(\frac{x-q}{s}\right) \xleftarrow{f_0^{-1}(\,)} \frac{x-q}{s} \xleftarrow{/s} x - q \xleftarrow{-q} x$$

This works whenever $r \neq 0 \neq s$.

Sometimes, it's useful to change other parameters in the formula that are not related to shifting and scaling, but to changing the shape in some other way.

We have a look at some functions that are often used in modeling. This is not a calculus textbook; I assume you have some basic knowledge about derivatives, but it's not absolutely necessary. I focus here on the properties that are useful for our modeling story, like monotony, peaks, periodicity, asymptotes, etc. and different possible parameterizations. And I introduce some interesting functions that you probably didn't encounter in your college years.

Recommended exercises: Enter the formula of a function in GeoGebra, Graphmatica, or your favorite math software, and play with the parameters, so you get a feeling of how they influence the shape of the curve. For example, if you enter a formula like "$y = ax + b/x$" in GeoGebra, it will automatically ask you if you want sliders to adjust a

and b. The allowed ranges can be adjusted by clicking right on them and selecting "properties".

Or, even better, study a formula and try to reason how the graph will look, and then check your ideas with software.

Most functions that are described here, can also be selected in FittingKVdm, and then you can change the parameters easily with the arrow-up/down keys in the parameter edit boxes.

1.2. Polynomials

A "polynomial of degree n" is a linear combination of positive integer powers of x, n being the highest.

1.2.1. Constant functions

The simplest polynomial possible is the "stupid" function that has the same result for every x: $f_0(x) = 1$.

Applying a linear transformation results in:

$$f(x) = s \cdot f_0\left(\frac{x-p}{r}\right) + q = c, \quad c = s + q$$

with c: a constant real number. Shifting and stretching a flat line horizontally don't change anything, and "stretching" it vertically has the same effect as shifting it vertically. Only one point is needed to know the entire course of the function.

1.2.2. Linear functions

Already more interesting is the simplest first degree polynomial: $f_0(x) = x$.

The graph $y = x$ is the bisector line between the x and the y axis in the first quadrant, see Fig. 1.3.

Applying a linear transformation gives us a more general form:

$$f(x) = s \cdot f_0\left(\frac{x-p}{r}\right) + q = \frac{s}{r}(x-p) + q = \frac{s}{r}x + q - \frac{sp}{r} = ax + b \tag{1}$$

with a and b some real numbers.

This is the equation of a straight line, because every time x goes up by 1, y goes up by a ($a < 0$ means it goes downward of course). Therefore, a is called the "slope". And $f(0) = b$; therefore b is called the "intercept" (with the y axis).

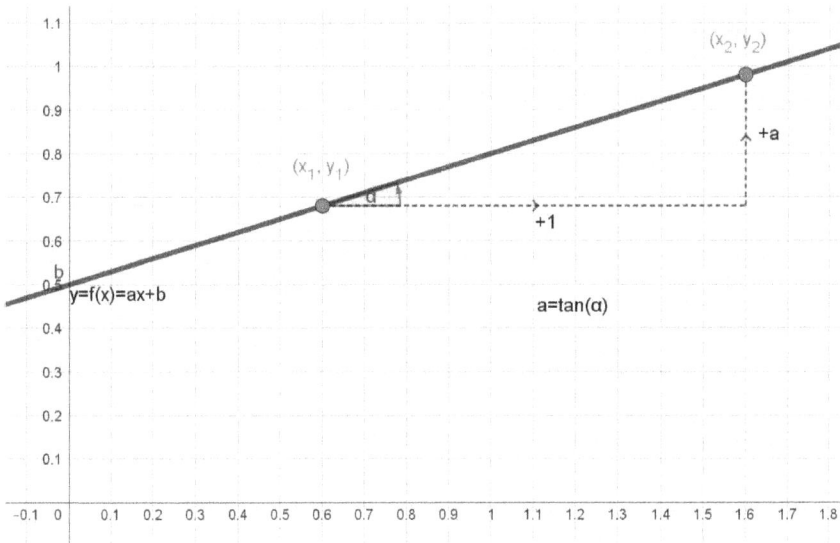

Fig. 1.3. A linear function.

The **inverse function** can be calculated as long as $a \neq 0$:

$$x \xrightarrow{\ \cdot a\ } ax \xrightarrow{\ +b\ } ax + b = f(x)$$

$$f^{-1}(x) = \frac{x-b}{a} \xleftarrow{\ /a\ } x - b \xleftarrow{\ -b\ } x$$

The equation can also be written as follows:

$$f(x) = a(x - c) \tag{2}$$

With $c = -b/a$, this is the x value that gives $f(x) = 0$ as the result, so it's called the "zero point".

So, the generalized form of this "linear" function can be written in two forms, two "parameterizations", and both of them have two independent parameters. Two of the four transformation parameters are obsolete, since the vertical shift of a straight line automatically causes a horizontal shift, and a horizontal stretch automatically causes a vertical stretch.

So, you need two points with different x values, (x_1, y_1) and (x_2, y_2), to determine f completely, because with two points you have a system of two equations and two unknowns. Using the first parameterizations, we get:

$$\begin{cases} y_1 = f(x_1) = ax_1 + b \\ y_2 = f(x_2) = ax_2 + b \end{cases}$$

Subtracting and moving gives:

$$a = \frac{y_2 - y_1}{x_2 - x_1} \quad \wedge \quad b = y_1 - ax_1$$

If you happen to know the intercept and one other point, you can calculate the slope immediately from $a = (y_1 - b)/x_1$.

If you happen to know the zero and one other point, you can use the second parameterizations to get $a = y_1/(x_1 - c)$.

Exercise: If f has these points: $(2,5)$ and $(6,7)$, find a, b, c.

A singular case is a line through the origin, meaning $f(0) = 0$, so $b = c = 0$, or $f(x) = ax$. This is also a singular case from the so-called "power" functions, see further.

Applications:
Any kind of relationship between two variables x and y can be *approximated* by a linear function, for example, x = age of spouse, y = age of husband; x = income, y = happiness; x = height of a person, y = shoe size; etc. Those approximations are often far from perfect; see the case studies.

Very often, natural laws are described quite well by the singular case $y = ax$, like: x = current through, and y voltage over a normal resistor; x = weight, y = friction; x = volume, y = mass; x = absolute temperature, y = pressure of a gas in a closed container; etc.

All too often, linear models are used out of laziness or habit, but there is a lot of other interesting functions that could be more accurate!

1.2.3. Quadratic functions

The next basic function in this series is: $f_0(x) = x^2$.
Its graph is a parabola with minimum in the origin and a vertical symmetry axis: $x = 0$.

Applying a linear transformation gives:

$$f(x) = s \cdot f_0\left(\frac{x-p}{r}\right) + q = \frac{s}{r^2}(x-p)^2 + q = a(x-p)^2 + q \qquad (1)$$

with $a = s/r^2$ a nonzero real number (Fig. 1.4).

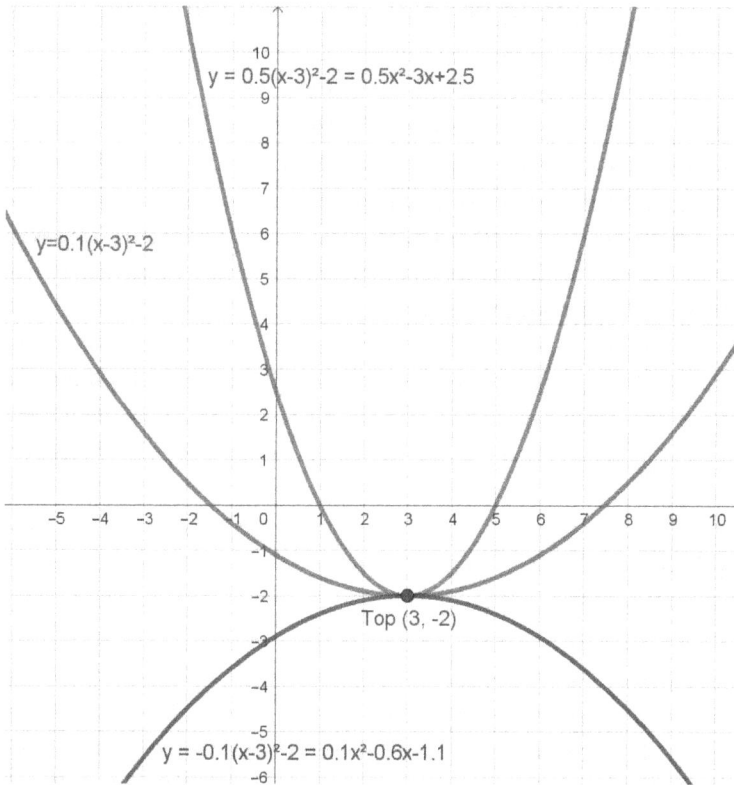

Fig. 1.4. Graphs of some quadratic functions (parabolas).

Here, one parameter is obsolete, because stretching the graph horizontally will automatically cause a vertical stretch too.

So, you only need three bits of information to determine the size and position of such a parabola: the co-ordinates of the top (p, q) and one other point, or three points with different x values that you put into the formula, so you can obtain the parameters by solving a 3×3 equation system.

You can expand the brackets to get:

$$f(x) = ax^2 - 2apx - ap^2 + q = ax^2 + bx + c \qquad (2)$$

This is the general form of any second degree polynomial. With a little algebra, you can easily calculate backwards, so any formula in this form can also be put in the first form.
If the parabola needs to go through the origin, set $c = 0$.

There is a third possible parameterization:

$$f(x) = a(x - x_1)(x - x_2) \qquad (3)$$

where x_1 and x_2 are the two zeroes (roots) of the function. But this can only be done with real numbers if $b^2 > 4ac$. Otherwise, the roots will be complex numbers, but we don't need that here. See any college algebra book for more about this.

So, if you have data, and you want to see which parabola connects them, you can always enter them in (2), but if you know the top and one point, it's easier to use (1), and if you know the roots and one point, it's easier to use (3).

Inverting f is the easiest if you start from (1):

$$x \xrightarrow{\ -p\ } x - p \xrightarrow{\ (\ )^2\ } (x-p)^2 \xrightarrow{\ \cdot a\ } a \cdot (x-p)^2 \xrightarrow{\ +q\ } a \cdot (x-p)^2 + q = f(x)$$

$$f^{-1}(x) = p \pm \sqrt{\frac{x-q}{a}} \xleftarrow{\ +p\ } \pm\sqrt{\frac{x-q}{a}} \xleftarrow{\ \pm\sqrt{(\ )}\ } \frac{x-q}{a} \xleftarrow{\ /a\ } x - q \xleftarrow{\ -q\ } x$$

We can reverse the calculation of $f(x)$ if $(x - q)/a > 0$. The problem is that we then have two possible outcomes (unless $x = q$), so f^{-1} is not a function!

Applications:
Many phenomena can be approximated by quadratic functions, but the most exact usage is probably the distance traveled (y) vs. time (x) by an

object that is accelerated by a constant force (gravitational, electric, motor, etc.), e.g., a falling apple, if the influence of friction can be neglected.

1.2.4. Cubic functions

If we linearly transform the graph of the basic third degree polynomial function, $f_0(x) = x^3$, we get:

$$f(x) = s\left(\frac{x - p}{r}\right)^3 + q = a(x - p)^3 + q$$

which can be expanded to

$$f(x) = ax^3 - 3apx^2 + 3ap^2x - ap^3 + q \qquad (1)$$

Now, the general formula for a third degree polynomial function is:

$$f(x) = ax^3 + bx^2 + cx + d \qquad (2)$$

Here is a difference with quadratic functions: not any third degree polynomial can be written in form (1) because (1) has only three independent parameters and (2) has four of them.

This means that their shape can differ from the basic shape. You can easily verify this by drawing the graphs of, for example: $f(x) = x^3 - x$ and $g(x) = x^3 + x$. Please don't use a computer, just calculate a few values, e.g., for $x = -2, -1, -0.5, 0, 0.5, 1$ and 2.

You'll see that they both have an inflection point in the origin, but only f has a (relative) maximum and a minimum. If there are extrema, the function is not invertible.

Since there are four independent parameters, you'll need four points with different x values to determine the function.

1.2.5. Higher polynomials

The general formula for a polynomial function with degree n can be written as

$$f(x) = a_nx^n + a_{n-1}x^{n-1} + \ldots + a_2x^2 + a_1x + a_0$$

(With $a_n \neq 0$. The parameters are also called "coefficients".)

It's easy to figure out that a linear transformation will always result in a polynomial with the same degree but with different parameters. If f needs to go through the origin, set $a_0 = 0$.

This kind of function can have up to $n - 1$ extrema, so it's generally not invertible.

Other parameterizations are used sometimes, because of their interesting properties, e.g., "Hermite" polynomials, very fascinating, but I refer to Mr. Google for more details.

Applications:
If you have at least n data points and you want to see a smooth curve connecting the dots (or coming close to them), you can always use a polynomial of degree $n - 1$ (see Fig. 1.5). So, you could use this for **interpolation** (estimating y

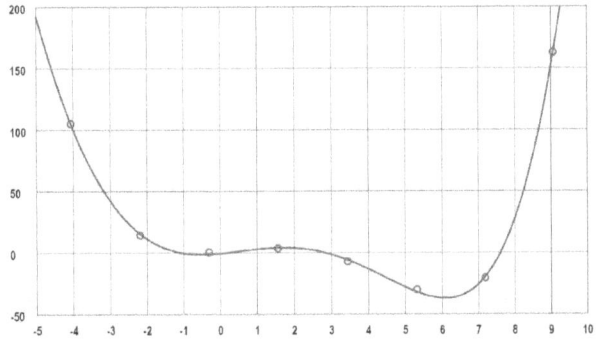

Fig. 1.5. Eight points can always be connected by a 7th degree polynomial function. (Yes, 7th degree! The part where it goes down on the left side is not visible here.)

values between known data points, for example, when studying complicated body movements), but you can't ever trust any *extrapolation*, one of the purposes of modeling. Not the slightest bit. We see examples in the case studies.

A very fascinating use of polynomials is the theory of so-called **Taylor series**. Many complicated functions like trigonometric and exponential functions can be approximated very well by well-designed high degree polynomials. Actually, that is how your computer calculates a sine or a cosine, etc.
If you are not familiar with this theory... you can find it in many calculus textbooks, and it's well covered in Wikipedia,
https://en.wikipedia.org/wiki/Taylor_series
It's a beautiful and very useful theory!

1.3. Power Functions

The basic power function looks like: $f_0(x) = x^b$.
Here, b can be any real number except 0.
Since x and $f(x)$ should be positive (strictly speaking, except in cases when b is an integer), and the practical applications require $f(0) = 0$, the only linear transformation that makes sense, is a multiplication with a positive number a. So, the general form can be written as:

$$f(x) = ax^b$$

The value of parameter b has a big influence on the graph. Especially the sign of b is important!
When $b > 0$, the graph is always ascending (Fig. 1.6); a special case is $b = 1$, because it's also a linear function. With $b < 0$, it's always descending toward 0 (Fig. 1.7), and it looks like a hyperbole (= special case $b = -1$, which is also a homographic function, see further), with $x = 0$ as a vertical asymptote and $y = 0$ the horizontal asymptote.

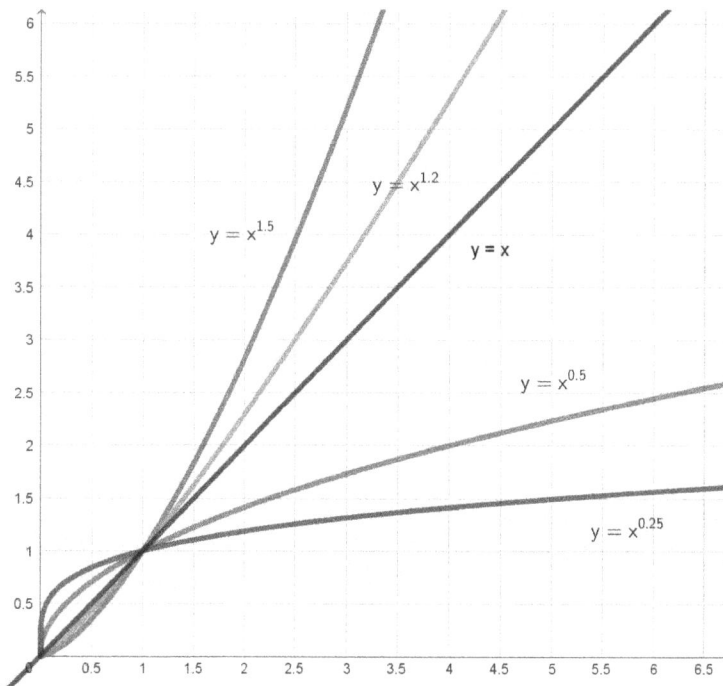

Fig. 1.6. Some power functions with positive exponents.

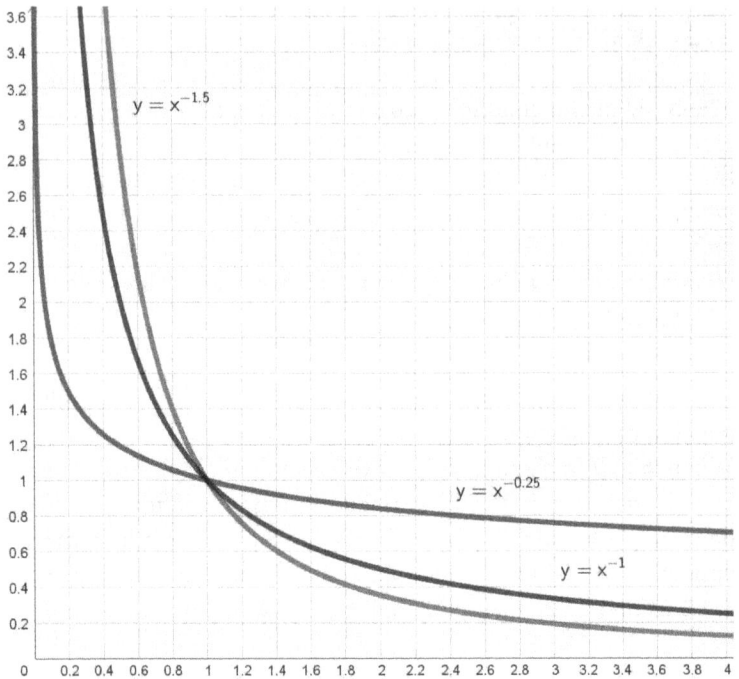

Fig. 1.7. Some power functions with negative exponents.

The inverse can easily be calculated:

$$x \xrightarrow{\;(\;)^b\;} x^b \xrightarrow{\;\cdot a\;} ax^b = f(x)$$

$$f^{-1}(x) = \sqrt[b]{\frac{x}{a}} = \left(\frac{x}{a}\right)^{\frac{1}{b}} \xleftarrow{\;\sqrt[b]{(\;)}\;} \frac{x}{a} \xleftarrow{\;/a\;} x$$

Applications:

Many relationships that have to do with scaling laws, the change of properties related to size, especially in biology, can be described with power functions.

For example, think of a cube with rib size z, volume V, surface A: $V = z^3$, $A = 6z^2$, so $A = 6V^{2/3}$; z could also be the size of an animal or a plant, A the wool production, and V proportional to the mass.

In the case studies, we discuss many other examples too.

1.4. Homographic Functions

The "seed function" here is: $f_0(x) = 1/x$, which is a special case of the power functions.

We might also do a linear transformation here, to get:

$$f(x) = s \cdot f_0\left(\frac{x-p}{r}\right) + q = \frac{rs}{x-p} + q \tag{1}$$

This can be expanded to:

$$f(x) = \frac{rs + q(x-p)}{x-p} = \frac{qx + rs - qp}{x-p} = \frac{ax+b}{x+c} \tag{2}$$

(with $rs - qp = b$, etc.)

Often you find definitions like $f(x) = (ax+b)/(cx+d)$, the general form of a **first-degree rational function**, but the fraction can always be simplified so that the coefficient of x in the denominator can be set to 1.

There are three independent parameters, so three points are needed to fix the function.
The vertical asymptote has now shifted to $x = -c$.
And if $x \gg 1$, $f(x) \approx ax/x = a$, so the horizontal asymptote is $y = a$.

(1) Can also be written as:

$$f(x) = a - \frac{b}{x+c} \tag{3}$$

Note that the parameters have different meanings here than in (2). They're a bit easier to interpret: $x = -c$ is the VA, $y = a$ is the HA, and b determines the steepness, see Fig. 1.8.

The function is invertible as long as $x \neq a$:

$$x \xrightarrow{+c} x + c \xrightarrow{(\ )^{-1}} \frac{1}{x+c} \xrightarrow{\cdot(-b)} -\frac{b}{x+c} \xrightarrow{+a} a - \frac{b}{x+c} = f(x)$$

$$f^{-1}(x) = \frac{b}{a-x} - c \xleftarrow{-c} \frac{b}{a-x} \xleftarrow{(\ )^{-1}} \frac{a-x}{b} \xleftarrow{/(-b)} x - a \xleftarrow{-a} x$$

Applications:
This form (3), with all the parameters positive, might be used to describe *approximately* some variable y that goes to a limit when x goes to in-

finity. See the case studies "Wine ratings vs. price" (p. 368), or "Happiness vs. income" (p. 347), etc.

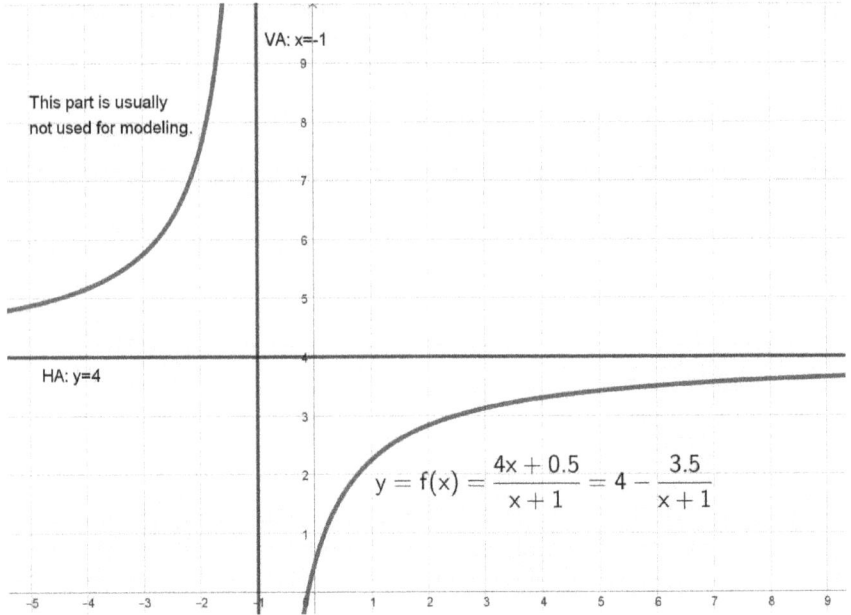

Fig. 1.8. A typical homographic function.

An interesting singular case of (2) is the one that goes through the origin ($b = 0$, Fig. 1.9).

That describes quite perfectly the voltage over a battery versus the load resistance (see case study p. 222), or the speed of a chemical reaction vs. the substrate concentration (Google: Michaelis-Menten kinetics).

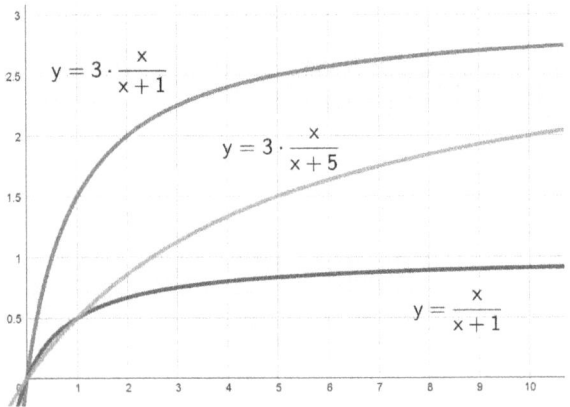

Fig. 1.9. Homographic functions through the origin (only the useful branch).

1.5. Other Rational Functions

Sometimes we could use a model with an oblique asymptote ($y = mx + q$). The first situation that comes in my mind, is the position of a parachutist: after he opens his parachute, the accelerated motion will become more and more linear with a constant speed.

The simplest functions that have oblique asymptotes, are rational functions where the degree of the numerator is one more than that of the denominator. And the simplest of those is a quadratic function divided by a linear one:

$$f(x) = \frac{ax^2 + bx + c}{x + d}$$

(No coefficient for x is needed in the denominator, since the fraction can always be simplified.)

The slope of the asymptote (m) can easily be calculated as

$$m = \lim_{x \to \infty} \frac{f(x)}{x} = a$$

and the intercept (q) as

$$q = \lim_{x \to \infty} \left(f(x) - ax \right) = b - ad$$

There is one problem (besides the vertical asymptote $y = -d$) why it is kind of "dangerous" to use them for modeling though: they are extremely **unstable**, just like higher order polynomials. By that, I mean: a small change in data points can have a dramatic influence on the parameters. Or in other words, you can change some of the parameters quite a lot and get almost the same curve. It follows from the formula of the intercept that it stays the same for example if you increase b and increase d with the right amount while the slope stays the same.

To illustrate this, Fig. 1.10 is an example that you can easily reproduce yourself using GeoGebra.
In the top graph, $b = 12.9$, $d = 10.9$, while in the graph, $b = 15.6$, $d = 15.1$, and they look very similar, and both go through the same point. (Both functions have $a = 0.3$ and $c = 0.$)

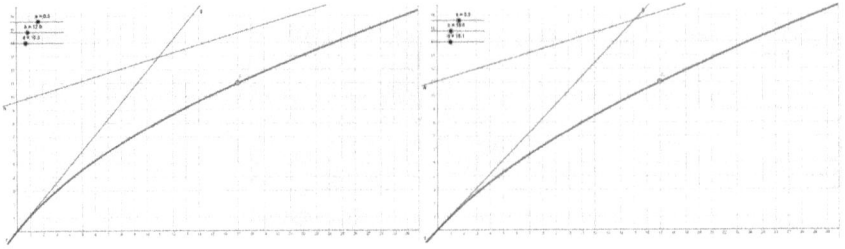

Fig. 1.10. Rational functions: a big change in parameters (see text) can give almost the same curve (GeoGebra screenshots).

1.6. Exponential Functions

The basic exponential function looks like the following: $f_0(x) = e^x$, e being Euler's number (2.71828...).

It is always positive and ascending and going through $(0,1)$. It never touches the x axis but it comes closer for x going to $-\infty$, so the x axis is a horizontal asymptote.

Applying a linear transformation, we get:

$$f(x) = s \cdot e^{\frac{x-p}{r}} + q \tag{1}$$

At first sight, it looks like the four parameters play an independent role, but... if you remember the property $a^{n+m} = a^n a^m$, we can also write this as

$$f(x) = s \cdot e^{-\frac{p}{r}} \cdot e^{\frac{x}{r}} + q = b \cdot e^{\frac{x}{r}} + q \tag{2}$$

since $e^{(p/r)}$ is just a constant that can be joined with s. (Of course, instead of writing "x/r", we could also use a multiplicative constant and write "ax", with $a = 1/r$.)

That means that a horizontal shift has the same effect as changing b, so we have only three independent parameters in the formula. So, three points are needed to calculate a, b and q, but the calculations get nasty when q is unknown. If q is known, a and b are easy to find from two points.

The parameter r determines how fast the graph goes up to infinity, if $r > 0$. But if $r < 0$, the graph is horizontally mirrored.

If $r > 0$, q is the limit for $x \rightarrow -\infty$. If $r < 0$, $f(x)$ goes down to q when x goes to $+\infty$. In any case, the horizontal asymptote is $y = q$.

Now, using the property $a^{nm} = (a^n)^m$, we can rewrite (2) as

$$f(x) = b \cdot \left(e^{\frac{1}{r}} \right)^x + q = b \cdot a^x + q \tag{3}$$

with

$$a = e^{\frac{1}{r}} \Leftrightarrow \ln(a) = \frac{1}{r}$$

This parameterization can be very useful since the parameter a tells you by which factor $f(x)$ is multiplied every time you add one to x, see Fig. 1.11. For example, if $y = 2^x$, then y doubles every time you add 1 to x ($2^{x+1} = 2 \cdot 2^x$). In general, the difference with q doubles every time you add $\ln(2)/\ln(a)$ to x; check this as an exercise.

The derivative of a^x is interesting: if you calculate it (do it as an exercise if you forgot it), you find that it is proportional to itself. The proportionality factor is a limit, called the "natural logarithm of a" ($\ln(a)$). In one special case, that limit is 1, and that a value is called Euler's number (e). The logarithmic function $\ln(x)$ is the inverse function of e^x.

For more details, see any higher grades' college calculus textbook.

The inverse can easily be calculated if $(x - q)/b > 0$:

$$x \xrightarrow{a^{(\ )}} a^x \xrightarrow{\cdot b} b \cdot a^x \xrightarrow{+q} b \cdot a^x + q = f(x)$$

$$f^{-1}(x) = {}^a\log\left(\frac{x-q}{b} \right) = \frac{\ln\left(\dfrac{x-q}{b} \right)}{\ln(a)} \xleftarrow{{}^a\log(\ )} \frac{x-q}{b} \xleftarrow{/b} x - q \xleftarrow{-q} x$$

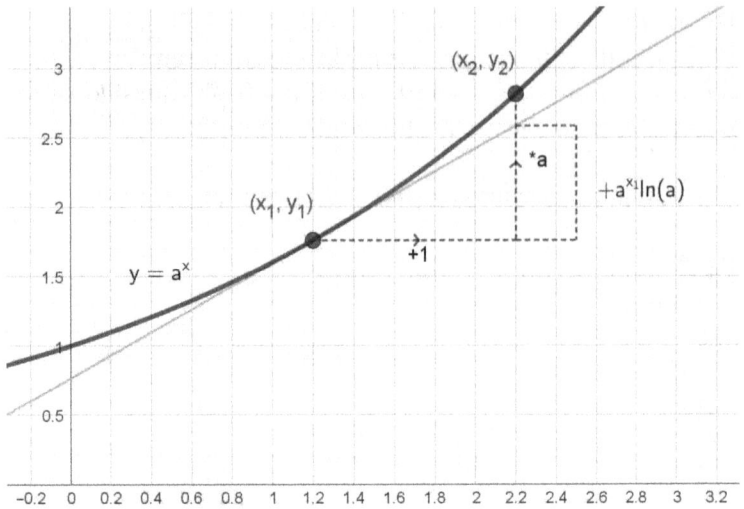

Fig. 1.11. Example of an exponential function: $a = 1.6$, $b = 1$, $q = 0$.

Applications:
Usually, x = time. If $r > 0$ (equivalent to $a > 1$), variable y is something that grows in an unhindered way, like bacteria with enough food and space, or money that gets interest on a savings account (once upon a time); the growth of y is proportional to the already existing value of y. If $r < 0$ (equivalent to $a < 1$), f can depict radioactive decay, the cooling of an object, absorption of light through a layer of material with thickness x, etc.

There is another parameterization which is useful for the description of the charging of a capacitor from empty to full. It requires some juggling, starting from (2), assuming $b < 0$, $r < 0$:

$$f(x) = b \cdot e^{\frac{x}{r}} + q = q\left(1 - \left(-\frac{b}{q}\right) \cdot e^{\frac{x}{r}}\right) = q\left(1 - e^{\frac{x}{r} + \ln\left(-\frac{b}{q}\right)}\right)$$

$$= q\left(1 - e^{\frac{x-c}{r}}\right) = q\left(1 - e^{-\frac{x-c}{d}}\right)$$

(4)

with

$$c = r \cdot \ln\left(\frac{-b}{q}\right) \wedge d = -r, d > 0$$

Note that $f(c) = 0$ here, so if $c = 0$, the graph starts steeply from the origin and approaches the limit q slower and slower. See the capacitor case study (p. 225).

What if we want the curve to be ascending and going through the origin? Setting $c = 0$ and changing the sign of the other parameters will do the job. Usually, this form is used (with b having a different meaning than previously):

$$f(x) = a \cdot \left(e^{bx} - 1 \right)$$

Of course there are only two remaining independent parameters now. This parameterization is used to describe the behavior of diodes; see the case study (p. 236).

1.7. Logarithmic Functions

The logarithmic function with base a is the inverse of exponential functions with base a: $^a\log(x)$ is the inverse of a^x; in particular, the "natural" logarithm $\ln(x)$ is the inverse of e^x.
For modeling, it doesn't matter which base you use, since they only differ by a factor:
$^a\log(x) = \ln(x)/\ln(a)$.
So we can use $f_0(x) = \ln(x)$ as a base function.

The derivative $f_0'(x) = 1/x$ is always positive when x is, so f_0 is an always ascending bijection with a vertical asymptote (VA): $x = 0$, no other asymptotes and domain $\mathbb{R}_0^+$.

Applying a linear transformation, we get the more general version:

$$f(x) = s \cdot \ln\left(\frac{x - p}{r} \right) + q$$

These functions have an annoying VA at $x = p$, which is not often liked for modeling.

However, if we set $p/r = -1$ and $q = 0$, we get an interesting subset of this function class, because they pass through the origin. We can also use this parameterization:

$$f(x) = \frac{a}{b} \cdot \ln(1 + bx)$$

If we use this for positive variables, the VA is always at a safe distance.

Remark: the division by b is actually not necessary here, but it makes parameter a easier to interpret as the slope in the origin, since:

$$f'(x) = \frac{a}{1 + bx} \Rightarrow f'(0) = a$$

Interesting to note is that this set of functions has similarities with power functions with an exponent between 0 and 1; see the case with $a = b = 1$ in Fig. 1.12.

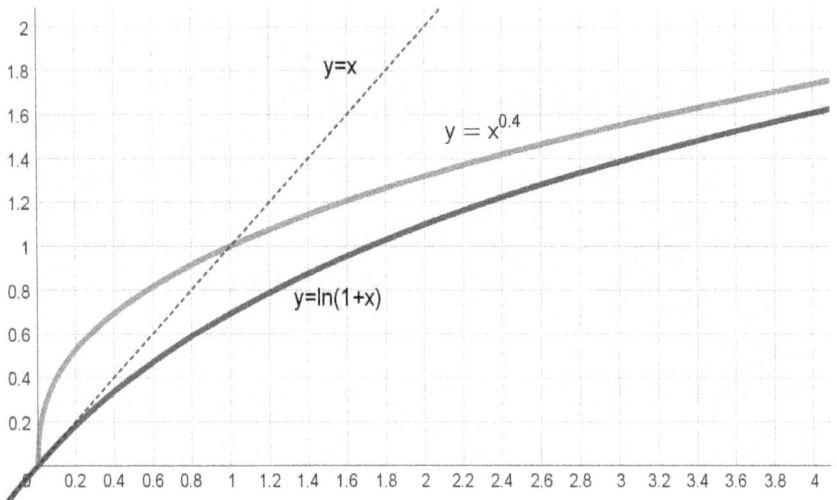

Fig. 1.12. Comparison of a power function and a logarithmic function.

If we only look in the first quadrant, they both go through the origin, where there slope is maximal, and then the slope becomes less and less without going to zero though. So, from far they look the same, but... the power functions have an infinite slope at the origin! And that might make a big difference if we use them as a model.

Applications:
See case studies about food expenditure (p. 367) and the vocabulary of writers (p. 354).

1.8. Sigmoid-like Functions

1.8.1. Introduction

The so-called "sigmoids" are curves that go monotonously from one value (y_-) at $-\infty$ to another (y_+) at $+\infty$. The shape looks like a flattened letter "s" (sigma), hence the name. So, they have two horizontal asymptotes: $y = y_-$ and $y = y_-$. Since the slope is going from "nearly flat" to "nearly flat", there must be an x where its absolute value is maximal, in other words, the derivative is a peak-shaped function (Fig. 1.13), and hence the second derivative has a zero, so the function has an inflection point.

Examples are the "logistic" function (see further), Arctan, the "Gompertz growth" function, and several others.
There are also similar functions that work only on positive numbers, like the Weibull function.

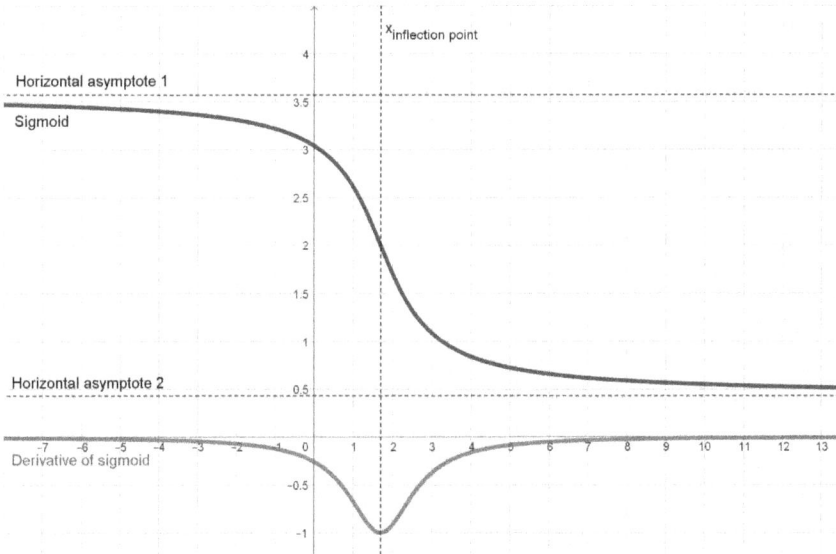

Fig. 1.13. A sigmoid shaped function (a linear transformation of Arctan) and its derivative.

1.8.2. The logistic function and tanh (hyperbolic tangent)

The most used sigmoid is the so-called "logistic" function, basically:

$$f_0(x) = \frac{1}{1+e^{-x}} \tag{1}$$

It ascends smoothly from 0 to 1 through an inflection and symmetry point (0, 1/2), and it has horizontal asymptotes: $y = 0$ and $y = 1$.

More generally, after a linear transformation, we get:

$$f(x) = \frac{s}{1+e^{-\frac{x-p}{r}}} + q \tag{2}$$

Here, the four parameters are indeed all independent: the inflection point is at $x = p$, the steepness is determined by r, and if $s > 0$, the lower and upper limits are $y_- = q$ and $y_+ = q+s$. If you want the curve to be descending, set s negative of course. Often, $k = 1/r$ is used.

Remark: (1) can also be written as:

$$f_0(x) = \frac{1+\tanh\left(\frac{x}{2}\right)}{2}$$

with tanh the "hyperbolic tangent" function:

$$\tanh(x) = \frac{e^x - e^{-x}}{e^x + e^{-x}}$$

which is ascending from -1 to 1 through the origin (symmetry point), and hence has horizontal asymptotes $y = -1$ and $y = 1$. So, the logistic function and tanh are linear transformations of each other (Fig. 1.14). I leave the proof as a little exercise for you; it's straightforward from the definition of tanh.

If you prefer a parameterization where you can set the average value and the "amplitude" (I mean the difference between the average and the top limit), you might like the tanh version better.

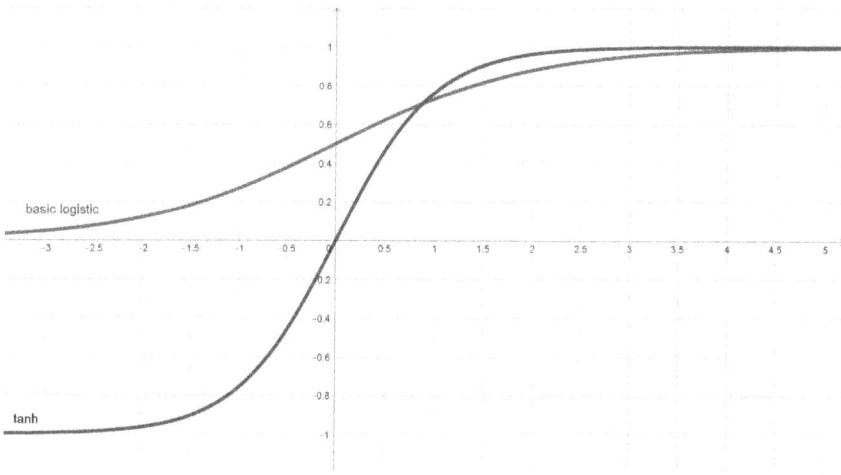

Fig. 1.14. The basic logistic and hyperbolic tangent function, only differing by a linear transformation.

The inverse function can easily be calculated:

$$x \xrightarrow{\cdot(-1)} -x \xrightarrow{e^{(\ )}} e^{-x} \xrightarrow{+1} 1 + e^{-x} \xrightarrow{(\ )^{-1}} \frac{1}{1 + e^{-x}} = f_0(x)$$

$$f_0^{-1}(x) = -\ln\left(\frac{1-x}{x}\right) = \ln\left(\frac{x}{1-x}\right) \xleftarrow{\cdot(-1)} \ln\left(\frac{1-x}{x}\right) \xleftarrow{\ln(\ )} \frac{1}{x} - 1 = \frac{1-x}{x} \xleftarrow{-1} \frac{1}{x} \xleftarrow{(\ )^{-1}} x$$

The domain of f_0^{-1} is obviously $]0, 1[$, and in general, it's $]q, q+s[$, or $]q+s, q[$ if $s < 0$.

Applications:
The logistic function is often used to model the chances for survival vs. age, for getting a disease/cure vs. some variable,... or an evolution from one to another situation.

1.8.3. Gompertz growth

This function is named after the British mathematician Benjamin Gompertz (1779–1865):

$$f_0(x) = e^{-e^{-x}}$$

It's a bit more complicated to understand than the logistic function, but it still has a sigmoid shape, ascending from 0 to 1.

The derivative

$$f_0'(x) = e^{-x - e^{-x}}$$

is maximal at $x = 0$, so f_0 has an inflection point $(0, 1/e)$. A big difference with the previous sigmoids is that this curve is *not* symmetrical around the inflection point: it's steeper on the left, and less steep on the right (Fig. 1.15).

A commonly used parameterization of the linear transformation is:

$$f(x) = a \cdot e^{-e^{b-cx}} + d$$

(With e still being Euler's constant ≈ 2.71.)

The inflection point is now at $x = b/c$, and the slope at this point is ac/e, so it will be *descending* from $a + d$ to d if a and c have different signs.

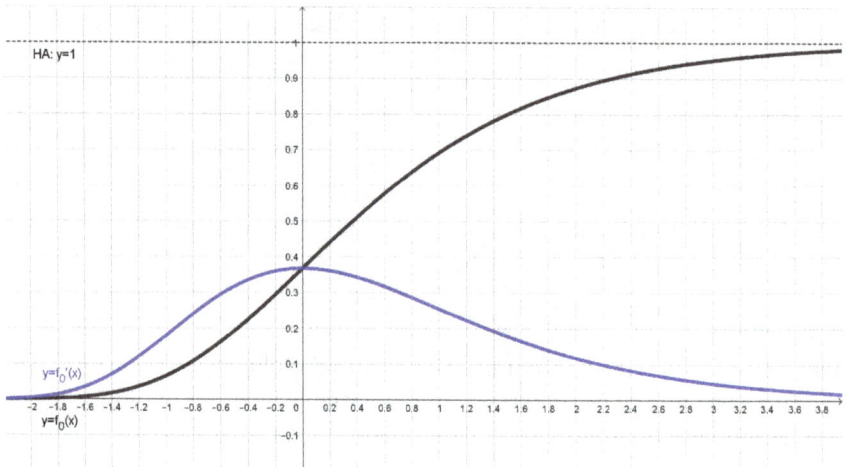

Fig. 1.15. The Gompertz function (black) and its derivative (blue).

Application:
Gompertz used it to describe mortality chances. It also seems to describe the growth of tumors quite well, or in general, population growth within a confined space. See case studies about tumor growth (p. 331), children's vocabulary (p. 345), and "The chance to be alive or dead" (p. 320).

1.8.4. A versatile "transition" function

Another function that has a sigmoid shape, with the same symmetry as tanh, is as follows:

$$f_0(x) = \frac{x}{\sqrt{x^2 + 1}}$$

The derivative:

$$f_0'(x) = \frac{1}{\sqrt{(x^2 + 1)^3}}$$

has a maximum for $x = 0$, so there is also the inflection point of f_0 (Fig. 1.16).

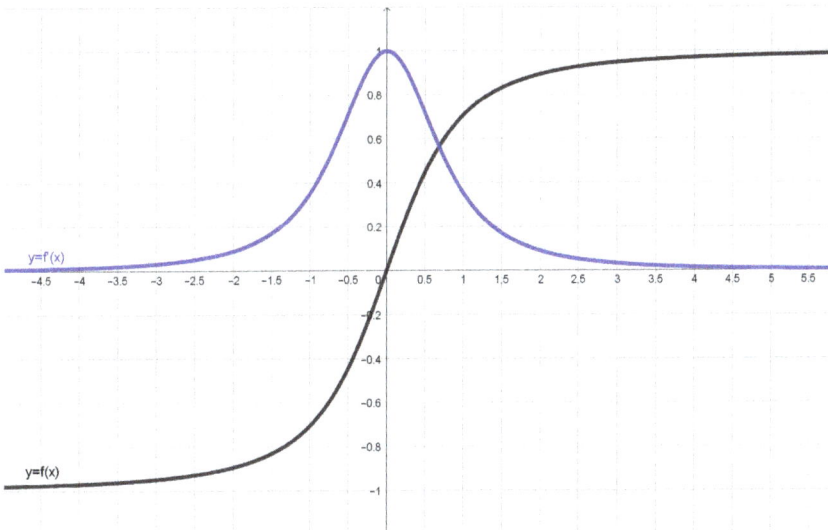

Fig. 1.16. $y = x/\sqrt{(x^2 + 1)}$ (black) and derivative (blue).

It can be made more versatile though, by changing the square and the square root into another positive exponent, say "k". To avoid trouble with some exponents, we also have to use absolute values.

So, the transformed version can be written as:

$$f(x) = \frac{s \cdot (x - p)}{\left| \left\| x - p \right\|^k + |r|^k \right|^{\frac{1}{k}}} + q$$

Note that the case $k = 2$ is just the linear transformation of f_0.

Fig. 1.17 shows what to expect when k changes.

Applications:
I haven't seen this function in the literature yet, but I'm sure someone will find situations where it might be more appropriate than other sigmoids.

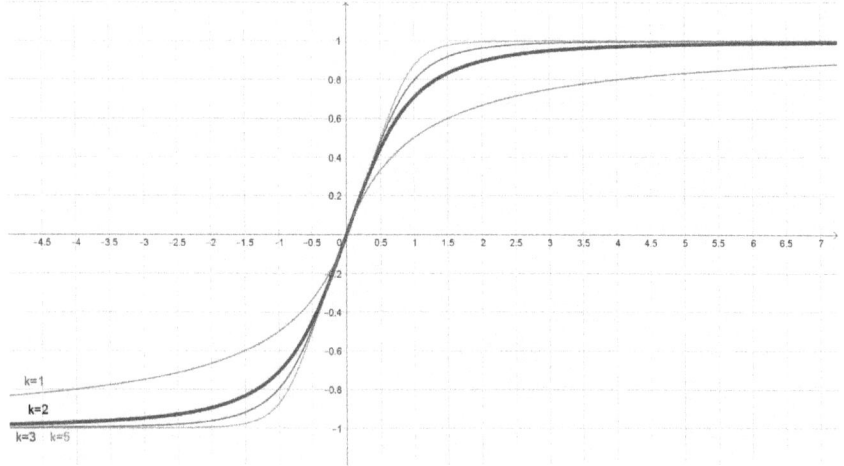

Fig. 1.17. The "transition" function with different *k* values.

1.8.5. Weibull growth and decay

By adding one additional parameter ($k > 0$) to an exponential function, you can make it more versatile:

$$f_0(x) = e^{-x^k}$$

If $k = 1$, it reduces to the basic exponential function. If $k > 1$, it looks like a sigmoid. With $k < 1$, it's like a "half sigmoid" (Fig. 1.18).

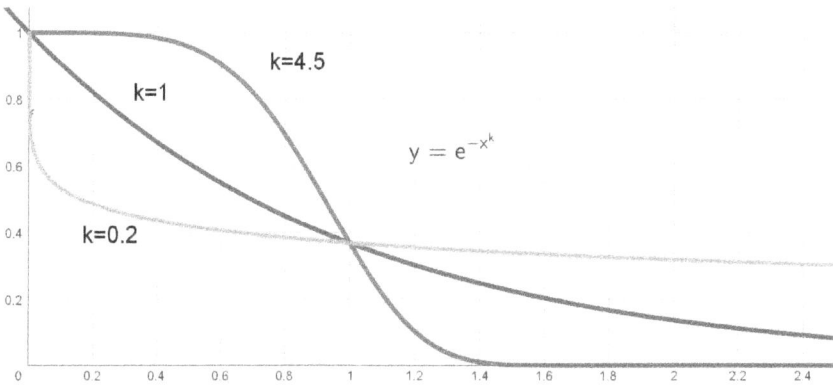

Fig. 1.18. The basic Weibull decay function with different k values.

One possible linear transformation (horizontal and vertical stretch only) is the following: it starts from (0, *a*) to descend to the asymptotic value 0, so it can be used to describe a **decay**:

$$f_d(x) = a \cdot e^{-\left(\frac{x}{b}\right)^k} \qquad a, b > 0$$

If we turn this upside-down, we get another version that starts from (0,0) to ascend to an asymptotic value (*a*), describing a **growth**:

$$f_g(x) = a \cdot \left(1 - e^{-\left(\frac{x}{b}\right)^k} \right)$$

This is also the cumulative distribution function for the so-called **Weibull distribution**.
See, https://en.wikipedia.org/wiki/Weibull_distribution.

Applications: f_d can depict the percentage of broken down items, deceased individuals, suspended matter, etc., where f_g depicts the opposite (% survival) versus time. See case studies "Lifetime of vessels", "Chance to be alive or dead" (p. 320) and "Does money make us happy" (p. 347).

1.9. Peak-Shaped Functions

1.9.1. Introduction

By "peak functions" we mean functions that have one extremum, and on the right side of that, they go to some base value, so they have one horizontal asymptote.

They can be seen as the derivative of sigmoid-shaped functions, in which case they go to the asymptote also on the left side of the peak. Or, they can have positive x values in their domain only, if they are the derivative of a "half" sigmoid.

Usually, the ones that have all real values in their domain are symmetric around the peak, but we can invent asymmetric ones too, and some that work on positive numbers only!

They can be used to depict distributions (y = counts), resonance phenomena (x = frequency or wavelength), evolutions of some quantity over time, etc.

1.9.2. Lorentzian peak

Probably the simplest possible peak-shaped function is basically this one, named after the Dutch physicist Hendrik Lorentz (1853–1928):

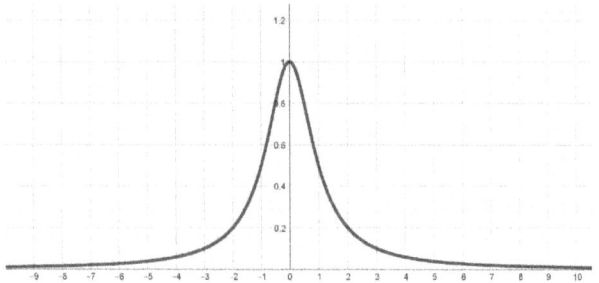

Fig. 1.19. The basic Lorentzian peak function.

$$f_0(x) = \frac{1}{1 + x^2}$$

The graph is a peak with height 1 and position $x = 0$ (Fig. 1.19). Note that $f_0(\pm 1) = 0.5$. If you have some calculus background, you will note that this is the derivative of the Arctangent function ($\tan^{-1}$).

More generally, after a linear transformation, we get:

$$f(x) = \frac{s}{1 + \left(\dfrac{x-p}{r}\right)^2} + q$$

The graph is now a peak centered around $x = p$, amplitude (height) s, baseline $y = q$; and r determines the width.

Other parameterizations are used too, sometimes one that is "normalized" to have an integral (area below the curve) of 1. In that case, it's usually referred to as "**Cauchy distribution**", after the French mathematician Augustin Cauchy (1789–1857).

This function shows up in the description of resonance phenomena (absorption or emission peaks in spectroscopy).

1.9.3. Gauss distribution

When you plot the graph of this function,

$$f_0(x) = e^{-x^2}$$

you will also see a peak with height 1, very similar to the previous one, but narrower and with a different curvature (Fig. 1.20).

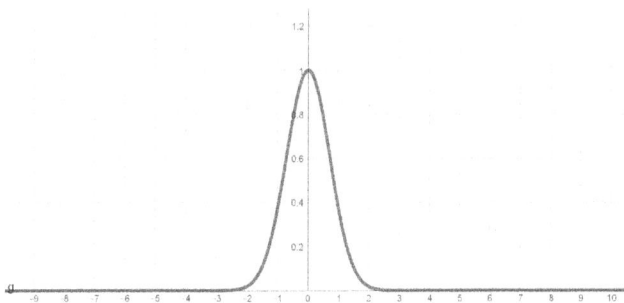

Fig. 1.20. The basic Gauss function.

It's the derivative of a sigmoid called "**error function**" (**erf**):

$$\frac{d}{dx}\left(\frac{\sqrt{\pi}}{2} \cdot erf(x)\right) = e^{-x^2}$$

More about this in https://reference.wolfram.com/language/ref/Erf.html.

We could generalize it as usual, but most often, this parameterization is used

$$f(x) = \frac{N}{\sigma\sqrt{2\pi}} \cdot e^{-\frac{1}{2}\left(\frac{x-\mu}{\sigma}\right)^2}$$

to make sure that the area under the curve is N.

It represents the **probability density** of the so-called **"normal" distribution around an average value** $x = \mu$ **and standard deviation** σ. This means that, if you pick one random x value from such a population, the chance that it has a value between two values x_1 and x_2 is equal to the area under the curve between x_1 and x_2.

1.9.4. The "double logistic" function

A logistic function describes the evolution from one level to another one. If we multiply two of these, one ascending from 0 to 1 and one descending from 1 to 0 and shifted over a distance d, and then scale it, we get

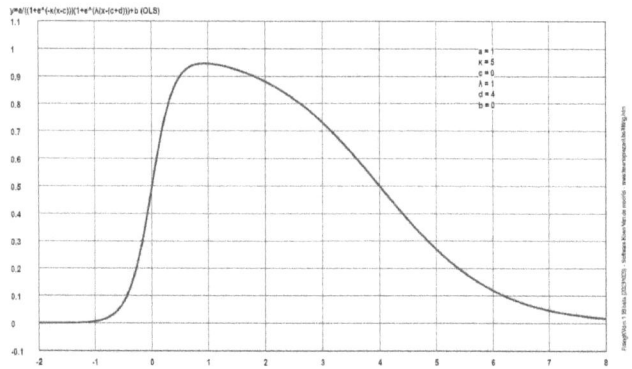

Fig. 1.21. A steep ascent ($\kappa = 5$) and a less steep descent ($\lambda = 1$); $d = 4$, baseline $b = 0$, $a = 1$ (peak height is almost 1).

$$f(x) = \frac{a}{\left(1+e^{-\kappa(x-c)}\right)\left(1+e^{\lambda(x-(c+d))}\right)} + b$$

which looks like a skewed peak. The bigger κ, the steeper it ascends; the bigger λ, the steeper it descends. The parameter a approaches the peak height if d is big enough. The baseline is $y = b$, see Fig. 1.21.

I refer to it as "double logistic" by lack of a better name... suggestions are welcomed!

Applications: If x = time, it might depict some event where an equilibrium state is disturbed and later restored. See case study "A hot stone in water" (p. 220).

1.9.5. Skewed peak functions

Could we find more interesting functions with an asymmetric peak?
Are you ready for a fine calculus cooking class?
Let's start with this simple function:

$$g(x) = \ln\left(e^{-ax} + e^{bx}\right) \qquad\qquad \text{with } a > 0, b > 0$$

What's special about it? If $x \gg 1$, e^{-ax} becomes very small, so $g(x) \approx bx$; if $x \ll -1$, e^{bx} is negligible, so $g(x) \approx -ax$. That means that it is approximately V-shaped with two different oblique asymptotes $y = -ax$ and $y = bx$.

The minimum can be found by finding the zero of the derivative:

$$g'(x) = \frac{-ae^{-ax} + be^{bx}}{e^{-ax} + e^{bx}}$$

$$g'(x) = 0 \Leftrightarrow x = \frac{\ln b - \ln a}{a + b} = x_{top}$$

In the singular case when $a = b$, the minimum is (0, ln2).

Now, all we need to do, is to turn this function "upside down" and "squeeze" the oblique asymptotes into horizontal asymptotes $y = 0$.

Two possible ways to do this, are inspired by the Lorentzian and the Gaussian peak:

$$f_{1,0}(x) = \frac{1}{\left(g(x)\right)^2} \quad \text{and} \quad f_{2,0}(x) = e^{-\left(g(x)\right)^2}$$

(No "1+" was necessary in the denominator of $f_{1,0}(x)$ since $g(x)$ is never zero.)

Both look like asymmetrical (if $a \neq b$) bell-shaped curves (Fig. 1.22).

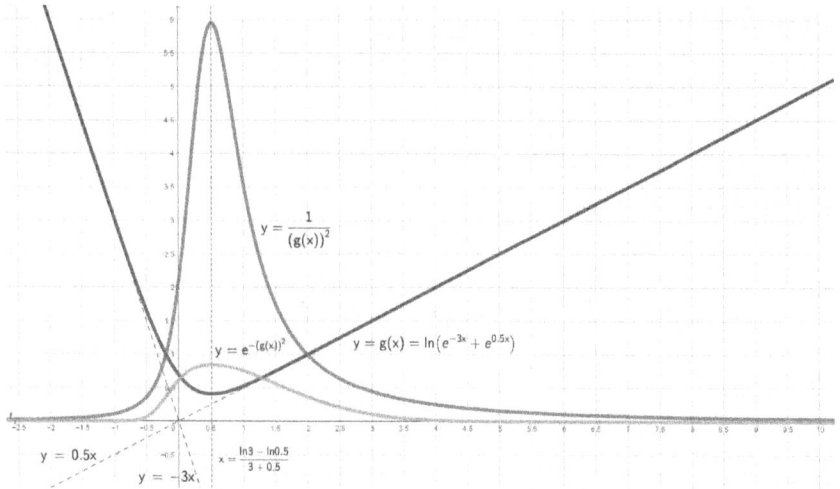

Fig. 1.22. A function with two different oblique asymptotes (g) and two possible transformations to turn it "upside down".

The height of the peaks depends quite a lot on a and b. We don't really like that. For modeling, we prefer to have one clear parameter that tells us how high the peak is. The trick to obtain that, is to divide $f_0(x)$ by $f_0(x_{top})$ so the height becomes 1, and then do the usual linear transformation:

$$f_1(x) = \frac{h}{m} \cdot \frac{1}{\left(\ln\left(e^{-a \cdot \frac{x-c}{s}} + e^{b \cdot \frac{x-c}{s}} \right) \right)^2} + d$$

with

$$m = f_{1,0}\left(x_{top}\right) = \frac{1}{\left(\ln\left(\left(\frac{a}{b}\right)^{-\frac{a}{a+b}} + \left(\frac{a}{b}\right)^{\frac{b}{a+b}} \right) \right)^2}$$

and

$$f_2(x) = \frac{h}{m} \cdot e^{-\left(\ln\left(e^{-a \cdot \frac{x-c}{s}} + e^{b \cdot \frac{x-c}{s}} \right) \right)^2} + d$$

with

$$m = f_{2,0}\left(x_{top}\right) = e^{\left(-\left(\ln\left(\left(\frac{a}{b}\right)^{-\frac{a}{a+b}} + \left(\frac{a}{b}\right)^{\frac{b}{a+b}}\right)\right)^2\right)}$$

Note that m is not an additional independent parameter; it's just an abbreviation to make the formula more readable.

So, the parameter h is now nicely the top height, d the baseline height, and c gives the top position. The attentive reader will note that we have actually one parameter that is not strictly needed: the horizontal stretch factor s. You can normally set this just to 1. But there might be cases where you want symmetry, and then you can just set $a = b = 1$, and then s becomes an indicator of the peak width.

Applications:
Just like the "double logistic" function, these functions might be tried for modeling events where an equilibrium is temporarily disturbed; see the "hot stone" case study (p. 220).
They might also be interpreted as probability distributions, since their integrals are finite, but they can only be calculated numerically.

1.9.6. Weibull distribution

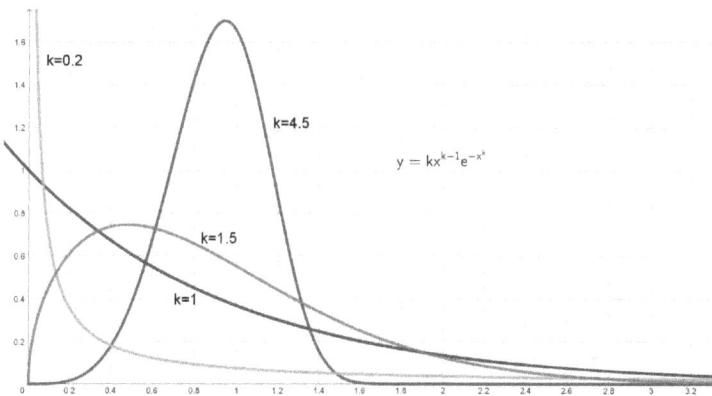

Fig. 1.23. Weibull distribution with $a = b = 1$.

This is the derivative of the Weibull growth function (see p. 33):

$$f(x) = f_g'(x) = \frac{ak}{b^k} x^{k-1} e^{-\left(\frac{x}{b}\right)^k}$$

so, it only works on positive x values too (see Fig. 1.23).

1.9.7. RLC Serial filter function

This looks like a weird function, but it comes from the calculation of signal transmissions in passive electronic filters, where only the positive half ($x > 0$) is used:

$$f(x) = \frac{1}{\sqrt{\left(1+r\right)^2 + \left(ax - \frac{b}{x}\right)^2}} \qquad a \geq 0,\, b \geq 0,\, r \geq 0,\, 0 < f(x) < 1$$

If $a > 0$ and $b > 0$, this curve (the used right half) ascends from the origin (in fact, the origin is the *limit* for x going to 0) to a peak at $x = \sqrt{(b/a)}$, then it slowly descends to 0 at infinity. The peak height is $1/(1 + r)$, so it will only reach the maximally possible value of 1 (= "perfect" transmission) if $r = 0$.

If $a = 0$ the peak moves to infinity and it turns into a horizontal asymptote $y = 1/(1 + r)$. If $b = 0$ it's in the origin (see Fig. 1.24).

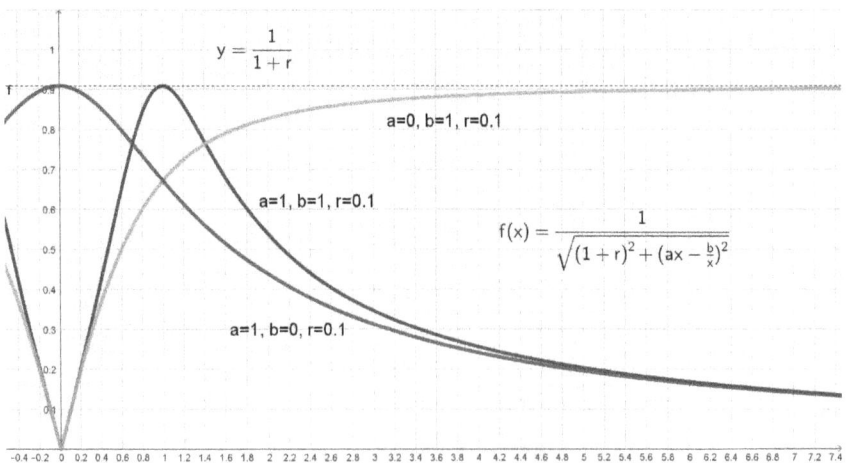

Fig. 1.24. The RLC transmission function with "normal" and extreme parameters.

Application:
This function describes the transmission of an RLC filter if a signal with frequency x passes. The parameters r, a and b are connected with the resistance, the capacitance, and the inductance in the circuit. See case study "RLC filters" (p. 230). The peak position is then the resonance frequency $1/(2\pi \sqrt{(LC)})$.

Of course, other applications might exist too! Whenever there is some "resisting force" that is proportional to x, competing with a resisting force *inversely* proportional to x, there will be an "optimal" x that causes the least resistance. The parameter r can be seen as a measure for "efficiency", $r = 0$ being ideal.

1.9.8. Dagum distribution

This weird function has a peak shape for some parameter values (Fig. 1.25):

$$f(x) = N \cdot \frac{ap}{x} \cdot \frac{\left(\dfrac{x}{b}\right)^{ap}}{\left(\left(\dfrac{x}{b}\right)^{a} + 1\right)^{p+1}} \qquad\qquad a, b, p, N > 0, x > 0$$

To be honest, I don't like it since the shape of it changes in a way that is very difficult to understand, but I mention it here since it is *supposed* to describe income distributions, but it usually doesn't do a good job!

More information is available at
https://en.wikipedia.org/wiki/Dagum_distribution.

See the case study "Income distributions" (p. 372).

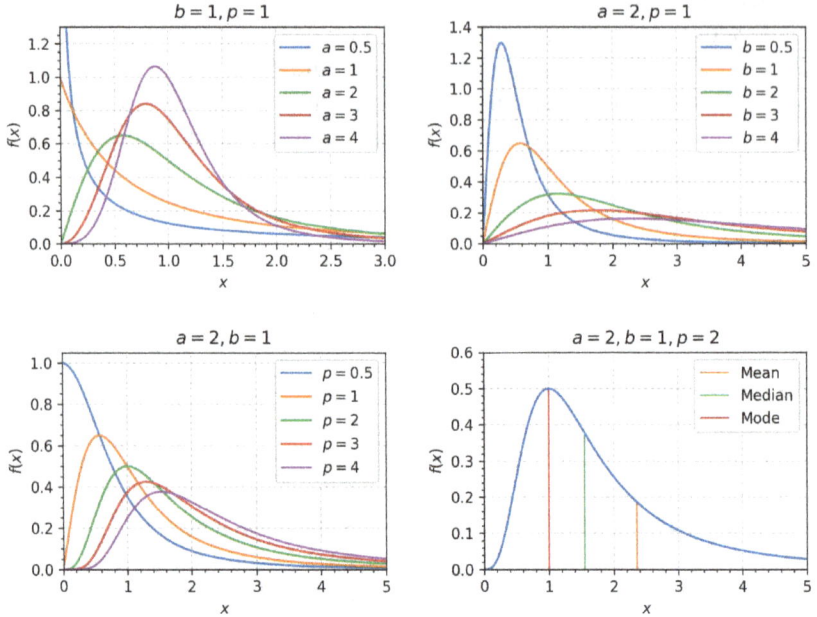

Fig. 1.25. Dagum distribution with different parameters. Image by Rafael Marcondes, https://commons.wikimedia.org/wiki/File:DagumPDF.png.

1.9.9. Power function multiplied with exponential decay

How would a function look that describes something that grows from zero to a maximum and then slowly decays back to zero? For example, a biological ability like strength or endurance during a lifetime? Or a property of a material that improves by adding something but not too much?

Something like the following might be a good candidate:

$$f_0(x) = x \cdot e^{-x}$$

This starts like a line from the origin and has a maximum for $x = 1$.

If we want some more variation in the start, we can pimp it up with an exponent in the first factor:

$$f_1(x) = x^n \cdot e^{-x}$$

The derivative

$$f_1'(x) = x^{n-1}e^{-x} \cdot (n - x)$$

shows that the top shifts to $x = n$, and its height ($n^n e^{-n}$) increases dramatically if n increases.

The curve starts to go up from the origin, vertically if $0 < n < 1$, linearly if $n = 1$ and flat if $n > 1$. In the limit for n going to 0 f becomes a simple descending exponential function (Fig. 1.26).
Dom(f_1) = $\mathbb{R}$ for integer n values, but in general, for any positive n, it's limited to $\mathbb{R}^+$.

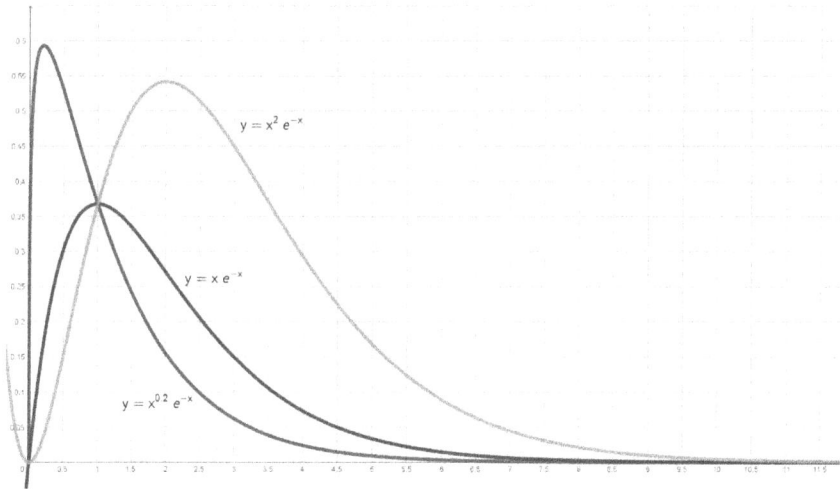

Fig. 1.26. Power functions multiplied by an exponential decay function.

If we want an easy-to-understand stretchable parameterization of this, we could divide it by the peak height and add vertical and positive horizontal scaling parameters a and b:

$$f(x) = a \cdot \left(\frac{x}{bn}\right)^n \cdot e^{-\frac{x}{b}+n}$$

The formula looks a bit frightening, but the peak point is now simply the point with coordinates (bn, a).

More "pimping" could be done by adding an exponent in the exponent, like in the Weibull function.

Applications:
See the case studies "Concrete" (p. 209) and "Ultra-marathons" (p. 291).

1.9.10. Other peaks

If you like to experiment, you can find more pretty peaks. Try these for example:

$$f(x) = \frac{x}{\cosh(x)} \quad \text{or} \quad f(x) = \frac{1}{ax^2 + bx + c} \quad \text{with } b^2 - 4ac < 0$$

1.10. Periodic and Semiperiodic Functions

1.10.1. Introduction

A function is called strictly **periodic** if there is some real number $T > 0$, so that $f(x+T) = f(x)$ for every $x \in \mathbb{R}$. T is called the "period" of f.

Some phenomena occurring in reality can be described quite well by a periodic function, like a pendulum clock, or light waves emitted by atoms when they change states, or the incident solar light intensity on a specific position on earth that varies with the seasons, etc.

Many phenomena can be described by a complicated mix of (sometimes changing) periodic functions, like the tides, a heartbeat, sounds, brain waves, animal populations, climate variations, solar spots, etc.

There are also functions that show something like a periodic pattern, but it changes over time, like a damped wave, for example. They are not strictly periodic, but we could call them "semiperiodic".

1.10.2. Sine and cosine

The simplest periodic function is the well-known sine (Latin and many other languages: sinus), $f_0(x) = \sin(x)$.
It looks like a wave that goes up and down between -1 and 1 around an average value of 0, and the period is 2π radians $= 360°$. The maximum nearest to the origin is at $x = \pi/2$, and the nearest minimum at $x = -\pi/2$.

The linearly transformed generalization looks like the following:

$$f(x) = s \cdot \sin\left(\frac{x - p}{r}\right) + q$$

Usually, p is called φ (Greek letter phi, "phase shift"), and ω (Greek letter omega, "pulsation") is used instead of $1/r$.

Or, often, it's written like

$$f(x) = m + A \cdot \sin\left(2\pi\frac{x - c}{T}\right)$$

with m the average value, A the amplitude, and T the period (usually a time or a distance, the same as x). T can also be replaced by the frequency $f = 1/T$. Parameter c (or sometimes, x_0) is now the phase shift in the same units as T (Fig. 1.27).

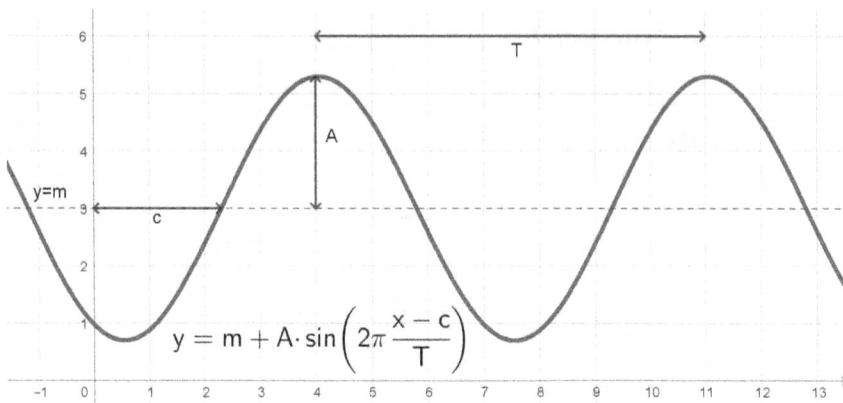

Fig. 1.27. A possible parameterization of the linear transformation of the sine function.

Note that the cosine function is one of those linear transformations of the sine function, since $\cos(x) = \sin(\pi/2 - x)$.

If you want an ascending inflection point in the origin, set $m = 0$ and $c = 0$. If you want a top at $x = 0$, the simplest is to replace sin by cos.

Applications: See the case studies "Temperature in Chatanga", "Temperature in one day" (p. 269), "Temperature vs. latitude" (p. 273), etc.

Remark: We have four independent parameters here, so you might think that four points will suffice to determine the shape of the sine that connects them. Well ... there is a problem here: if you put those four points in the equation $y = f(x)$, you get a system with an infinite number of solutions, because of the periodicity. This is called the **"aliasing"** problem.

Example: Take the four points $(0, 0)$, $(1, 5)$, $(3, 2)$ and $(4, -2)$. Fig. 1.28 shows two possible sine waves that connect them. The one on top has the longest possible period ($T \approx 5.104$). The one below has a period of approximately 1.244.

Test aliasing y = m + A·sin(2π(x-c)/T) S: 3.353920482E-034 (OLS)

A = 4.35129291130809
T = 5.1042993121954
c = 0.419840205322781
m = 2.15

Test aliasing y = m + A·sin(2π(x-c)/T) S: 1.772984008E-024 (OLS)

A = 4.35129291130786
T = 1.24364694773763
c = 0.51953068930429
m = 2.150000000000171

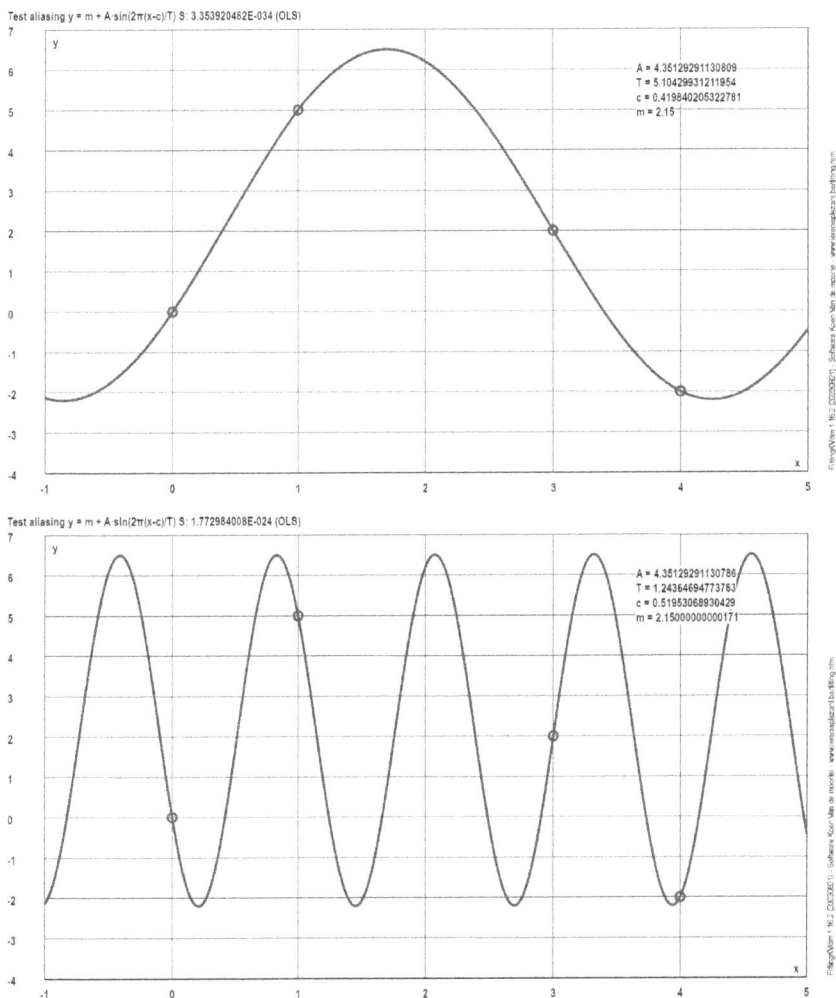

Fig. 1.28. The "aliasing" problem: not just one sine function can connect 4 dots.

1.10.3. A sum of two random sines

A sum of sines with periods T_1 and T_2 is only strictly periodic if the ratio of both periods is rational, and the periodicity is only clearly visible if that ratio is a simple rational number, like $T_1 = 2$ and $T_2 = 3$. In that case, the period of the sum will be $2 \cdot 3 = 6$. Think of wall papers with pattern sizes of 2 m and 3 m hung below each other; the total pattern will have a size of 6 m (Fig. 1.29).

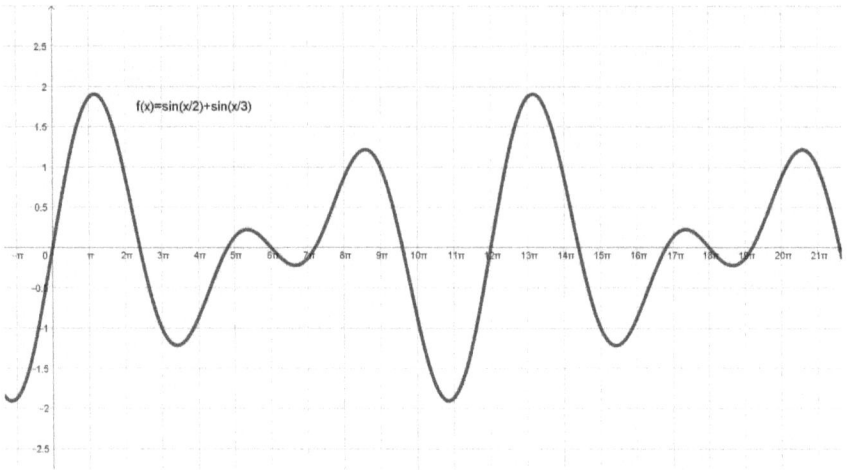

Fig. 1.29. $f(x) = \sin(x/2) + \sin(x/3)$, period $T = 12\pi$.

If $T_1 = 2$ and $T_2 = 2.001$, you will need a lot of both periods to see the total period. Find out how many, as an exercise. Tip: find the smallest integers k and l so that $k \cdot T_1 = l \cdot T_2$.

Applications: See the case studies about the tidal currents (p. 283).

1.10.4. A sine wave + harmonics

If you want to describe a phenomenon that repeats itself with a period T, but with a more complicated pattern than the ordinary sine wave, you might add some "harmonics", meaning sines with multiples of the frequency, or in other words: period $T/2$, $T/3$, $T/4$, etc. (Fig. 1.30).

Examples: A heart beat, sounds of instruments, etc.

You can parameterize this as a sum of sine or cosine functions with phase shifts, or a sum of sines and cosines without phase shifts. They are all equivalent, but depending on the situation, one might be more convenient. If you want a shape that is symmetrical around the y axis (even), you can use cosines with phase shifts of zero, and if you want an odd symmetry, you can use sines with phases zero, and an average term of zero if you want it to go through the origin.

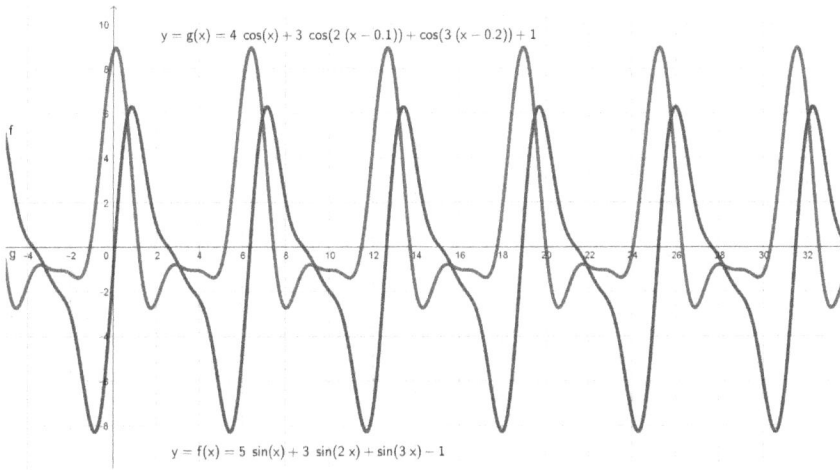

Fig. 1.30. Two trigonometric functions with a possible second and third harmonic; period $T = 2\pi$.

1.10.5. Periodic peaks

Is it possible to modify a wave-shaped function into regularly spaced sharper peaks, not by adding a lot of harmonics but by adding just one parameter? Sure!

How about starting from the following one?

$$f_0(x) = \frac{1}{1 + \sin^2(x)}$$

This function varies between 0.5 and 1 with a period of π, and it still looks very "wavy", but by adding a parameter k, we can influence the sharpness.

We can put it on two places:

$$f_1(x) = \frac{1}{k + \sin^2(x)} \quad \text{or} \quad f_2(x) = \frac{1}{1 + (k\sin(x))^2}$$

In the f_1 formula, making k smaller will result in sharper peaks, but also the height will increase. In f_2, increasing k will sharpen the peaks, and the top of the peaks will always have $y = 1$ (Fig. 1.31).

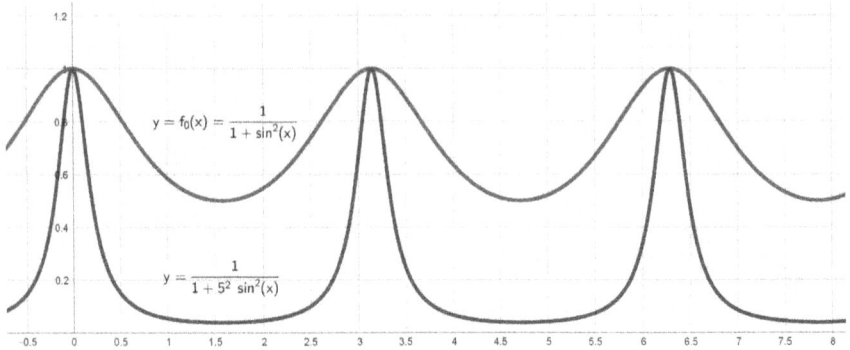

Fig. 1.31. A peak-shaped periodic function.

If you want to manipulate the formula so that the period is T, the height of the peaks A, and the base level m, you could do this linear transformation of f_1 (check as an exercise):

$$f(x) = A \cdot \left(\frac{k(k+1)}{\sin^2\left(\pi \cdot \dfrac{x-c}{T}\right) + k} - k \right) + m$$

You might also start a linear transformation from f_2:

$$f(x) = \frac{A}{1 + k^2 \sin^2\left(\pi \dfrac{x-c}{T}\right)} + m$$

but here the base line is $y = A/(1+k^2) + m$ ($\approx m$ if $k \gg 1$) and the height of the peaks is less than A.

Remark: A negative value of A would make sense here, if you want "upside down" peaks.

Applications, at least approximately: Prey-predator cycles and sunspot cycles (see case studies p. 339 and 262).

1.10.6. A skewed wave

What would be the recipe for a periodic function that looks like a simple sine wave but with asymmetrical peaks?

According to Fourier's theory, any periodic function can be approached by a sum of sine waves: one basic wave that determines the period: $\sin(\omega(x - \varphi))$ plus a number of harmonics (double, triple, quadruple frequency etc.) with adjustable amplitudes and phases.
For example, a so-called "sawtooth" function can be obtained by adding all the terms $\sin(nx)/n$ with $n = 1$ to... infinity! See
https://mathworld.wolfram.com/FourierSeriesSawtoothWave.html.

Now, in the computer world, we don't really like "infinity". It takes quite a long calculation time to get there... But okay, maybe the first 10 or 100 terms will be a good enough approximation? Hmm... that would still be time consuming, and worse, it would still look "wobbly", and you might have to change a lot of parameters to move the peaks to the place you would like them to be.

Nice to have would be a smooth function with one parameter that determines the skewness. I challenge you to find one yourself! This is not an easy one! But, somewhere in the hidden corners of the internet, someone mentioned the existence of such a beauty:

$$f_0(x) = \frac{1}{k} \cdot \operatorname{Arc\,tan}\left(\frac{k \cdot \sin(x)}{1 - k \cdot \cos(x)}\right)$$

As long as $-1 < k < 1$, this is a nice, continuous, and strictly periodic function with period 2π (Fig. 1.32).
If $k > 0$, the ascending slope is steeper than the descending slope.
If $|k| = 1$, the function becomes a "sawtooth".

With $k = 0$, strictly, the function is undefined (0/0), but fortunately, using de l'Hôpital's rule, we can find the limit:

$$\lim_{k \to 0} f_0(x) = \lim_{k \to 0} \frac{d}{dk}\left(\operatorname{Arc\,tan}\left(\frac{k \cdot \sin(x)}{1 - \cos(x)}\right)\right) = \lim_{k \to 0}\left(\frac{\sin(x)}{k^2 - 2k\cos(x) + 1}\right) = \sin(x)$$

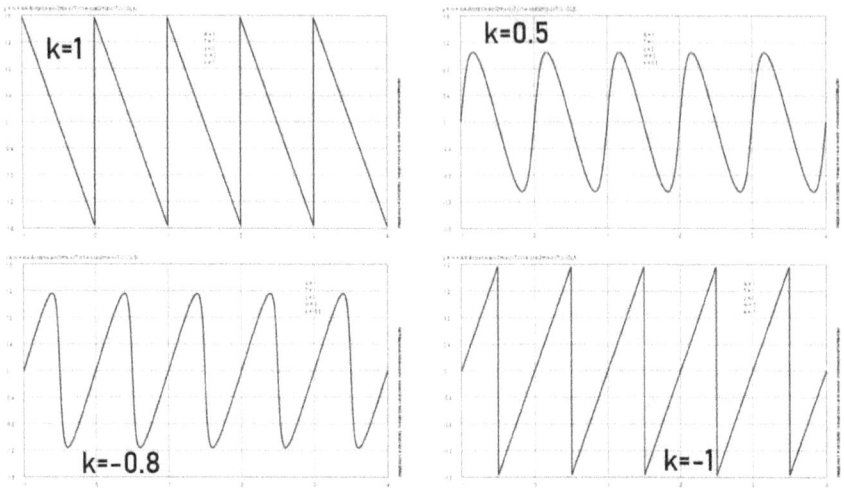

Fig. 1.32. The "skewed wave" function with some different values for *k*

Where are the peaks? To find out, we calculate the derivative:

$$f_0'(x) = \frac{df_0(x)}{dx} = \frac{\cos(x) - k}{k^2 - 2k\cos(x) + 1}$$

$$f_0'(x) = 0 \Leftrightarrow x = \pm\operatorname{Arc}\cos(k) \qquad (\text{+ multiples of } 2\pi \text{ of course})$$

(+: maximum, –: minimum)

So, the amplitude is

$$f_0(\operatorname{Arc}\cos(k)) = \frac{1}{k}\operatorname{Arc}\tan\left(\frac{k}{\sqrt{1-k^2}}\right)$$

varying between 1 ($k = 0$) and $\pi/2$ ($|k| = 1$).

Allowing linear transformations we get the more versatile function (with *T*, *c*, *A*, and *m* having the same meaning as in the generalized sine function):

$$f(x) = m + \frac{A}{k} \cdot \operatorname{Arc}\tan\left(\frac{k \cdot \sin\left(2\pi\dfrac{x-c}{T}\right)}{1 - k \cdot \cos\left(2\pi\dfrac{x-c}{T}\right)}\right)$$

This function is implemented in FittingKVdm as the "Skewed wave" model.

Theoretically, there is no objection to having negative A or T values, but it's custom to keep them positive, since changing the sign of A is equivalent to adding half a period to c, and changing the sign of T is like changing that of k.

This function is periodic with period T, and the first peak position is at
$$x_{top} = T \cdot \text{Arc} \cos(k) + c$$
and the amplitude is
$$y_{top} - m = \frac{A}{k} \text{Arc} \tan\left(\frac{k}{\sqrt{1 - k^2}} \right)$$
Amplitude $\approx A$ when $|k|$ is small, $\pi/2 \cdot A$ when $|k| = 1$.

Applications:
I have not seen this function being used in the literature yet, but there must be situations where it can be applied. For example, when looking at weather cycles (daily/yearly), it seems that the warming up can be slower than the cooling down; see case study p. 268. Also, in electronics, sawtooth-like functions are used.

1.10.7. A damped sine wave

In real physical situations, where friction cannot be neglected, oscillations that are not fed with energy will eventually fade out.
Their amplitude will decrease exponentially to zero, so they can be described basically by

$$f_0(x) = e^{-\lambda x} \sin x$$

with $\lambda > 0$. Usually, $x =$ time (Fig. 1.33).
The bigger the value of λ, the quicker the oscillation will be damped.

Applying a linear transformation, we get:

$$f(x) = m + A \cdot e^{-\lambda x} \cdot \sin\left(2\pi \frac{x - c}{T} \right)$$

Of course, the function is not strictly periodical anymore, unless $\lambda = 0$. Once λ is too big, there are no ups and downs anymore, but otherwise T is still the distance between two tops.

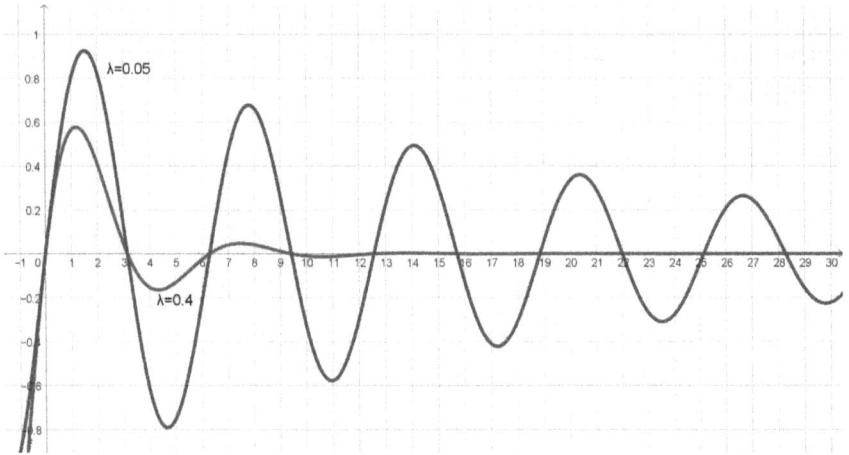

Fig. 1.33. Two different damping parameters.

See the case study about pendulums (p. 194).

1.11. Miscellaneous Functions

These are some functions that are useful for specific situations. The domain can be limited to make them invertible.

1.11.1. "Parallax" function

This is not how this function is officially called, but I refer to it with this name, because it can be used for measurements that are related to the so-called "parallax", the difference in observation angle if you look at a point from different viewpoints:

$$f(x) = \text{Arc} \tan\left(\frac{h}{x+d}\right)$$

If $h > 0$, $d > 0$, $x_i \geq 0$ then this is a descending function toward a horizontal asymptote: $y = 0$. It looks a bit like an exponential decay, but the curvature is different (Fig. 1.34).

As long as the conditions are fulfilled, the vertical asymptote at $x = -d$ is out of sight.

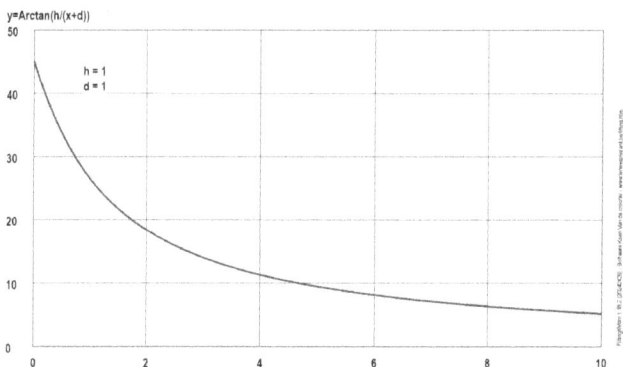

Fig. 1.34. Example with $h = 1$, $d = 1$.

Applications: $x + d$ is the distance to an object, h its height, and $f(x)$ the observation angle; see case study "Measuring the height of a building" (p. 183).

1.11.2. "Refractive index"

This function is needed to model optical refraction. In that case, x is the angle of the incoming beam, n the refractive index, and $f(x)$ the angle of the refracted beam:

$$f(x) = \text{Arc}\sin\left(\frac{\sin(x)}{n}\right)$$

Constraints: $n \geq 1$; $0 \leq x < 90°$.

"Arcsin" (with capital A) is that part of $\sin^{-1}$ that produces results in the first and fourth quadrant ($-90° \leq \text{Arcsin}(x) \leq 90°$).

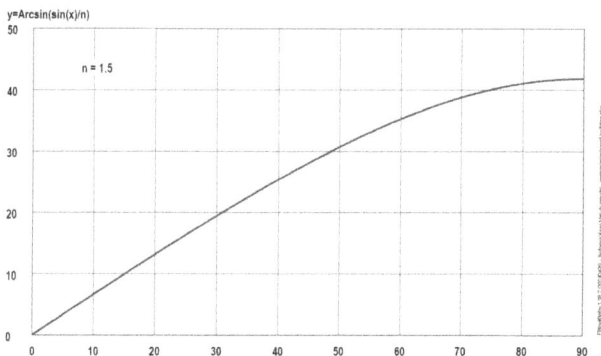

The graph is a curve through the origin, ascending and ending flat at $x = 90°$, at least if $n > 1$, otherwise it's just a straight line.

Fig. 1.35. Example with $n = 1.5$.

Remark: If the trigonometric functions on your computer expect radians and their inverses produce them, don't forget to multiply x with $\pi/180$ and $f(x)$ with $180/\pi$.

See case study "The refractive index of a CD box" (p. 252).

1.11.3. Cosh (the hyperbolic cosine)

The hyperbolic cosine is defined as:

$$\cosh(x) = \frac{e^x + e^{-x}}{2}$$

Its graph looks similar to a U-shaped parabola, with top $(0,1)$.

A useful linear transformation is the following:

$$f(x) = a \cdot \left(\cosh\left(\frac{x-c}{a}\right) - 1 \right) + b$$

The top is now (c, b); it is U-shaped if $a > 0$, and upside down if $a < 0$; it reduces to a flat line $y = b$ if $a = 0$ (Fig. 1.36).

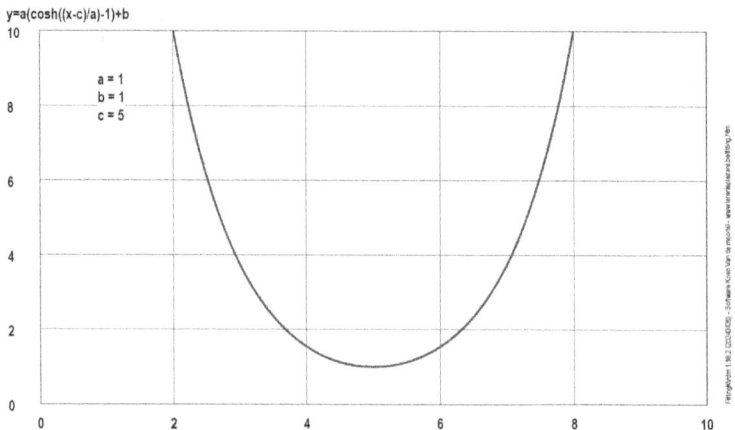

Fig. 1.36. $y = \cosh(x - 5)$, top at $(5,1)$.

This function describes a hanging cable $(a > 0)$ or an ideally constructed arch $(a < 0)$. See case study "Hanging chain" (p. 213).

1.11.4. Power-Möbius function

Changing x into $x/(1-x)$ is sometimes called a "Möbius transformation" (after the German mathematician August Ferdinand Möbius (1790–1868).

Taking a power of this result, produces an interesting function, which we might baptize "power-Möbius":

$$f(x) = a \cdot \left(\frac{x}{1-x}\right)^b$$

The domain is limited to $[0, 1[$ (0 to 100%).

If $b > 0$, the curve starts in the origin and ascends toward a vertical asymptote at $x = 1$.

If $b < 0$ the curve is mirrored horizontally.

If $|b| < 1$ it starts or ends vertically (see Fig. 1.37).

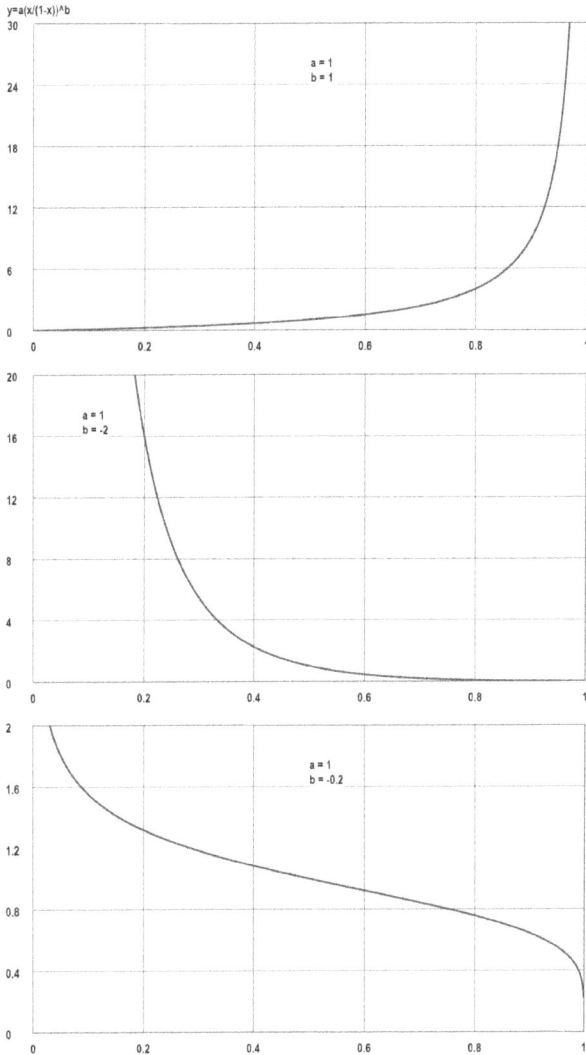

Fig. 1.37. Power-Möbius function with $a = 1$ and different exponents $b = 1, -2, -0.2$.

Where could it possibly be used?

Suppose you add X units of cement to 1 unit of sand and the strength of the concrete $S \sim X^b$, then the percentage of cement is $x = X/(1 + X)$, so $X = x/(1 - x)$, and $S = a(x/(1-x))^b$.

Also, if x is a probability, $x/(1 - x)$ is the odds.)

2

Harvesting Data

Not all numbers are sacred.

What happens with imprecise data if you do operations on them?

Learn to minimize errors in your hunt for data.

2.1. Measurement Uncertainties

I sometimes ask my (male, hetero) students: "Would you rather spend an evening with a group of females who have an average age of 18, or 40?" As you can expect, they always answer: 18 of course! But then I tell them: group 1 consists of 36-year-old mothers and their newborns, while group 2 consists of 18-year-old single girls and their grannies; and then of course they regret their choice!
The moral of the story: the average doesn't tell you much; it's important to know also the standard deviation or at least some kind of dispersion indicator!

Every number published should be accompanied by an "error margin" or a so-called "confidence interval". At least, that was the first thing our professors taught us when I started my physics studies! To my surprise, this rule is very often neglected. For example, in the popular (still sold) *A Handbook of Small Data Sets* by D.J. Hand *et al.* (Springer, 1994), with 510 interesting educational sets, not a single one has any indication of measurement precision!

In better books or articles, you'll see results published like, e.g., "$x = 123.4 \pm 5.6$ cm", which is much better, but even then, it's not always clear what they mean exactly. The "123.4" and the "5.6" might be for example:

(1) The average μ and standard deviation σ of a series of n repeated measurements of x values (e.g., heights of a number of randomly chosen 10-year old girls). In that case it means that about 68% of all the measurements were between $123.4 - 5.6 = 117.8$ and $123.4 + 5.6 = 129.0$ cm, if they are normally distributed.

(2) The average and the expected uncertainty on that average, which is $\sigma/\sqrt{n}$.

(3) The result of one measurement and an approximate error given by the manufacturer of the used instrument. In that case, it usually means that the real value of the measured quantity is within these limits with a probability of about 95%.

(4)   One measurement and a maximal deviation, estimated by common sense.

Examples:
(1)   Suppose you want to study the pebbles on a certain beach. You grab five of them and weigh them with a scale that gives results rounded to 1 g: 20, 26, 27, 37, 40 g.
Using the formulas,

$$\mu = \overline{x} = x_{avg} = \frac{\sum\limits_{i=1}^{n} x_i}{n} \qquad \sigma = \sqrt{\frac{\sum\left(x - \overline{x}\right)^2}{n-1}}$$

or using any spreadsheet software, you find: $\mu = 30$ g, $\sigma = 8.276$ g, which means that you can expect that **68%** of all the pebbles on the beach weigh between 30 − 8.276 and 30 + 8.276 g
if the measurement method allows distinguishing different x values in the sample (seems OK here, although not optimal),
if the measurement device is well calibrated (we have to have some trust in the manufacturer...),
if the distribution of all the x values is normal (probable, if the pebbles all came from the surface, since the smallest sink lower),
if the sample was randomly chosen (probable, if you picked one by one with your eyes closed),
if the sample is "large enough" (clearly not here, but for the simplicity of the example...).
That's a lot of "ifs", isn't it?

(2)   From the same pebble weights, you can conclude that the average mass of all the pebbles is between 30 − $8.276/\sqrt{5} \approx 26.3$ and $30 + 8.276/\sqrt{5} \approx 33.7$ g with a probability of 68%, again if the same conditions are valid.

(3)   You measure a voltage of 4.56 V and the manual of the voltmeter says the precision is 0.03 V. That means that the real voltage is "most probably" between 4.56 − 0.03 and 4.56 + 0.03 V. How probable? Well... who knows?

(4)   You measure a length with a ruler, and you round to the nearest mm, so: "123 mm" means that the real length is between 122.5 and 123.5 mm.
How certain can you be? 80%? 90%? 99%? 100%? This actually de-

pends on your eyesight! How well can you see where the middle is between two lines 1 mm apart, to decide if you need to round upwards or downwards? If you take ± 1 mm as your error estimation, you'll have a 100% confidence interval, but that doesn't "exist" in the normal distribution theory.

Remark: Your data can have an asymmetric distribution, for example, the income of people in a certain area. You can still calculate μ and σ of course, but the interpretation with the 68% will just not be valid anymore.

2.1.1. The standard deviation in some special cases

• Counts

If the variable you are measuring is a count, for example, the number of crows you see in one day in your garden, how much should you take as the "measurement error"? Obviously, if you counted 7, it could not have been 7.1 or 6.7. You may have missed one or counted one double, yes, but in principle, it is possible to have an error of zero. But, usually, what you want to know is the following: how many crows can we expect to see in a day (in this area in this season)?

Then, following from the theory of normal distribution (a good approximation for the Poisson distribution if the numbers are big enough), usually, the square root of the number is taken. This is not the error in the counting, but an estimation of the possible mistake you can make if you generalize the observed number for the whole population.

So, we can expect to see, in 68% of the days, between $7 - \sqrt{7}$ and $7 + \sqrt{7}$ crows in a day, based upon this one observation.

Actually, this estimation is only good for large numbers and the range is not precisely symmetrical. And especially if the count is zero, there is a problem for which there is no general agreement, it seems. I would vote to use an error of 1 rather than 0, if you can assume it was "just a bad day". If you were observing elephants at the North Pole, you would have an average and standard deviation of 0 with very high confidence.

If you want to think more about this, there is this interesting article by Tommaso Dorigo (2011), *Those Deceiving Error Bars*, see
www.science20.com/quantum_diaries_survivor/those_deceiving_error_b ars-85735.

● Proportions (fractions)

If you measure a proportion p, like for example the percentage of people driving while using their phone, there is also a useful rule of thumb to estimate the standard deviation.

If your total count was n, and the count of the target subgroup was m, $p = m/n$ (between 0 and 1 of course), then

$$\sigma \approx \sqrt{\frac{p\left(1-p\right)}{n}}$$

The approximation is better if n is bigger, and if p is not too close to 0 or 1.

By the way, I did such a count in Serbia, August 2023, and out of 203 drivers, I saw 19 with their phone in their hands, so $p = 19/203 \approx 0.0936$, $\sigma \approx 0.0204$. This means the "real" p (for all the drivers in that area) is probably between 0.073 and 0.113.

Besides the "rules", use common sense! The following are examples:

● In the case of the phoning drivers, I would estimate the error on p a bit bigger. Why? In some cases, I could see that both hands were not on the steering wheel, but it wasn't clear if they had a phone in their invisible hand. I didn't count these observations, which certainly has an influence on the realistic uncertainty.

So, don't just trust "the numbers", but always ask yourself how they "appeared" out of the blue!

Also, maybe p would be different in a neighborhood 5 km away where more relaxed farmers live instead of nervous business people? So, the generalization for "the whole population" is always difficult.

● How much is the measurement error here?

Suppose you want to know the relationship between someone's mass and how high he/she can jump.

So, you measure someone's mass as precisely as possible, say 81.32 kg and you measure the jump height, and repeat this with many test persons. We can assume a precision of $\pm$ 0.005 kg on the mass *if* the scale is properly calibrated. Fine!

Now, we want to study the relationship between a person's mass and how much beer he or she consumes per week. We use the same scale, so we have the same precision, right? No! We can't assume that this person weighs the same all the time, before and after eating or going to the toilet, etc. So, in this case we must assume a bigger "measurement error", probably in the order of ± 0.5 kg. Ideally, we should weigh the person many times in this week, and calculate the average and standard deviation, but that would be impractical.

Remark: Often, the letter "*s*" is used for sample standard deviations or measurement errors, but to avoid all confusions with parameters or quantities named "*s*", in this book, **I will use σ_x for the estimated uncertainty on *x*,** but depending on the situation, it might mean the literal standard deviation, or a wider confidence interval.

In graphs, this uncertainty is shown as "error flags". Boxplots can also be a good way to do this if your variable is a median.

2.2. Resolution Versus Accuracy

If you read a value from the display of a digital measuring instrument, you might assume that it has a maximal deviation of half of the least significant digit. For example: 123.4 g on a weighing scale should mean that the real mass of the object is between 123.35 and 123.45 g. But... that is only is the instrument were perfectly calibrated. A resolution or precision (smallest measurable difference) of 0.1 g doesn't mean that the accuracy is the same. The latter is usually worse.

Measurements themselves are done with analog sensors that produce a voltage within a continuous range, which is then turned into a number using an Analog to Digital Converter (ADC). If the ADC has 8 bits, the result will be a number between 0 and 255 (2^8 different values). If it has 14 bits, the result can have $2^{14} = 16384$ different values, which is a lot better. Anyhow, the analog measurement has to be rounded, and that causes some error. On top of that, the analog value can be disturbed by a lack of calibration (can be solved if you have money or time), non-linearity, or nonideality of the sensor, the influence of temperature, humidity, pressure, nearby electric or magnetic fields, wind, vibrations, age, etc.

The technical specifications of most multimeters these days mention how these two add up as a percentage of the reading plus a number of times the least significant digit. Let's have a look at a typical multimeter with "6000 counts", i.e., the display can show "0000" to "5999" and a decimal point somewhere.

Accuracy Specifications		
DC millivolts	Range/resolution	600.0 mV / 0.1 mV
	Accuracy	± ([% of reading] + [counts]): 0.5% + 2
DC volts	Range/resolution	6.000 V / 0.001 V 60.00 V / 0.01 V 600.0 V / 0.1 V
	Accuracy	± ([% of reading] + [counts]): 0.5% + 2
AC millivolts[1] True RMS	Range/resolution	600.0 mV / 0.1 mV
	Accuracy	1.0% + 3 (DC, 45 Hz to 500 Hz) 2.0% + 3 (500 Hz to 1 kHz)

Fig. 2.1. Specifications of a typical multimeter.

How do you use this?

Suppose you read 5.000 V on the display, using the 0–6 V range. The resolution is 0.001 V in that case. So you can detect if the voltage goes up or down by 0.001 V.

The accuracy should be 0.5% of the reading + 2 "counts" (sometimes the word "digits" is used).

That means: $\pm (0.5/100 \cdot 5.000 + 2 \cdot 0.001) = \pm 0.027$. The percentage of the reading is obviously the most important here.

But what if the reading is 0.700 V? Then the error margin becomes $\pm (0.5/100 \cdot 0.700 + 2 \cdot 0.001) = \pm 0.0055$. Those "2 counts" have more impact now, and the actual relative error is now $0.0055/0.700 \approx 0.0079 \approx 0.8\%$. If you measure voltages below 0.6 V you should switch to the lower range (0–600 mV) if you want optimal accuracy.

If you can't go down anymore, let's say the display shows "5.0 mV" in the 0–600.0 mV range, then the error is much worse: $\pm (0.5/100 \cdot 5.0 + 2 \cdot 0.1) = \pm 0.225$ mV, that is a relative error of $0.225/5 = 0.045 = 4.5\%$.

Remark: Also with analog instruments, you can have a similar kind of troubles.

Beware: Since this error margin calculated as above consists normally mainly of a **systematic error**, much bigger than the "noise" of the instrument, it can be too big in some situations, and the resolution *can* be used as an uncertainty measure; see the capacitor case study (p. 225).

You might have experienced this with a classical thermometer with an alcohol or mercury tube: if the tube shifted by pressing it accidentally, the resolution will still be the same, but the absolute error will be several degrees. You could recalibrate it if you have melting ice at your disposition, but if not, you can still use the thermometer to measure temperature differences!

2.3. "Difficult" Quantities

Not all "precise" numbers that are published *are* precise!
Some quantities are just difficult to measure "precisely" because they are difficult to define or to quantify, the technical challenge is big, they can only be measured indirectly, or they are sensitive to subjectivity, or a combination of these factors. Some examples will make that clear, I hope:

• A well-known example is the following: How do you measure the length of a **coastline**? And what would be the error?
You might drive a stake into the ground on the water line every kilometer and then add the lengths of all the line segments between those stakes, but you could also place the stakes every hundred meters or every meter, and each time the length will get bigger. You can even measure from grain to grain of sand, and then it becomes very complicated to know exactly the total length, especially if you still have to take the tides into account!

• In the scale range of atomic particles, there are fundamental **quantum mechanical oddities**, like if you try to measure the position of a particle more precisely, you lose precision on the speed measurement (Google: "Heisenberg uncertainty principle").

• The **hardness of a substance** is measurable, but there are more than

100 different definitions and procedures to do it (scales of Mohs, Brinell, Rockwell, Vickers, Leeb, etc.), and they are not very nicely correlated!

• **How much energy can we get from an amount of food?** All the "calories" written on a package of chips or a soda, where do they come from? From calorimetric measurements, meaning, burning the dry substance and then trying to catch all the heat (which is by itself quite imprecise)? And who says how much of this energy is actually "used" by the body? A lot has to be "assumed" to produce those numbers!

The following is a similar question: "**How much energy does a living being consume?**". It requires a complicated setup to make an estimation here, and of course, it makes a difference if the animal was sleeping or moving a lot during the experiment! See more in the case study about this (p. 309).

• What is or was "**the average temperature**" on the planet? First, you have to define what you mean: air, ground, or sea? How deep, how high? Nowadays, we have satellites that can measure the whole earth simultaneously, but a 100 years ago, we only had a number of measuring stations, mostly on the northern hemisphere and not really randomly distributed. And if we want to know the temperature from a 1000 years ago, we can only deduce it from indirect measurements such as tree rings, pollen in glacier ice, or corals that require complicated calibrated models!

• All teachers and professors know the following: to judge if a student can pass, you have to do a few measurements, called **exams** and assignments. What is measured exactly? Does the teacher focus on memorization, on problem solving, on theoretical concepts, etc.? Did the student have a good or a bad day? Was the teacher very harsh or forgiving for small mistakes? The same student might have very different grades for the same subject, depending on the teacher, the books used, the circumstances, etc. So, school results are never "absolute numbers"! The same reasoning can be made for any cognitive test, like IQ tests, for example.

• What is "**the value**" of a house, a painting, or in fact, anything that **can be traded**? I don't have to tell you how subjective this is! You might define it as the money that was finally offered to buy it, but then the value of unsold items remains a bit like the situation of Schrödinger's cat! And how about the value of a human life? The government, the insurance company, and the victims probably all have a different opinion about that!

• On social media, you often see things like "Your post/advertisement had xxx **views**". Well, how do they actually know that? If someone scrolled over it for 10 milliseconds, do they count it as "viewed"? Maybe it was on your screen for half an hour but you were doing something else? If it was an advertisement, the platform has every reason to lie and exaggerate!

• How do you count the **number of ants** in a nest? Without destroying it please...

• Organizations like the World Health Organization (WHO) produce numbers like "4.2 million **premature deaths** worldwide per year in 2019", see:
https://www.who.int/news-room/fact-sheets/detail/ambient-(outdoor)-air-quality-and-health.
Now, ask yourself the following:
(1) What do they even mean by that? When does a death qualify as "premature"? Below the average or the median? If so, half of the population qualifies.
(2) How on earth can they count this? How do they know you died of air pollution? This is only obvious in very specific cases.
It's very clear that these numbers are made up from models and hence from very questionable assumptions!
(3) Suppose they did have "the correct" numbers: does a 90-year-old who dies of pollution count as much as a baby? The quantity "premature death" actually means totally nothing and it is utterly misleading and only to be used for political purposes. The only statistic that would be meaningful here, would be the number of **qualitative life years lost** ("qaly"), which is also not a very precise quantity, but at least it is more meaningful. See, e.g.,
https://en.wikipedia.org/wiki/Quality-adjusted_life_year.

• What is the **age of a fossil, a rock, a star, or the universe**? Methods exist, but they are all indirect and based on many assumptions. It might very well be that in 50 years from now, the universe is estimated a few billion years less or more.

• How are **character traits** measured? Suppose you want to know if someone is introvert or extravert. How do you even define that, precisely, independent of age, culture, education, etc.? Then, how do you measure it? With questionnaires that might be misunderstood or filled out by a not so honest person? Or should you ask friends and family how they see this person? See case study (p. 363).

• How do you quantify the **relative importance of a city**? By the number of inhabitants, or by the traffic going in and out? Maybe, but London is probably not very "important" in the life of a farmer in the Carpathians. By the surface? Then Rovaniemi (Finland) would be the most important place in Europe (check it out!). Or maybe, by the distance the first road signs pointing to this place show up? Using that criterion, Narvik (Norway) must be quite important, since just north of Bergen, there is a sign saying "Narvik 900km". Narvik *is* indeed one of the most important places in Northern Scandinavia, even if it has only 10,000 inhabitants.

• How do you quantify "**happiness**" or "**health**"? Some questionnaires ask people to give themselves a rating of 0–10 (a so-called ladder score), which is certainly meaningful, but still quite "fuzzy". A psychologist or a medical doctor might also judge differently. See also the case study about happiness and income (p. 347).

• How do you quantify "**overweight**"? Using the body mass index (BMI), the corpulence index (CI), the body roundness index (BRI), etc.? See the case study about the BMI (p. 293).

• How do you quantify the **quality of a road connection** between A and B? By taking the ratio (R) of the road distance and the distance as the crow flies? In that case, the connection between the center of Brussel and the center of Oostende (Belgium) is quite good ($R \approx 112/107.4 \approx 1.043$) and that between the Swiss village of Binn and the Italian Goglio is very bad ($R \approx 108/9.82 \approx 10.998$) since you need to make a huge detour to go around the mountains that separate both places. But you might also use time or cost as a criterion of course.

• How do you quantify the **quality of a song, a movie, a wine, a restaurant, etc.**? That's highly subjective of course, but if we want "a number", should we let the masses decide, like it is done on www.ratingraph.com or www.imdb.com for example? Or should we let "experts" decide? Should Madonna or Elvis have a higher rating than

Bach or Mozart? See the case study about wine ratings (p. 368) and the Eurovision song contest (p. 341)!

● How reliable are **"cause of death" numbers**? During the COVID-19 crisis (2020–2022), many people were counted as "COVID victims" while they were weakened by some other disease and then diagnosed to have COVID (with an error-prone PCR test) in their last days, because some industries wanted the numbers to be as big as possible. If you need a measure of the severity of an epidemic, it's probably more reliable to use the so-called **"excess mortality"**. But even that is also an estimation based upon models that calculate the "normal" mortality for each season. And which data do you use to calibrate this model? The last 5 years? The last 20 years? Are demographic changes taken into account? You see, there is a lot of room to "stretch" this "excess" according to your preferences...

● How reliable are **"adverse effect from medication"** numbers? Many complaints from people after having a medicine or a vaccine are not registered, because usually doctors are not obliged to do so, or people don't report it to their doctor. Of course? not all complaints have to be causally related to this medicine, but if they occur to an otherwise healthy person, they probably are. So, such numbers are often too optimistic.

Why am I giving these examples? I just want you to be cautious, when-ever you do statistics and modeling: **don't assume that every number you get is a solid rock you can build on**, especially if there are stakeholders who favor higher or lower numbers! Always ask yourself the following: "How was it measured?", "What were the assumptions and the conditions and the instruments?", "Would anyone benefit from twisting these numbers?".

2.4. Disturbance by Measurements

The act of doing a measurement might influence the quantity you are trying to measure, not only in quantum mechanics. Following are some examples:

● Every electronic engineer or hobbyist with practical experience knows that **measuring a voltage** over a component in a circuit can lower the voltage and change the behavior because of the (too low) input impedance of the voltmeter. Even approaching a high frequency circuit can disturb it seriously because your body works as an antenna!

● More general, if you measure any kind of **field strength**, be it a gravitational, an electric, or a magnetic field, you have to put a small mass, or charge, or magnet sensitive object in the field, so you can measure the force upon that object. But doing that, you always influence the field, in the best case just a little bit.

● But also in life science, you have to be aware that just you being there as an **observer** can influence your subjects: you can't expect that animals in captivity act exactly the same as in the wild. And I guess human behavior observed in a "reality show" might differ from "reality" when no one is there.
If you interview people, they might give "desirable" answers. Or, if there is anything "suggestive" or "intimidating" in your questions, the answers might very well be changed. So, always ask yourself: do I actually measure what I am supposed to measure?

● If you have an RBC nuclear scan done, a small amount of **marker** radioactive material is injected in your blood to track your red blood cells. There is always a risk this may cause side effects.

● If you measure a **pH** with electrodes, you always cause some ionization in the fluid, which can have a small influence on the result.

My point is to always think about the possibility that your measurement can be a disturbing factor in the system you study.
Interesting to Google: "nondestructive testing".

2.5. Error Propagation

What happens when we do some operations with measurements that have uncertainties?
Why do we need to bother about this anyway?

● Well, suppose you need to know how much you used from a certain fluid, and you see 78 ml of it in your measuring jug at the start of your experiment, and 33 ml at the end. So the answer is 78 − 33 = 45 ml. But what is the precision of this result, if all the measurements are rounded to 1 ml?

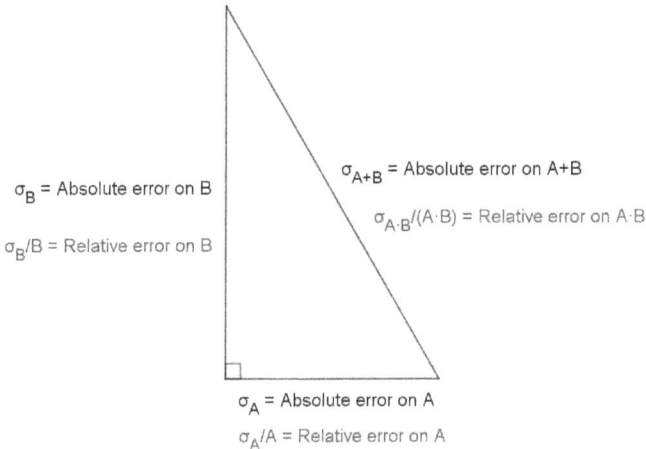

σ_B = Absolute error on B

σ_B/B = Relative error on B

σ_{A+B} = Absolute error on A+B

$\sigma_{A\cdot B}/(A\cdot B)$ = Relative error on A·B

σ_A = Absolute error on A

σ_A/A = Relative error on A

Fig. 2.2. Errors on sums and products can be calculated with the Pythagorean theorem.

The reading "78" means, the real value could be between 77.5 and 78.5; and "33" means: between 32.5 and 33.5. So, the result could be 44 or 46 if we have bad luck. Statistically, the probability that the "worst cases" happen is small. From the theory of the normal distribution, it can be proven that the errors from two independent results add like "orthogonal" vectors if you calculate their sum or difference. In other words: the error on the sum $a + b$ is the diagonal of the rectangle with sides σ_a and σ_b, which is calculated using the theorem of Pythagoras.

● Suppose you measured temperatures in °F, precise to 0.5 °F, but you need to convert them to °C. What is the precision then?

$T_C = (T_F - 32)/1.8$, so a change of 1 °F means a change of $1/1.8$ °C; an error of 0.5 °F is an error of $0.5/1.8 = 0.27777...$°C. So, 80 ± 0.5 °F becomes: 26.67 ± 0.28 °C.
(Usually, the error is rounded to two significant digits, since it doesn't make much sense to show more.) You should never write 26.266666667 since that could give the wrong impression that the temperature was measured with extreme precision.

• In general, if you do some operations with a number x, say you calculate $y = f(x)$, a small change in x (Δx) will usually cause a small change in y (Δy). You can know how much, for each specific x value, by calculating $f(x + \Delta x)$ and $f(x - \Delta x)$, but it can be estimated generally using the derivative: $\Delta y \approx f'(x)\Delta x$, so $\sigma_y \approx |f'(x)|\sigma_x$ because we only need the absolute value of the error.

Example: You measure an angle α, precise up to 1°, but you need $y = \tan(\alpha)$.

$$\sigma_y = \frac{d \tan \alpha}{d\alpha} \sigma_a = \frac{\sigma_\alpha}{\cos^2 \alpha}$$

Attention: This is if α is in radians; if it is in degrees,

$$\sigma_y = \frac{\sigma_\alpha}{\cos^2 \alpha} \cdot \frac{\pi}{180}$$

So, if $\alpha = 10 \pm 1°$, $y = \tan(10°) \pm \pi/180/\cos^2(10°) \approx 0.176 \pm 0.018$. You can check that $(\tan(11°) - \tan(9°))/2$ is also 0.018.
But if $\alpha = 85 \pm 1°$, $y \approx 11.4 \pm 2.3$. The same uncertainty in α causes a much bigger uncertainty in y now!

• If there are several quantities (x_1, x_2,...) influencing the quantity you need (y), σ_y becomes

$$\sigma_y = \sqrt{\left(\frac{\partial y}{\partial x_1} \cdot \sigma_{x_1} \right)^2 + \left(\frac{\partial y}{\partial x_2} \cdot \sigma_{x_2} \right)^2 + ...}$$

We use the partial derivatives and the "Pythagorean" sum here since we assume the errors are independent. (Calculating the partial derivative $\partial(\ )/\partial x$ is the same as calculating the normal derivative $d(\ )/dx$ assuming all the other variables in the formula are constants.) Exercise: use this to prove that the relative errors on a product add up like the absolute errors on a sum.

Example: Suppose you want to find the relationship between the angle of ascent of a road and your walking speed. So you need to find the angle (α) first by measuring the road distance of the route (d), using the odometer of a bike for example, and the height difference (h), using a topographic map or a GPS. Then $\alpha = f(h,d) = \text{Arcsin}(h/d)$.

An error in h (σ_h) will cause an error in α (in degrees):

$$\sigma_{\alpha,1} = \frac{\partial \alpha}{\partial h} \cdot \sigma_h = \frac{180°}{\pi} \cdot \frac{1}{\sqrt{1 - \left(\frac{h}{d}\right)^2}} \cdot \frac{1}{d} \cdot \sigma_h$$

You could also estimate this numerically as
$\sigma_{\alpha,1} \approx (f(h + \sigma_h, d) - f(h - \sigma_h, d))/2$.

An error in d (σ_d) will cause an error in α:

$$\sigma_{\alpha,2} = \frac{\partial \alpha}{\partial d} \cdot \sigma_d = \frac{180°}{\pi} \cdot \frac{1}{\sqrt{1 - \left(\frac{h}{d}\right)^2}} \cdot \left(-\frac{h}{d^2}\right) \cdot \sigma_d$$

which you could also estimate numerically as
$\sigma_{\alpha,2} \approx (f(h, d + \sigma_d) - f(h, d - \sigma_d))/2$.

So, the total error becomes:

$$\sigma_\alpha = \sqrt{\sigma_{\alpha,1}^2 + \sigma_{\alpha,2}^2}$$

By the way, in these days, you might also just use an app (like "Angle Meter" for Android) on your smartphone to measure the angle directly...

This knowledge will come in handy if you would do nonlinear transformations of your data; see "Why non-linear transformations should be avoided", p. 145.

More examples can be found in this text by Richard Daley,
https://foothill.edu/psme/cascarano/images/error_propagation_summary_rd.pdf.

Also recommended: John R. Taylor (1997), *An introduction to Error Analysis*, 2nd ed., University Science Books, Sausalito, CA, USA.

2.6. Minimizing Errors

2.6.1. Some obvious things...

Bad measurements cause bad conclusions, so of course you should use **good instruments and quality samples**, whenever possible, but that might be costly, I know. If you are on a budget, make the best use of what you have, using common sense. For example, if you have doubts about your cheap instrument, try and compare it with a better one whenever you have a chance, and use it to correct yours. Or measure some known reference quantities to (re)**calibrate** it.

Try to **stick to the same correct procedure** for all your measurements. Just a few well-known but sometimes forgotten examples are as follows:

*   Measure temperatures in the shade.
*   Read the value from an analog instrument perpendicular to the needle.
*   Read the level of a liquid in a graduated cylinder at eye level and at the center of the "meniscus" (the curve of the surface, produced by surface tension); those cylinders are calibrated taking this curvature into account (at least for water).
*   Do the necessary cleaning and maintenance of instruments.
*   In human sciences, ask your questions as neutrally and clearly as possible, to avoid bias and noise in your data from people who misunderstood them!

Fig. 2.3. The meniscus on a liquid surface. 20.0 is the correct reading.
Photo: PRHaney, Wikimedia Commons.

If you want to analyze the evolution of a quantity over time (a so-called **time series analysis**), and you want to be precise, please don't use months or years as a time unit, since they have varying lengths! Instead, use the number of days starting from a certain reference day (like the "Julian day", something like the "stardate" in Star Trek, but more serious). Spreadsheet programs have functions to do this calculation without effort. If you have to convert days to years, remember that 1 year is not 365 days, but 365.25 days in a good approximation.

If you have to collect **data from different sources**, measured with different instruments, etc., that's no problem, but make sure you know their precisions, so you can give all the data the appropriate weights if you want to create a model from them. That's another reason why measurement errors are so important.

In the following paragraphs, I give you some tips for better measurements without a higher cost. They might not be applicable literally in your field, but they are meant as examples to cultivate good habits.

2.6.2. Example: Height of a building

Suppose you want to determine the height of a building (x) by measuring the length of its shadow (s). If you know the length of the shadow (r) of a known vertically standing stick with a height of 1 m, then $x/s = 1/r$, so $x = s/r$, assuming the surface is horizontal and the building is not the tower of Pisa.

Now, we can ask ourselves, does it matter whether you do this measurement while the sun is high or low?

The relative uncertainty on x depends on those of both r and s. The absolute errors stay the same if you measure distances with a tape measure (e.g., 1 cm or 1 mm), which means that the relative errors are smaller when r and s are bigger. So, yes, you will have a higher precision if you measure when the sun is low!

Now, you want to determine the same height, but by measuring the angle (α) between the surface and observation direction from a distance d. That angle varies from almost 90° (observing close to the building) to almost 0° (far away). The height can easily be calculated as:

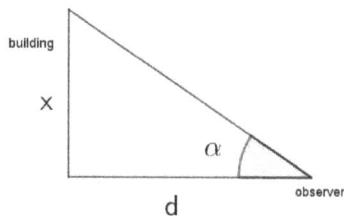

Fig. 2.4. Measuring d and α to calculate x.

$$x = d \cdot \tan(\alpha).$$

Now again, does it matter if we do this measurement from close or from far? Let's assume the absolute errors on α and d stay the same (e.g., $\sigma_\alpha \approx 1°$ and $\sigma_d \approx 1$ cm).

If we calculate the error propagation:

$$\sigma_x^2 = \left(\frac{\partial x}{\partial d}\sigma_d\right)^2 + \left(\frac{\partial x}{\partial \alpha}\sigma_\alpha\right)^2 = \left(\tan(\alpha)\cdot\sigma_d\right)^2 + \left(\frac{d}{\cos^2(\alpha)}\cdot\sigma_\alpha\right)^2$$

we see that the first term goes to infinity when α approaches 90°; the second one does that when α approaches 0°!

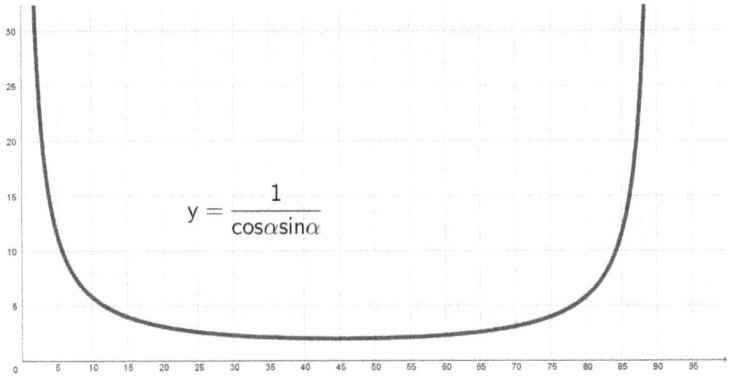

$$y = \frac{1}{\cos\alpha\sin\alpha}$$

Fig. 2.5. Variation of σ_x/x vs α.

The relative error (squared) becomes

$$\left(\frac{\sigma_x}{x}\right)^2 = \left(\frac{\sigma_d}{d}\right)^2 + \left(\frac{\sigma_a}{\cos(\alpha)\sin(\alpha)}\right)^2$$

Now, the first term is not bad as long as we stay away at least a few centimeters from the building, but the second becomes catastrophic when α is near 0 or 90°, and it is minimal when $\alpha = 45°$. So, the answer is now: if possible, measure from a position where $\alpha \approx 45°$ for optimal precision! (Somewhere between 20 and 70° looks fine according to the graph.)

2.6.3. Example: Measuring the density of a fluid

Suppose we have a liquid whose density we would like to know, and we have a measuring cup with an accuracy of ± 1 ml and a scale with an accuracy of ± 1 g.

Okay, we put the empty measuring cup on the scale, press the "tare" button to set it to zero, and then we pour 11 ml (= 11 cm³) of the liquid into it. The scale says: 13 g. This means that the density ρ = 13/11 = 1.181818... g/cm³ ("ρ" = the lowercase Greek letter "rho", not to be confused with "p").

Those many digits after the decimal point are of course not all reliable. At school (at least in my area), the pupils learn some rules of thumb for the "correct" rounding: "when multiplying or dividing two measurements, you should retain the smallest number of significant digits of both". In this case, you would be allowed to have two digits left. So we would have to write: ρ = 1.2 ± 0.1 g/cm³.

Now, suppose we repeat the measurement, using the same measuring cup and scale, but this time we fill the cup a little fuller. We now see that 77 ml weighs 91 g. The calculation of the density gives the same result: ρ = 91/77 = 1.181818... g/cm³.
According to our rule of thumb, we should round to two significant figures again, because the numerator and denominator also have two, so again ρ = 1.2 ± 0.1 g/cm³.

Someone with a good scientific intuition will now have an uncomfortable feeling. The second measurement should actually be more accurate, because the *relative* error on both measurements (the inaccuracy of the measurement in relation to the measured value) is much smaller: 1/77 ≈ 1.3% on the volume and 1/91 ≈ 1.1% on the mass, while with the first measurement, they were respectively 1/11 ≈ 9.1% and 1/13 ≈ 7.7%! So, there is something wrong with that rule of thumb.

How do we get an idea of the true accuracy?
In measurement 1, in case of the worst reading errors (denominator larger and numerator smaller or vice versa), we could obtain a density of 14/10 = 1.4 or 12/12 = 1.0, which means that ρ = 1.2 ± 0.2 g/cm³.
For measurement 2, this becomes 92/76 = 1.2105..., or 90/78 = 1.1538..., so ρ = 1.18 ± 0.03 g/cm³.

Our rule of thumb gave the same accuracy for the density in both cases: 0.1 instead of 0.2 and 0.03 respectively, which was too optimistic in the first measurement and much too pessimistic in the second.
If we use the statistical error propagation rules, we get a more realistic estimation. For the first measurement,

$$\frac{\sigma_\rho}{\rho} = \sqrt{\left(\frac{\sigma_V}{V}\right)^2 + \left(\frac{\sigma_m}{m}\right)^2} = \sqrt{\left(\frac{1}{11}\right)^2 + \left(\frac{1}{13}\right)^2} = 0.1190... \approx 11.9\%$$

and for the second,

$$\frac{\sigma_\rho}{\rho} = \sqrt{\left(\frac{1}{77}\right)^2 + \left(\frac{1}{91}\right)^2} = 0.01701... \approx 1.7\%$$

This gives us the *absolute* error for the first measurement: $\sigma_\rho \approx 1.1818 \cdot 0.1190 \approx 0.14$ g/cm³, and for the second, $\sigma_\rho \approx 1.1818 \cdot 0.01701 \approx 0.020$ g/cm³, which is a serious difference!

One usually writes (with two significant figures for the error):
in case 1, $\sigma_\rho = 1.18 \pm 0.14$ g/cm³, and
in case 2, $\sigma_\rho = 1.182 \pm 0.020$ g/cm³.
We see that the statistically expected errors are indeed smaller than those of the previously estimated "worst case scenarios".

The lesson to be learned is that by using the same material, sometimes you can obtain more precise results if you just use it wisely.

2.6.4. Example: The Wheatstone bridge

Suppose you want to measure some quantity that can be translated into an electrical voltage. Many kinds of sensors exist to do this, for example, light dependent resistors (LDR), temperature dependent resistors with positive and negative correlation (PTC & NTC), or strain gauges to measure the strain on an object.

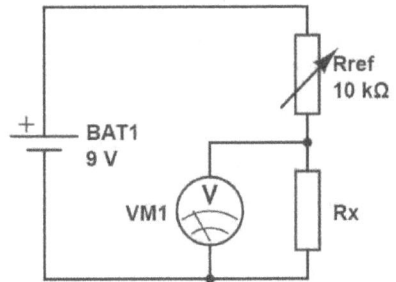

Fig. 2.6. Simple measurement of the voltage over an unknown resistor (sensor).

You could simply connect the sensor in series with another resistor, connect both with a power source, and measure the voltage over the sensor to detect changes in the quantity you want to measure. A problem is that the changes you want to detect may be very small.
For example, suppose a PTC has a resistance of 5 kΩ at a certain ref-

erence temperature. If you put it in series with a fixed 5 kΩ resistor, and add a 9 V battery, the voltage over the PTC will be 4.5 V. Now suppose the temperature rises and the resistance of the PTC rises to 5.1 kΩ. The voltage will now be 5.1/10.1·9 V $\approx$ 4.5445 V; that is a small relative rise of about 1%.

Now, add two other resistors as shown in the circuit (Fig. 2.7) and put the voltmeter as a bridge between the two sides. The same small rise in resistance will now cause a huge relative rise from "0" (= unmeasurably low) to 0.0445 V. The same voltmeter can now be set to a smaller range, e.g., 0–200 mV instead of 0–20.0 V and detect very small differences in R_x! The extra cost of two resistors is negligible.

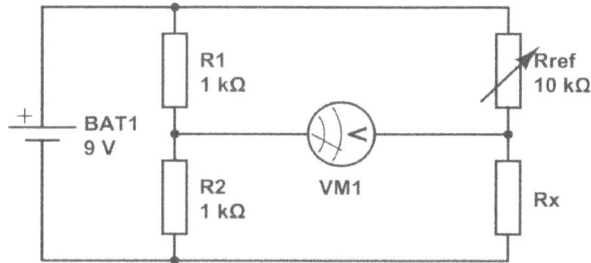

Fig. 2.7. A Wheatstone bridge. R_x can be a sensor.

Another advantage is that the calibration of the zero point is independent of the battery voltage!

The circuit is named after Sir Charles Wheatstone (1802–1875).

Completely analogous is the use of a *scale* based on weight versus a *balance* and a set of calibrated masses, to determine the mass of an object. A balance is very sensitive to small differences and works with unchanged accuracy if it moves to a place with a different gravitational field strength. A torsion balance, like the one Henry Cavendish (1731–1810) used to determine the strength of the gravitational force between two objects, is extremely sensitive; see any college physics textbook, or en.wikipedia.org/wiki/Cavendish_experiment.

2.6.5. In human sciences...

Questionnaires produce lots of garbage data. People sometimes don't really know what to answer or what they are "expected" to answer (not many people would dare to say on a test that they are "racist" or "virgin" or "pedophile", for example, especially if the test is not anonymous); the mood of the moment plays a role, etc.

For example, if the Belgian traffic institute VIAS concludes from a survey (May 2021) that 61% of Belgians "sometimes" forget to stop at a stop sign, while the European average is 43%, does that mean that this is really the case, or that Belgians are more honest, or that they understand the word "sometimes" differently than in other languages?

Another example is from a survey by the University of Antwerp regarding COVID-19 (May 2021): "Do you find scientific experts (e.g., virologists) in the media (e.g., TV, newspapers, news websites,...) reliable?". Well, what do they actually measure with this? What if you find certain experts reliable but not those on TV? Such questions are therefore completely worthless.

The **unambiguity of questions** is really a prerequisite to produce valuable data. For example, you might ask "Are you against abortion, yes/no?". The respondent might have an answer in mind like "Yes, but..." or "No, but...", so what should he answer? The answer is as good as worthless. You might improve the question by using a so-called Likert scale (named after the American psychologist Rensis Likert, 1903–1981) like "yes!/rather yes/don't know/rather no/no!", but that still doesn't give you much of a clue. You might ask more specific questions such as: "Does your religion prohibit abortion?". But then again, the respondent might answer "yes" and yet he might disagree with that point in his religion.
If you want reliable answers, you need to ask clear unambiguous questions, such as: "Are you always against abortion, because your religion tells you so?" or "Would you allow abortion if the life of the mother is in danger?", or "Would you allow abortion before week 10 if the woman was pregnant because of rape?".

In April 2024, in an election survey from the Belgian newspaper *De Standaard*, they asked "Hunting should be allowed only for wildlife management; agree/don't agree?". When you answer "don't agree", do you mean that hunting should not be allowed at all, or that it should be allowed also just "for fun"?

A last example is a survey from the university of Ghent for math teachers in February 2024. They asked: "How often do you use a graphic calculator in your classes? Never/a few times per year/a few times per month/weekly/multiple times per week/every day/multiple times daily".

Now, what do they want to know? What should I answer when I use it in half of my classes but I only teach math 2 days per week? Do they want to know what percentage of the time I use it, or do they want absolute numbers? Why would they want absolute numbers?

It would be better to ask the following: "In which percentage of your teaching hours do you use...?". The answers will still be a rough estimation of course, but at least you won't cause noise by confusing the respondents.

In my opinion, the American organization *PEW Research Center* does a good job: they usually ask very clear questions, and in their reports, they show exactly which questions they asked to whom and the percentages of the different answers, so that the reader can interpret for himself.

See, www.pewresearch.org.

Another reason to question questionnaires is the following: *who answered them?* Is the sample representative? And most of all, are all the answers real? You don't want to know how many surveys, done by some students who wanted to earn some extra money, were "creatively completed" by those students themselves! I've witnessed it!

2.6.6. Avoid information loss

Suppose you want to collect data to answer the question "Do bigger men have bigger sons?".

The simple option would be to separate the fathers in two groups ("big" & "small"), calculate both averages and standard deviations of the heights and those of their sons, and do some statistical test (e.g., Student-t test) to compare them. This will give you some answer, and a "significance level", but you wasted a lot of information. And the significance will depend on the quite arbitrary choice of "What is big?".

With the original raw data (x = father's height, y = son's height) you could as well do a regression analysis to see a more detailed pattern; see the next chapter!

Similar: to answer the question "Do city people vote more for party X?", you could divide your test persons in two groups (city/rural), or on the other hand you could plot the percentage of X voters versus the number of inhabitants of the place where they live. Or, you might make a plot with x = number of inhabitants, y = 1 if the person voted for X, and y = 0 if not. Maybe you'll see a pattern, or regression will help you to see it.

If you are gathering data related to people's political preferences, don't ask: "For which party will you vote?". Better let them give a rating in a Likert scale for each party, like "never", "probably not", "don't know", "likely", "certainly", but then you still have a problem to quantify those into numbers. You could use a rating like 0..10 to avoid that though. That way you get much more information for the same price and effort!

Suppose you want to find out how some property (health or whatever) is related to age. Very often, data from the test persons are then put together in **bins**, e.g., age 0–10, 10–20, 20–30, etc. I don't know why they do that, since you lose a lot of precious information. There might be a big change between zero and ten, for example. If you want to see the pattern, you would have to replace each bin by the average x value of each bin: $x_1 = 5 \pm 5$, $x_2 = 15 \pm 5$, $x_3 = 25 \pm 5$, etc. In this case, ± 5 means a confidence interval of about 100%. The y values would become the bin averages $\pm$ their standard deviations.

Especially, bins like "80+" are just unusable since you don't know the middle of the class! So, if possible, use the raw data!

For the study of distributions, unfortunately, bins are necessary, unless you use cumulative distributions; see further (p. 144).

3

Finding the Pattern

Finding patterns is like seeing things in clouds.

Patterns can be captured in numbers and formulas in different ways.

Regression is a powerful tool but it has to be used wisely.

Besides the traditional regression, there is a new method!

Monte Carlo is not just the casino part of Monaco!

3.1. Some Rudimentary Tools

3.1.1. Pearson r and Kendall tau

So, we have collected some data $(x_i, y_i)_{i=1...n}$, that can be visualized in a scatterplot.

Before we search for a very specific pattern in our data, we might want to get a global impression related to the shape of the cloud of data points, and we would like that impression solidified in one characterizing number.

Well, three such numbers are often used: the so-called "Pearson r", "Kendall τ" (tau), and "Spearman ρ" (rho). All three are designed to be between -1 and 1.

The first can be calculated right away:

$$r := \frac{\sum \left(x_i - \bar{x} \right)\left(y_i - \bar{y} \right)}{\sqrt{\sum \left(x_i - \bar{x} \right)^2} \sqrt{\sum \left(y_i - \bar{y} \right)^2}}$$

(The stripes above the letters signify the averages, and the sums are over all the measurements of course.)

If you try with some values, you will soon find out that $|r| \ll 1$ if the points are scattered out in all directions around one central point. The more the cloud looks like a long cigar, the closer the $|r|$ comes to 1. A perfectly ascending aligned set of points will give $r = 1$, and a perfectly descending one: $r = -1$. A catastrophe occurs when they are perfectly aligned horizontally, oops! A flaw in the design, but okay... There are more reasons why r can't always be trusted very well... Google, for example, "Anscombe's quartet" to see what I mean!

Now, there might be a nice monotonous pattern in your set, not forming a line, but more something like an exponential or a quadratic relationship for example. The number that will give a good score for this, is Kendall τ.

The way it is calculated is actually very simple: connect all the dots with lines; for each *ascending* line, add 1; for each *descending* line, subtract 1; if a line is vertical or horizontal, do nothing. Divide the sum by the total number of lines, which is $n(n-1)/2$, and that's it.

In the set shown in Fig. 3.1, for example, there are 11 descending (red) lines, 2 ascending (green), and 2 that don't count (black). There are 15 lines in total, so $\tau = (2-11)/15 = -0.6$. This number says there is a general tendency of "going down" in the set, but it's not perfect.

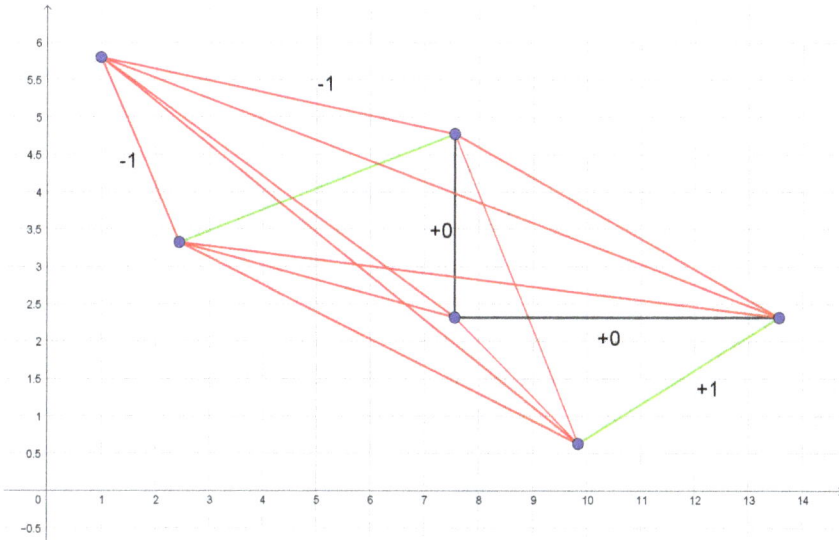

Fig. 3.1. Visualization of the Kendall τ calculation: green = ascending, red = descending.

It can also be put in a formula:

$$\tau = \frac{2}{n(n-1)} \sum_{i=1}^{n-1} \sum_{j=i+1}^{n} \text{sgn}\big((x_i - x_j)(y_i - y_j)\big)$$

with "sgn": the signum function (-1, 0, or 1 if the argument is negative, zero or positive).
(This is the simplest and most logical version; there are other versions for people who like to complicate things.) See, e.g., https://en.wikipedia.org/wiki/Kendall_rank_correlation_coefficient.

A value of $\tau = 1$ means, if x increases, y increases too, always. If $0 < \tau < 1$, it means the same but not always. If $\tau < 0$, y tends to decrease when x increases.

Spearman ρ is a bit similar to τ, also based on ranking. If you want to know more details,... literally tons of books have been written about these numbers, so you'll find more than you need using Mr. Google.

For example, this nice presentation "Rank Correlations: Spearman *r* and Kendall *τ* (FRM T5-06)" on the "Bionic Turtle" channel should give you some insight: https://www.youtube.com/watch?v=g DNmhEBZAO8

The main thing to remember is that *r* is **a measure for linearity and *τ* for monotony**. They are both positive if there is an ascending trend in the data, and negative when it's descending.

Remark: There are rules to estimate the probability to have a certain *r* or *τ*, the so-beloved "p values", but I would take those with a very serious grain of salt! One of the reasons is that they don't take into account any uncertainty in the data. Consider r and τ as useful indicative values, not much more.

3.1.2. Moving averages

If you have "noisy" data, going up and down, without a clear pattern, it might be useful to somehow "press the data together" in one curve to see a pattern, especially for time series of variables that may be very complex like cloud coverage or prices.

Many filtering and smoothing techniques have been invented to do this. I'll just show one that is very basic, and that's why I like it.

It's a kind of "moving average" graph that is calculated as follows:

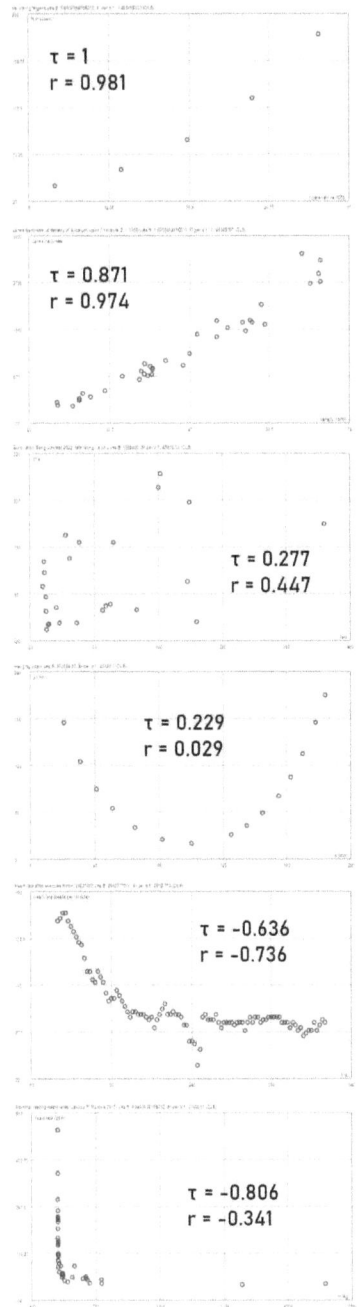

Fig. 3.2. Examples of *τ* and *r*. The last one is interesting: most points are descending (good *τ*) but they are not aligned (bad *r*).

- First, you choose the number of points you want to average; this can be any number between 1 and n. Let's call it m.
- Sort the data by increasing x-value.
- Take the first m data points and calculate the weighted average of their x and y values; that is the first dot.
- Now, do the same with data points 2 to $m + 1$ to get the second dot, then with points 3 to $m + 2$, etc. until you reach the end of your data set. You should have $n - m + 1$ dots now.
- Connect all the calculated dots with lines, and you'll see the moving average pattern.

If you choose $m = 1$, you will just see all your data points connected. If you choose $m = n$, you will see just one dot, showing the "center of gravity" of your data.

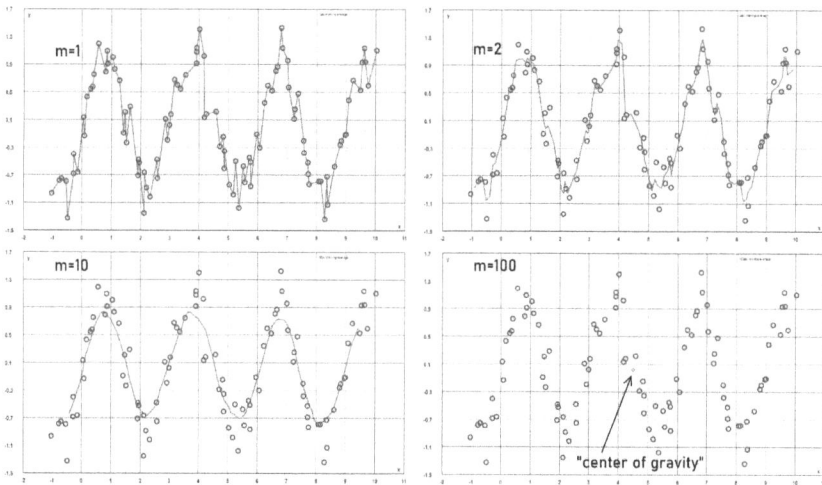

Fig. 3.3. Moving average graphs (100 data points).

Remark: If it doesn't matter whether you choose x or y as "independent" variable, you might switch them and follow the same procedure, and compare the results, see further.

3.2. Regression Analysis – What is it?

3.2.1. Connecting the dots, or flowing between the dots?

If you ask people: "What is the next number in the sequence: 1 2 3?", most will answer: 4. Why? That is the simplest pattern they see: they assume the n'th number y_n must be n. But in fact, the question makes no sense, since you can choose *any* number you like as the next one. You just have to find the rule. For example, if you want the next number to be 0, use the following rule:

$$y_n = (n-4)\left(-\frac{2}{3}n^2 + \frac{4}{3}n - 1\right)$$

It's as valid as any other rule, just a bit more complicated (illustrated in Fig. 3.4). So, it's very important to be aware that any recognized pattern is *based on an assumption*! Of course, it's nice to look for the simplest pattern, but there is no such thing as "the" pattern. By the way, that is an objection you could make against IQ tests, because they judge you on what the designer saw as "the pattern", but you might see another valid pattern...

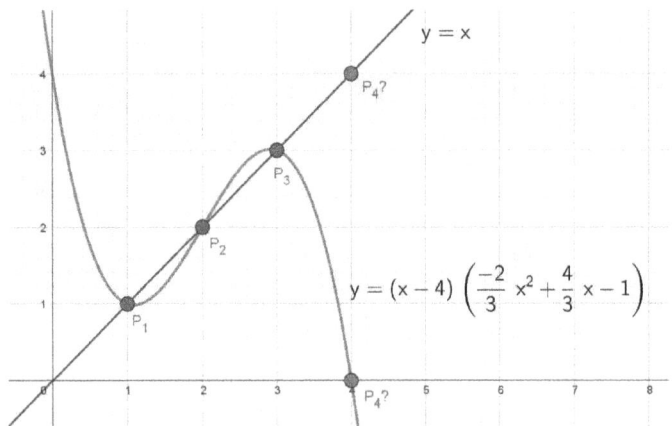

Fig. 3.4. What is the next dot? There is an infinite number of solutions!

Now, suppose we have done some *measurements*, i.e., some x and y values, for example x = time and y = temperature, or x = voltage and y = current, or whatever. The additional difficulty here is that measurements are not infinitely precise; they have a certain fuzziness.

We put them in a Cartesian coordinate system (a "scatterplot"), and now we want to draw some curve that represents the "idealized" completed pattern that connects the dots, more or less. Again, this can be done in an infinite number of ways! In Fig. 3.5, you see a few of the possibilities:

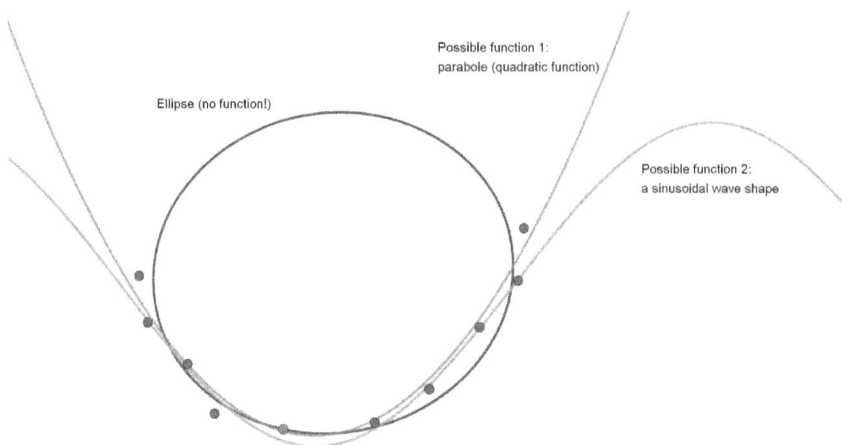

Fig. 3.5. Dots and curves that might depict "the pattern".

Which curve you choose as "the best", depends entirely on your *expectations*: you see what you want to see, just like you can sometimes recognize a cauliflower or a feather or a face in the clouds. Or, in the words of the German physicist Werner Heisenberg (1901–1976), "What we observe is not nature itself, but nature exposed to our method of questioning.".

Usually, we want to see the graph of a mathematical *function*, which means that for every x value there can only be one y value. That excludes the ellipse in the example, because that's not a function. If you expect the measurements to be a part of a repeating pattern (the temperature might go up and down every day, for example), you might choose the sine wave; if you expect to see one minimal y value, you'll choose the parabola or another function. Most of the time, you'll have some idea what to expect. Each kind of function should have some parameters in its formula that you can adjust to shift and stretch and bend the curve until it "fits". Once you decided what kind of function you *want* to see, you can go ahead and play with the parameters.

Let's say we have a good reason to assume it should be a parabola, and we try some values. Changing the numbers in the formula will make the parabola a bit different (see Fig. 3.6).

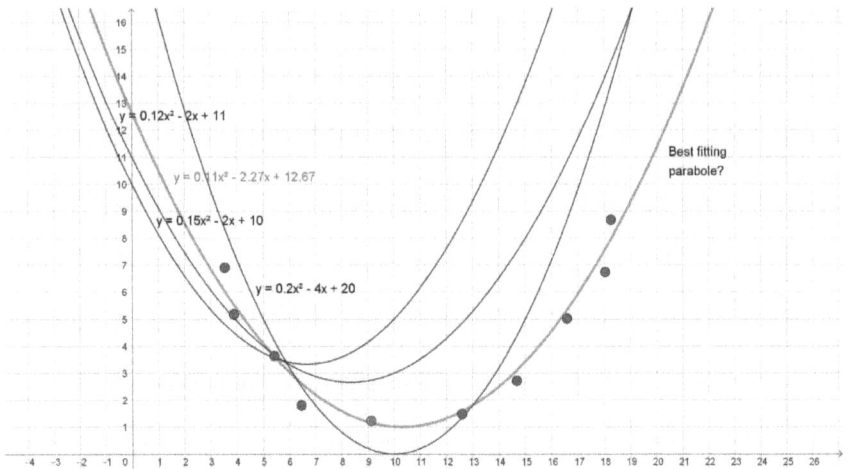

Fig. 3.6. Possible parabolas through the dots.

One of them clearly fits better than the others. We can see that with our naked eyes and our brain. Recognizing patterns is one thing our brain (and that of many mammals and birds and some other animals) is quite good at. It has given us an advantage in the evolution: those who were not able to recognize the shape of a snake or the movement of a lion in the grass quickly enough, didn't survive it.

Now, to find those exact parameter values... that seems more like a job we want to be done by a computer! But how can a piece of software "recognize" the same as we do?
Well, we'll have to calculate a number that reflects how well the points fit with the curve, some kind of "average distance" between them, and a method to minimize that number. This can be done in many ways. A prerequisite is that all the methods should give the same outcome if the points lie exactly on the curve for a specific set of parameters, of course. But that is seldom the case, since all measurements have imprecisions and functions that depict reality perfectly are also rare.

3.2.2. "Ordinary Least Squares" regression (OLS)

Traditionally, the distance between one measurement (nr. i) and the model curve $y = f(x)$ is measured vertically, as the absolute value of the difference between the measured value y_i and the "predicted" value by the model, i.e., $|y_i - f(x_i)|$.

This reflects the assumption that there is no error on the x value, only on the y. The x is supposed to be "chosen" independently, and for this x, a corresponding y value is measured. Often, x is seen as the "cause" and y as the "effect", or x is the time that unavoidably passes, and y is something that changes in time, like the percentage of cloud coverage, or the house prices, or the size of a tree, or whatever.

In many cases this assumption is not valid, we'll come back to that later, but it does make the calculations easier, which was an issue in the pre-computer days.

We could now use the average value (or the sum, let's call it S) of these vertical distances as an indication of the fit, and then fiddle with the parameters to make that number as small as possible. Zero is only possible if all the points happen to lie exactly on the curve.

There are at least two problems with that:

(1) Optimization comes down to derivation of S with respect to each parameter, and finding the zeroes of those derivatives. But if S has absolute values in it, the derivative will have discontinuities. We don't like that.

(2) Suppose, with some starting parameters, the distance of point 1 to the curve is 10, and that of point 2 is 1. If f is shifted or bent by a change of parameters and these distances now become 5 and 6, the sum of distances is still 11. Yet, we somehow feel that in the second situation, the distances are more "fairly" distributed than in the first; see Fig. 3.7.

If we only demand the sum of the distances to be minimal, the algorithm will not know how to choose between the first and the second parameter sets, since it makes no difference for S. So, we need to add another requirement: that the *distribution of all the distances be as small as possible*. That sounds like a minimal "**variance**" (or "standard deviation") but not from a point but from a curve... For a variance, the squares of distances are calculated, so that's what we should do.

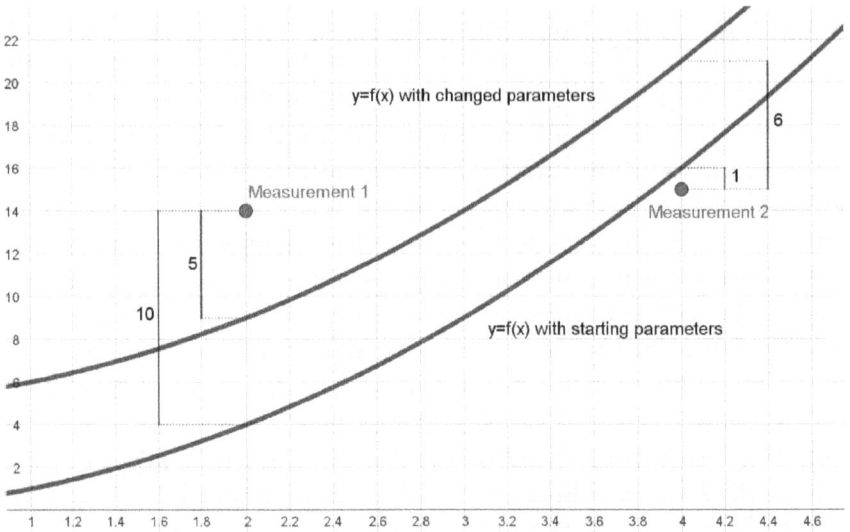

Fig. 3.7. Changing parameters causes a change in the distances between measurements and the model function.

In the so-called **"ordinary least squares regression"** (OLS), the sum (S) of the squares of the distances (also called **"residuals"**) is to be minimized (n = number of measurements):

$$S = \sum_{i=1}^{n} \left(y_i - f(x_i) \right)^2$$

This is actually n times the variance of the residuals.

To see what this means, let's look back to the above example: with the starting parameters, the sum of squares was $10^2 + 1^2 = 101$; with the changed parameters it became $5^2 + 6^2 = 61$, which is much lower. So now, the algorithm has an incentive to change the parameters toward the second situation!

Note: The variable "x" can as well be a vector of many independent variables, but in this book, we limit ourselves to the simplest case of a single one.

So, if we want function with a number of parameters in its formula, say a, b, c..., to be used as a pattern through a set of measurements, "all we have to do" is to find the parameter values that cause S to have a relative minimum (preferably the absolute minimum if possible). That implies we have to calculate the partial derivative of S with respect to each parameter and seek the zero thereof, i.e., solving

$$\frac{\partial S}{\partial a} = 0, \quad \frac{\partial S}{\partial b} = 0, \dots$$

With basic knowledge of calculus, this is a relatively easy task in the case f is linear ($f(x) = ax + b$). In that case, there is an exact solution, so you can obtain the best parameters with a simple straightforward calculation. I am leaving it as an exercise for you because you can find the solution in literally thousands of textbooks, or just by googling "linear regression".

In all other cases, the minimum of S can only be found by numerical approximation. In some cases, a non-linear model function can be "linearized" by substituting the variables with their logarithms, inverses, or by doing some other transformation. But that can cause trouble! I come back to that later.

The OLS algorithm is implemented in tons of popular software, like GeoGebra, Graphmatica, Excel, Quattro, R, Stata, SPSS, GraphPad Prism, Desmos, etc. and that of the TI-84 calculator.

I personally prefer to use a method that is universally applicable to any kind of model function: finding the minimum of S by **iteration**. You start with an initial (intelligent) "guess" for the parameters and then you change them step by step until S doesn't get significantly lower anymore. This is illustrated in the screenshots from FittingKVdm v.1.6 (Fig. 3.8).

I explain later in detail how this process works.

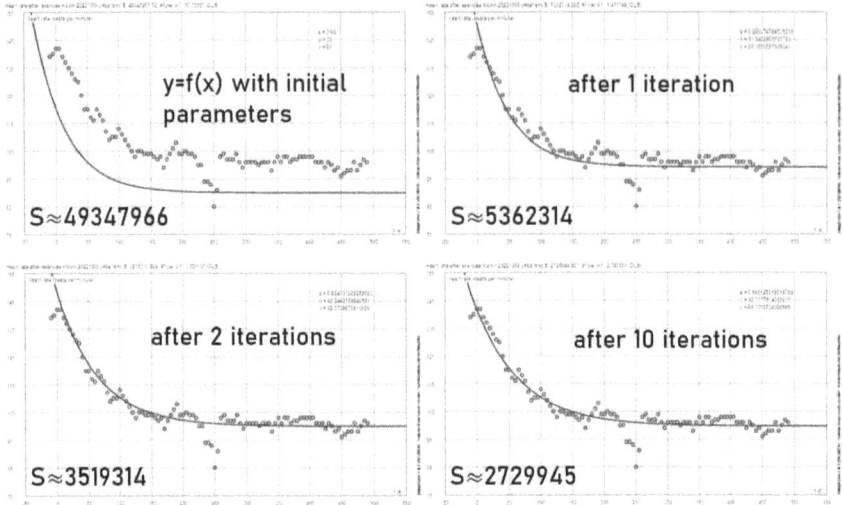

Fig. 3.8. The iteration process – an example (screenshots from FittingKVdm 1.6). (Don't try to read the small letters; just look at the dots and the curves.)

3.2.3. Weighted OLS

Suppose you want to measure the electric resistance of some component, using a voltmeter and an amperemeter. Strictly, you only need to do one measurement, but you want a more precise result, so you do one more. The first time you set your power source to $U = 7.90$ V, and you measure $I = 1.95$ mA. The second time you set $U = 9.3$ V, and you see that the current exceeds the range of 0–2 mA, so you have to switch the meter to the next range: 0–20 mA. It reads 2.2 mA.

You put the U and I values in a regression software with the model $y = ax$ (here $x = U$, $y = I$ and $a = 1/R$ with R the resistance according to Ohm's law).

You will get: $a = 0.240866$, or $R \approx 4.15$ kΩ.

Now, the attentive reader will remark that the second current measurement had only two significant digits, while the first had three. Hence, the second should be considered as "less precise", thus less reliable, so it seems fair that this should be taken into account while doing the regression, doesn't it?

How exactly will we do this? Well, the New Zealand mathematician Alexander Aitken (1895–1967) devised a statistical rule for this: the weight of a measurement with a (random, non-systematic!) measurement error $\sigma_{y,i}$ on the value of y_i is proportional to $1/(\sigma_{y,i})^2$. A point that is measured twice as accurately counts four times more. In comparison, to halve the noise in a photo, you also need to expose four times as long. Any difference between calculated and measured y value is therefore multiplied in the sum S by this weight factor w_i:

$$S = \sum_{i=1}^{n} w_i \left(y_i - f(x_i) \right)^2 = \sum_{i=1}^{n} \left(\frac{y_i - f(x_i)}{\sigma_{y,i}} \right)^2$$

If all measurement precisions are equal, you can of course take the weights all equal to 1. If the y values are *sample* counts, $\sigma_{y,i}$ is usually taken as $\sqrt{y_i}$. This follows from the theory of the normal distribution.

This formula is known, but not used very often. In most scientific articles or software programs nowadays, measurement uncertainties are not even considered. The same is true for student textbooks like these, for example:

- Bijma, Fetsje; Jonker, Marianne; van der Vaart, Aad (2016). *An Introduction to Mathematical Statistics*, Epsilon.
- Cohen, Harold (2011). *Numerical Approximation Methods*, Springer, p. 18.
- Dukkipati, Rao V. (2010). *Numerical Methods*, New Age International Publishers, chapter 6.

Only the last one mentions weighted OLS briefly.

In fact, if the error on y determines the weight of a data point, so should the error on x. So the calculation of S could be like this

$$S = \sum_{i=1}^{n} \left(\frac{y_i - f(x_i)}{\sigma_{x,i}\sigma_{y,i}} \right)^2$$

If we apply this to our example measurements, we obtain $a = 0.246695$, so $R \approx 4.05$ kΩ. It's clear that the influence is significant!

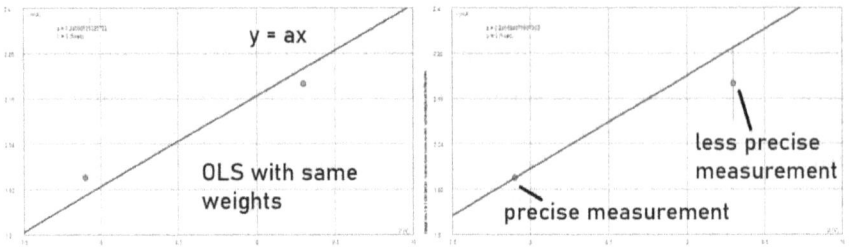

Fig. 3.9. Comparison of non-weighted and weighted regression.

3.2.4. Multidirectional least squares analysis (MDLS)

Even with weighted OLS, there is yet another big problem.
Suppose we measure two supposedly related variables x and y, e.g., voltage and current, mass and size, distance and time needed, temperature and volume, wine quality and price, whatever... Now, if we want to find a formula that connects them, even if it is only approximately, we have to choose one variable to be the "independent" one, and the other becomes the "dependent" one. Very often, this is totally arbitrary. If you can estimate someone's mass from his size, you should also be able to estimate the size from his mass, *using the same formula, inverted.* If you can estimate a wine rating from its price, you should also be able to guess how much you will probably pay for a wine with a 95/100 rating, using the same formula, yes?

Well, let's test this... You can easily repeat my steps with your favorite software.
Let's use this small dataset: {(16,7), (19,8), (20,14), (21,22), (24,25)}.
For the sake of simplicity, assume the errors in x and y are all 1.

Now do a traditional linear regression with any program you like.
You will get: $y = ax + b$, with $a \approx 2.529411823$, $b \approx -35.38823646$.

So you think you found "the" linear relationship that connects x and y.
Now, you want to calculate x from y, so, you use your first grade math:
$x = (y - b)/a \approx 0.3953488281y + 13.99069782$.
Okay?

Now... you swap the x and the y values in your dataset and do the regression again.
You obtain: $x \approx 0.3272450533y + 15.02587519$.

or:
$$y \approx (x - 15.02587519)/0.3272450533 \approx 3.055813953x - 45.83720930.$$

That's not what we expected! The graphs (Fig. 3.10) are clearly different.

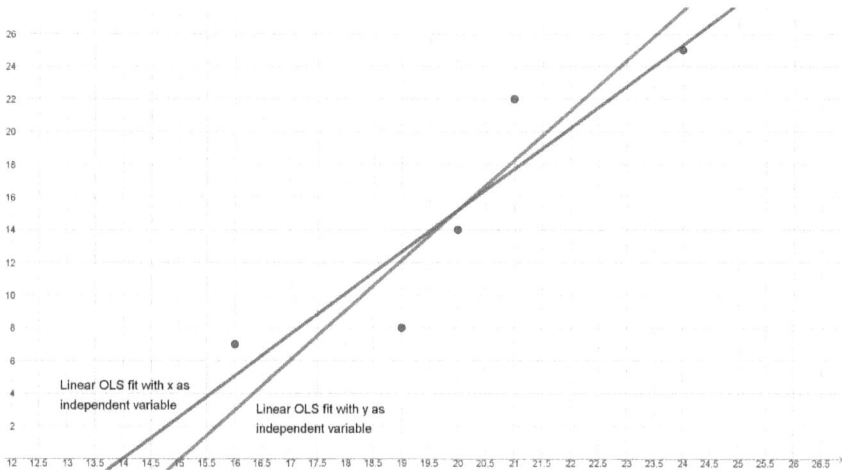

Fig. 3.10. The "best" fitting line changes when you switch the "dependent" and "independent" variables.

How is that possible?

The same happens when you want to see a different kind of pattern in this dataset, for example $y = f(x) = ax^b$, a power function.
Put them in a regression program, and you get $a = 0.0006332270576$, $b = 3.346564615$.

Reversing the formula, we get $x=(y/a)^{(1/b)} = 9.031054588y^{0.2988138928}$

But, if you switch x and y in your data and then do the regression, you obtain
$$x = 10.86186344y^{0.2325124702}.$$

That is again quite different!

The problem is that there is a fundamental asymmetry in the regression method! The classical algorithm that everybody uses, minimizes the sum of the (weighted) *vertical* distances between the measured (y_i) and the predicted y values $f(x_i)$. It is *assumed* as if this is the most obvious thing in the world, that only the y measurements have an uncertainty, and the x measurements are fixed like solid rocks. The whole theory is based upon this "dogma".

Apparently, it can make a big difference if we use the sum of the *horizontal* distances since that is basically what we did by switching x and y.

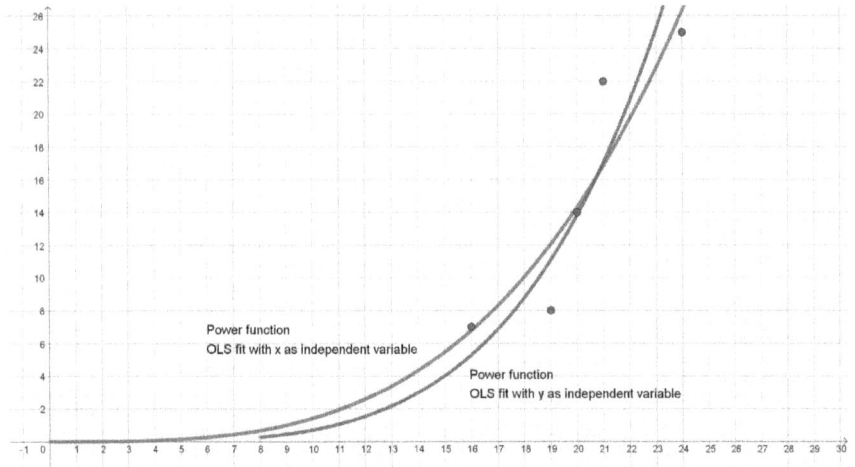

Fig. 3.11. Also when fitting power functions, the choice of "independent" variable matters.

The reason is simple: the method tries to pull the curve as flat as possible. If we switch x and y, the curve is pulled as vertical as possible. If the measurements are very close to the "ideal" curve, it doesn't make much difference, but if they form something like a cloud shape, like the data used for the Body Mass Index, for example (see case study p. 293), it can make a big difference. I challenge you to try this with your own data.

Is it possible that nobody ever noticed this? Or that nobody cared? Well, after a long search, I found some people who made the same observation, like Sebastian Kranz, economist at the University of Ulm (Germany), but he concludes his text with this strange little "poem": "Don't make you course a mess, but just be sly, and never simultaneously regress, y on x and x on y." [Kranz 2018]

Anyway, not many people seem to care, since in all the college textbooks I read, the numerous YouTube lessons about regression, and all

the software programs I ever tried, I never saw any remark about this phenomenon.
In fact, among economists, it is known that a fitted curve is pulled downward if there are measurement errors on the x values, or the so-called dependent variable has an influence in the "independent" variable, and they call it the problem of "**endogeneity**", and they invented a lot of tests and "remedies" to cure it, very far fetched artificial remedies that make things extremely complicated in my opinion! See for example, https://en.wikipedia.org/wiki/Endogeneity_(econometrics).

Why don't we just give x and y "equal rights" if they can be interchanged? Well, some people did give that a try, by minimizing the sum of the "perpendicular" distances between the points and the curve. This is called "**orthogonal regression**", or more generally "**Deming & York regression**", and "**total regression**" (applicable for nonlinear models). At first sight, that seems to make sense, but I see at least two problems here:

(1) That distance is easy to calculate if the curve is a straight line, but it becomes much more time consuming for other curves. In many cases, there is no exact solution, so an iterative approximation is needed for every data point. Yes, some nonlinear regression problems can be linearized, but you'll see I'm not a fan of that (p. 145).

(2) And, how do you define orthogonality if x is height and y is mass, or x is time and y is voltage? It's kind of strange to me if a 90° angle is not 90° anymore if you change units. This problem is "avoidable" if you always divide the distances by the measurement errors, to get a dimensionless distance, but then the best fit will depend on the value of those errors even if they are all the same. Is that desirable?

Isn't there a simpler way of "symmetrizing" the least squares method? Well, I think there is, by just multiplying the (weighted) vertical distances with the horizontal distances in the sum to be minimized. So, the "sum of squares" from weighted OLS changes into

$$S = \sum_{i=1}^{n} \frac{\left(y_i - f(x_i)\right)^2 \cdot \left(x_i - f^{-1}(y_i)\right)^2}{\sigma_{y,i}^2 \cdot \sigma_{x,i}^2} = \sum_{i=1}^{n} \frac{r_{y,i}^2 \cdot r_{x,i}^2}{\sigma_{y,i}^2 \cdot \sigma_{x,i}^2}$$

(See Fig. 3.12.)

I called this "**multidirectional least squares regression**" (MDLS) since it can be generalized with more than two variables, by just adding distances in other directions to the product.

For example, if we fit a relationship like $z = f(x,y) = ax^b y^c$, the three distances to a data point (x_i, y_i, z_i) would be:
$z_i - ax_i^b y_i^c$,
$x_i - ((z_i y_i^{-c})/a)^{1/b}$ and
$y_i - ((z_i x_i^{-b})/a)^{1/c}$.

I implemented this in my software program "FittingKVdm" (in version 1: only two-directional).

The parameters you obtain by minimizing this S, are somewhere in between the ones you obtain with the usual OLS and OLS with switched variables.

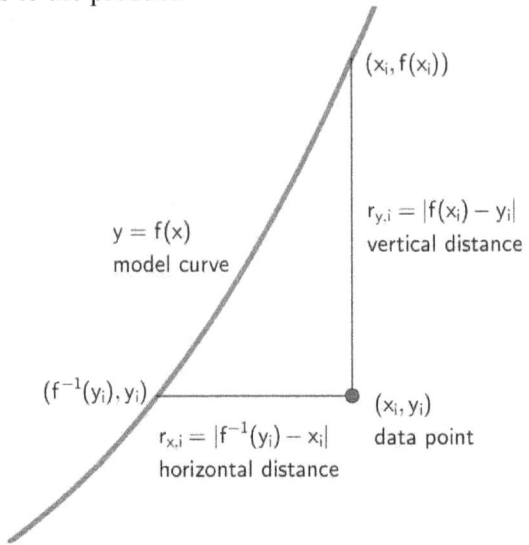

Fig. 3.12. Vertical and horizontal distances from a data point to a curve.

For example, applying this to our example five point dataset from above, it produces:
$y = 3.009650896x - 44.83942201$ if you want to see a straight line through the points, and
$y = 0.00005233923024x^{4.167235189}$ if you want to see a power function.

Limitations:
The main disadvantage of this method is that the model function should be *invertible* (the inverse relation should be a function), so it can't be used for periodic functions or functions that have a peak shape. For each y value, there is more than one possible x value, so you can't know the horizontal distance, unless you know to which slope of the curve the data point belongs.
Also if f has horizontal asymptotes, and some y values are above the highest asymptote or below the lowest, it's also impossible to calculate

the horizontal distance.

So, for monotonous functions, if the data are not too close to possible horizontal asymptotes, MDLS should work well.

It might be possible to define a better "sum of squares" that doesn't have these limitations, based upon something like the "total attraction force" between the whole curve and each point, calculated like the gravitation force or something similar, but that would require enormous calculation times.

Remark: how about generalizing the orthogonal regression?
Actually, if you calculate the (shortest = orthogonal, and weighted, meaning: divide each vertical or horizontal distance by the error in y and x) distance from the data points to a linear function f, you get the following (check this as a geometry exercise; you'll get there with the theorem of Pythagoras):

$$S_{ortho} = \sum_{i=1}^{n} \frac{r_{y,i}^2 \cdot r_{x,i}^2}{r_{x,i}^2 \sigma_{y,i}^2 + r_{y,i}^2 \sigma_{x,i}^2}$$

This is only the sum of the orthogonal distances if f is linear and the σ's are all equal to one. In case one of the r's is zero, you get $0/0$, but the limit is 0. Note that if all the errors on x are zero, you get the formula for OLS, as expected. If f is not linear, each term will be too small or too big, but it will still usually be in the same order of magnitude as the real distance. So, we might give it a try to test it with some functions, which have to be invertible too of course, since the horizontal distances from the points to the curve have to be calculated.

Remark: an apparent problem
Skeptics pointed me to this phenomenon, which was a big problem in their opinion. In some cases, the MDLS sum of squares (S) might have two minima, while this problem "was not encountered if you use OLS". For example, take four data points: (0,0), (0,1), (1,0) and (1,1).
The two solutions that produce a minimal S value, are the two diagonals ($y = x$ and $y = 1 - x$), that's true indeed (Fig. 3.13)! The iteration algorithm will lead you to one of both, depending on the initial parameter values.

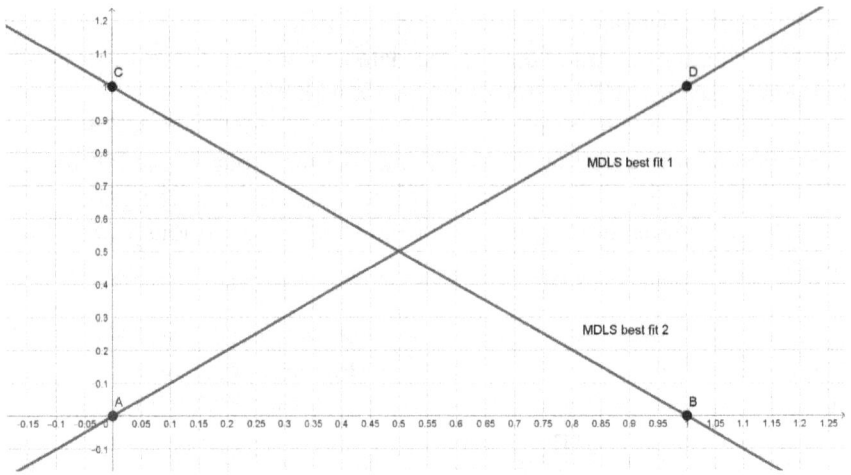

Fig. 3.13. The two "best" MDLS fits through these four points.

The OLS sum of squares has one minimum, when the line is horizontal, halfway between the points ($y = 0.5$). But... that is only because we crowned x as "the independent variable"! If we would do the regression with switched variables, we would get the vertical line between the points ($x = 0.5$)! See Fig. 3.14.

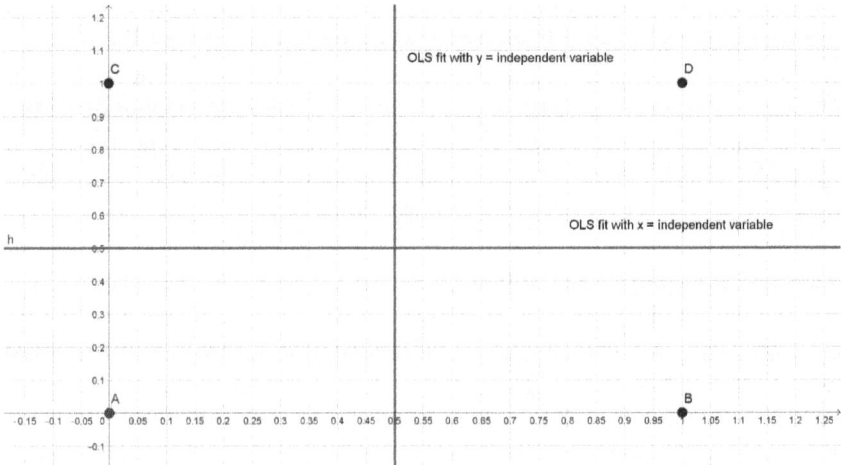

Fig. 3.14. The two "best" OLS fits through these four points.

By the way, Deming/orthogonal regression has exactly the same "problem" of course!

So, this is not a reason for not using MDLS! In fact, if you encounter such a strange case, it just means that it doesn't make much sense to choose a "best fitting line" anyway! In all "normal" cases where we can see a clear pattern, S will have 1 most pronounced minimum. By the way, even with OLS, many minima are possible, for example if f is (semi-)periodical, or if it has more than one minimum or maximum. If f is truly periodical, it doesn't matter which one you choose.

Reference:
- Kranz, Sebastian (Oct. 2018): *About a curious feature and interpretation of linear regressions* on his Economics and R Blog: https://skranz.github.io//r/2018/10/29/Curious_Regression.html.

3.2.5. Comparative simulation tests

To illustrate the difference between MDLS and OLS, and the "expanded orthogonal" regression, let's look at some simulations, illustrated with screenshots from FittingKVdm. Don't worry about the small character size; you don't need to read them; just look at the dots and the curves. First, the two-point example with the current vs. voltage measurements from above.

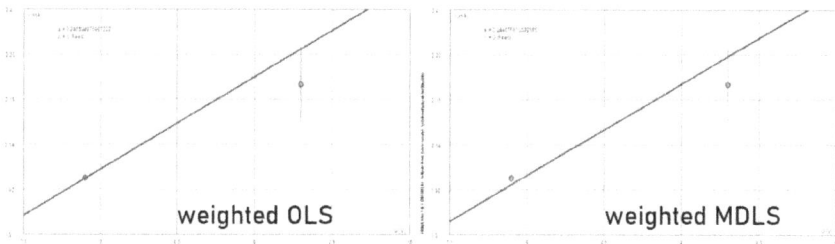

weighted OLS   weighted MDLS

Fig. 3.15. Comparison of OLS and MDLS in a simple two point situation with $f(x) = ax$.

While the weighted OLS seemed to give all the power to the most precise measurement, MDLS is a little bit "milder": the line in Fig. 3.15 is still closer to the left point, but you can see that the right point also had some "pulling power". That seems to make sense.

I did some more extensive tests too. For a few different functions, 100 data points were randomly generated, evenly spread in a given x range, and then to both the x and y values some gaussian noise was added with standard deviations $\sigma_{x,\,noise}$ and $\sigma_{y,\,noise}$. Then the fitting was done with the three methods. What we hope for, of course, is that the fitted parameters are close to the ones we started with.

● **Linear function model** $y = f(x) = ax + b$

$\sigma_{x,\,noise} = 1$, $\sigma_{y,\,noise} = 1$, measurement errors (flags) the same.

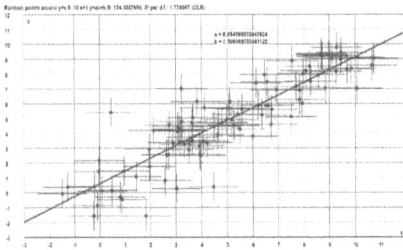

Fig. 3.16. Linear OLS fit.

Fig. 3.17. Linear MDLS fit.

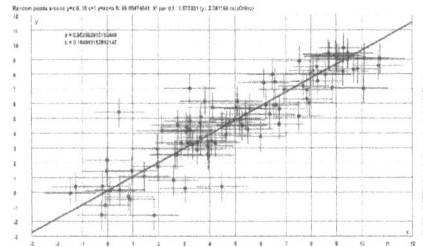

	a	b
Generated:	1	0
Fitted with...		
OLS:	0.854769	0.596989
MDLS:	0.976554	0.0930741
Ortho:	0.952583	0.104981

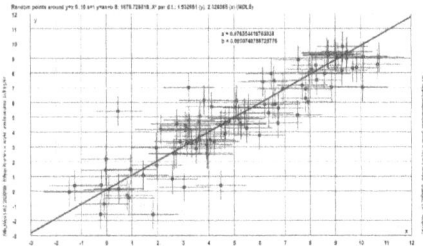

Fig. 3.18. Linear orthogonal fit.

● **Power function model** $y = f(x) = ax^b$

Here we added a royal amount of noise, proportional to the x and y values, as it is often the case when this kind of model is used (e.g. in biometry).

$\sigma_{x,\,noise} = 10\%$ of x_i, $\sigma_{y,\,noise} = 10\%$ of y_i, measurement errors (flags) the same.

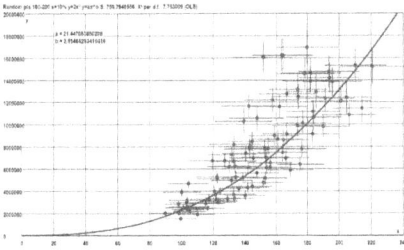

Fig. 3.19. Power OLS fit.

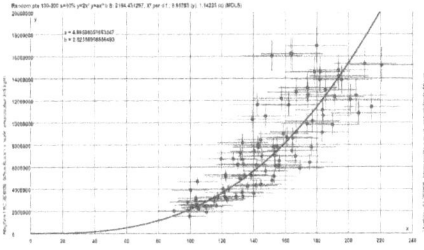

Fig. 3.20. Power MDLS fit.

	a	*b*
Generated:	2	3
Fitted with...		
OLS:	21.4471	2.51484
MDLS:	4.89508	2.82356
Ortho:	0.520229	3.27977

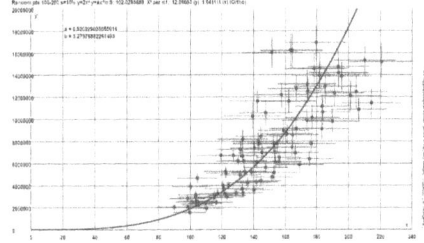

Fig. 3.21. Power "orthogonal" fit.

• **Exponential function** $y = f(x) = ba^x + c$

$\sigma_{x,\,noise} = 0.1$, $\sigma_{y,\,noise} = 2$, measurement errors (flags) 10% of each σ.

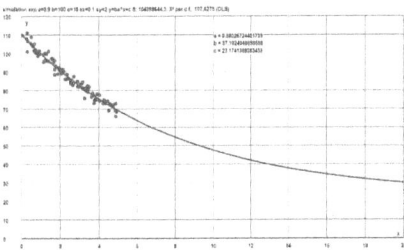

Fig. 3.22. Exponential OLS fit.

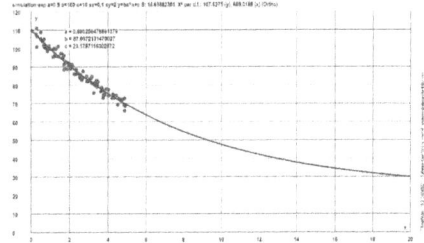

Fig. 3.23. Exponential MDLS fit.

	a	*b*	*c*
Generated:	0.9	100	10
Fitted with...			
OLS:	0.880267	87.1024	23.1741
MDLS:	0.895937	98.8407	11.2938
Ortho:	0.880256	87.0972	23.1797

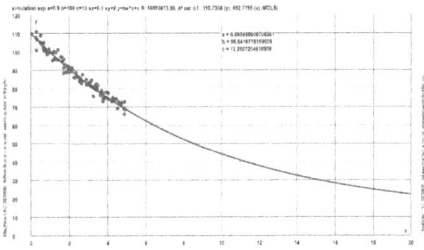

Fig. 3.24. Exponential "orthogonal" fit.

- **Shifted logarithmic function** $y = f(x) = a/b \cdot \ln(1 + bx)$

$\sigma_{x, noise} = 10\%$, $\sigma_{y, noise} = 10\%$, measurement errors (flags) 10% of each σ.

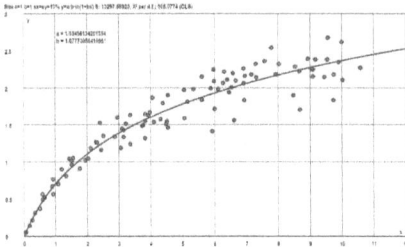

Fig. 3.25. Logarithmic OLS fit.

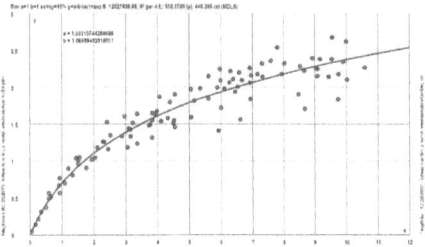

Fig. 3.26. Logarithmic MDLS fit.

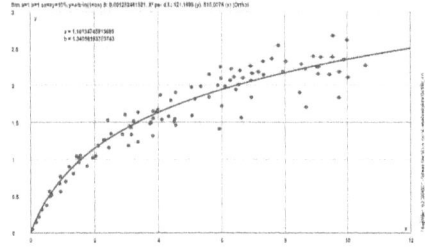

	a	b
Generated:	1	1
Fitted with...		
OLS:	1.03456	1.07773
MDLS:	1.03211	1.06659
Ortho:	1.18135	1.34056

Fig. 3.27. Logarithmic "orthogonal" fit.

- **Logistic function** $y = f(x) = a/(1 + e^{-k(x - c)})$

$\sigma_{x, noise} = 0.01$, $\sigma_{y, noise} = 0.02$, measurement errors (flags) 10% of each σ.

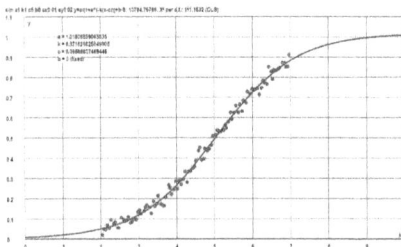

Fig. 3.28. Logistic OLS fit.

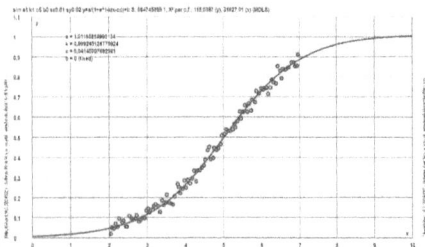

Fig. 3.29. Logistic MDLS fit.

	a	*k*	*c*
Generated:	1	1	5

Fitted with...

	a	*k*	*c*
OLS:	1.01809	0.971830	5.06569
MDLS:	1.01166	0.999240	5.04141
Ortho:	1.01809	0.971830	5.06569

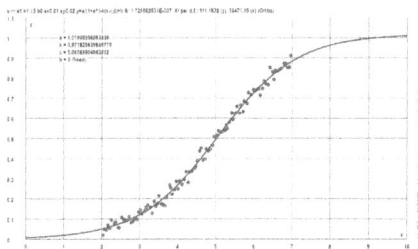

Fig. 3.30. Logistic "orthogonal" fit.

● **Homographic function** $y = f(x) = a - b/(x + c)$

$\sigma_{x,\,noise} = 0.1$, $\sigma_{y,\,noise} = 0.1$, measurement errors (flags) 10% of each σ.

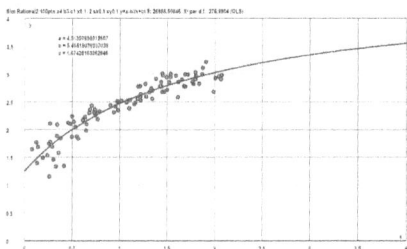

Fig. 3.31. Homographic OLS fit.

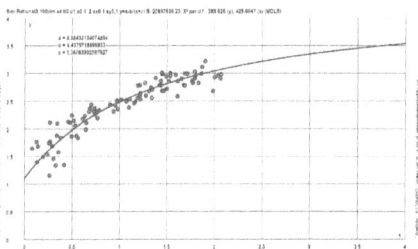

Fig. 3.32. Homographic MDLS fit.

	a	*b*	*c*
Gen.:	4	3	1

Fitted with...

	a	*b*	*c*
OLS:	4.51398	5.46519	1.67420
MDLS:	4.36432	4.43797	1.36064
Ortho:	4.51398	5.46520	1.67420

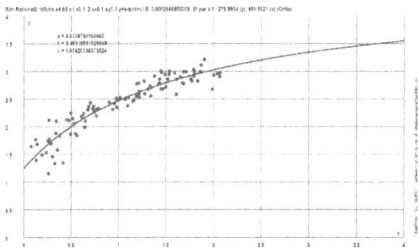

Fig. 3.33. Homographic "orthogonal" fit.

Note that

- in all cases, the OLS function is always horizontally biased,
- the "orthogonal" regression performs only reasonable in the linear case, as could be expected,
- MDLS always produces the results closest to reality!

Conclusion: It seems defendable to use MDLS whenever possible!

3.2.6. Regression or average?

In some cases, there is only one unknown parameter in the relationship between x and y. For example, $y = ax$, where x, y and a could be volume, mass and density, or voltage, current and resistance; or $y = ax^2$, with x, y and a being time, distance an object fell, and half of the gravitation field strength; or $y = \text{Arcsin}((\sin x)/a)$, where x and y are angles of a light beam on a surface and $a = n = $ the relative refraction index. You'll find them among the case studies.

One might ask, in all such cases, you actually only need *one* good measurement, and you can calculate the parameter a; if you want to improve the precision, you can do several measurements and take the average, so why would you need regression analysis?

Let's look at a very simple example: suppose we have two measurements $(5,10)$, $(11,15)$ with absolute errors on x and y all equal to one, and we assume the relationship $y = ax$.
We can make two estimations for a:
$a_1 = 10/5 = 2$.
$a_2 = 15/11 = 1.36364…$
The "real a" is probably (although not certainly) somewhere between those two values. Anyway, the "best" value we can derive from just these two measurements must be $a_2 < a < a_1$.

The simple arithmetic average of both, $a_3 = (a_1 + a_2)/2 = 1.68182…$, might be a better approximation for a, but it is certainly not the best. Why? It gives equal weights to both measurements. The second measurement should have more weight since it the measurements have smaller relative errors: $1/15$ and $1/11$ are smaller than $1/10$ and $1/5$.
The absolute error on a can simply be calculated with the general rule (see p. 71):

$$\sigma_a = \sqrt{\left(\frac{\sigma_x}{x}\right)^2 + \left(\frac{\sigma_y}{y}\right)^2} \cdot a$$

So, $\sigma_{a1} \approx 0.45$ and $\sigma_{a2} \approx 0.15$.
If you thought that a_1 and a_2 were suspiciously different, you see that the difference is not that unlikely with this kind of uncertainty: there is still an overlap between the intervals $]2 - 0.45, 2 + 0.45[$ and $]1.36 - 0.15, 1.36 + 0.15[$.

Now, we can calculate the classical weighted average:

$$a = \frac{\sum w_i a_i}{\sum w_i} \ , \quad w_i = \frac{1}{\sigma_{a_i}^2} \qquad \text{(sums over all measurements, obviously)}$$

With our example numbers, this becomes $a_4 = 1.43088...$, which is closer to the "probably better" value a_2. Good!

Now, how does this compare to different regression methods?
OLS makes the average vertical distance between the measured y_i values and the calculated values ax_i (the vertical residuals) zero, so it is believed. But that is true, only if the intercept is *not* kept fixed to zero! In this case, if we *demand* that the sum of the vertical residuals is zero, we get

$$\sum (y_i - ax_i) = 0 \Rightarrow \sum y_i - a \sum x_i = 0 \Rightarrow a = \frac{\sum y_i}{\sum x_i}$$

This makes a lot of sense, since it gives a bigger leverage to the second measurement, and at the same time the horizontal residual sum is zero too. This gives $a_5 = 1.5625$.

The OLS calculation is different!

$$\text{From} \quad \frac{\partial}{\partial a} \left(\sum (y_i - ax_i)^2 \right) = 0 \quad \text{follows:} \quad a_{OLS} = \frac{\sum x_i y_i}{\sum x_i^2}$$

So we get another approximation $a_6 = a_{OLS} = 1.47260...$
And MDLS produces $a_7 = 1.53664...$, quite close to a_5.

Summarized:

method	value	sum of vertical residuals	sum of horizontal residuals
a_1 (measurement 1)	2	-7	3.5
a_2 (measurement 2)	1.36364	3.18182	-2.33333
a_3 (average)	1.68182	-1.90909	1.13514
a_4 (weighted avg.)	1.43088	2.10586	-1.47172
a_5 (zero res.)	1.5625	0	0
a_6 (OLS)	1.4726	1.43836	-0.97674
a_7 (MDLS)	1.53664	0.41368	-0.26921

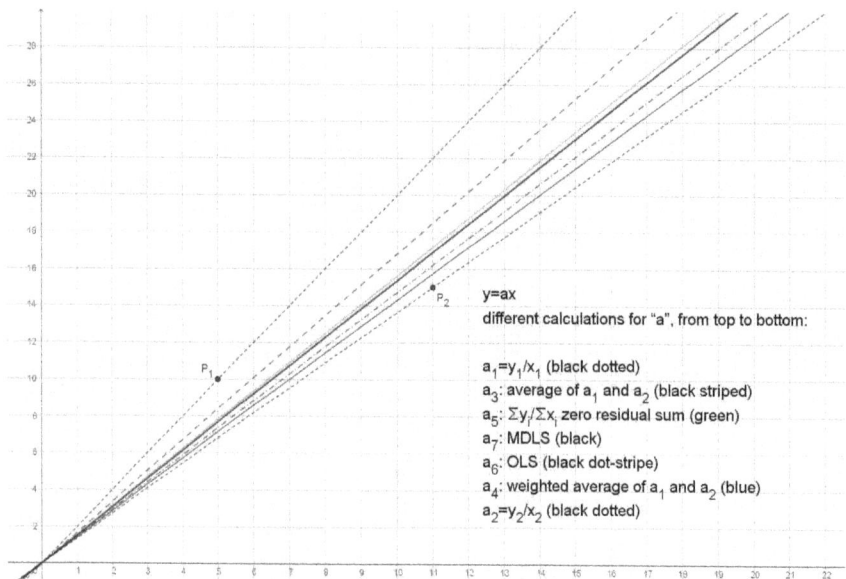

Fig. 3.34. Different ways of finding the "best" line through the origin, fitting with two measurements.

Which "a" is "best"? It depends greatly on your assumptions: OLS is definitely biased if the errors on x in reality are not zero. The weighted average seems to give even more weight to the second measurement, which is strange if we compare it to the fifth estimate which is elegant and simple.

One of the assumptions that have top priority in many cases, especially if the goal is some kind of (instrument) calibration, is that the found relationship between x and n stays the same if you apply the method with x and y switched: in the case "$y = ax$", it should produce $x = by$, with $b = 1/a$.

This assumption is obviously fulfilled if you use the ratio of the sums (a_5) or MDLS, not with OLS.

But how about the averages?

$$a_3 = \frac{1}{2}\left(\frac{y_1}{x_1} + \frac{y_2}{x_2}\right) = \frac{x_2 y_1 + x_1 y_2}{2 x_1 x_2}$$

$$b_3 = \frac{1}{2}\left(\frac{x_1}{y_1} + \frac{x_2}{y_2}\right) = \frac{x_2 y_1 + x_1 y_2}{2 y_1 y_2}$$

$$a_3 b_3 = \frac{\left(x_2 y_1 + x_1 y_2\right)^2}{4 x_1 x_2 y_1 y_2} \neq 1$$

So, the arithmetic average (and hence also the weighted average) does *not* qualify for this test! The attentive reader might remark that the geometric mean ($\sqrt{(a_1 a_2)}$) qualifies, but that doesn't work with negative numbers, and the sum of the residuals is also not zero.

If we demand symmetry and a zero residual sum and a fair weighing of the measurements, the winner is clearly the ratio of the sums (a_5)! MDLS is the only competitor, and it comes very close.

Now, making the average of the residuals zero, is generally not possible with nonlinear models, so we'll have to use a method that tries to *minimize* them, a regression.

MDLS might yield a bigger sum of the vertical residuals than OLS, but the horizontal sum will usually be smaller (except in the case $y = ax + b$), and the symmetry requirement is fulfilled.

Whatever parameter estimation you make, there will always be some disadvantage and some unknown uncertainty...

3.2.7. How precise are the parameters?

So, regression will determine the parameters that will make the function of a chosen class fit through your data, as good as possible. But, any numeric value in science doesn't mean much if you don't have an idea about its *precision*. If a software program produces a number $a = 1.23456789$, how many of those digits are reliable?

This is a very complex question! One thing is for sure, garbage in = garbage out. Unreliable data produce unreliable models. Even good data can be badly modeled. Even well-measured data can be incomplete, or biased, or influenced by some unknown factors or whatever.

I'm not going to discuss every possible theory that has been written about this problem. I will just explain one heuristic method that makes sense to me, and therefore I implemented it in my software.

Let's first start with a very simple example.
Suppose we want to make a model that describes the relationship between the height (h) and the shoe size (s) of men. There are biological reasons to believe that this relationship could be more or less linear ($s = ah + b$), so we need at least two measurements. I collected many (see case study p. 287), but I'll just use a couple of typical values, say (171, 41) and (187, 45). The heights are in cm and the sizes are European. If you ask people about their sizes, they usually give rounded values, which means the heights could be 0.5 cm off and the shoe sizes 0.5 unit. Anyway, you can't buy shoes with a size 44.3; if your feet would be most happy with that size, you will probably buy 44 and have the shoes stretched a bit.

The line through these points can be calculated directly:
$s = 0.25h - 1.75$.
Now, in the worst case scenarios, the real values (precisely measured heights and customized shoe sizes) could be (170.5, 40.5) and (187.5, 45.5), or (171.5, 41.5) and (188.5, 44.5).
In the first scenario, the line equation would become: $s = 0.294h - 9.65$, and in the second: $s = 0.176h + 11.2$.
The difference between the two a values is huge: 0.118, and between the two b values: 20.85, so we might conclude that we can be quite sure that $a = 0.25 \pm 0.059$, $b = -1.75 \pm 10.425$.
This is bad!

But,… worst case scenarios are rare. In reality, the real foot sizes of people who buy size 45 will be normally distributed around 45 with a standard deviation in the order of 0.5 (or 0.4 or 0.6, nobody knows), and in many cases the errors on the two points will not influence the parameters in the same direction.

So, the trick to get some idea of the deviations that can be expected realistically is, add some random noise to the data (with the measurement uncertainty as standard deviation) and reiterate to get new "optimal" parameters; then repeat this many times, and calculate the average and standard deviations of all the results. This is a so-called "**Monte Carlo method**". Such algorithms are often used when a probability is practically as good as impossible to calculate exactly.

This is implemented in FittingKVdm, with 100 different noise addings. A million would be better, but that would increase the calculation time too much.

For our data, we obtain:
$a = 0.25 \pm 0.012$, $b = -1.75 \pm 2.1$

This is visualized in FittingKVdm with dotted lines around the fitted curve. For all the combinations of extreme parameters, a graph is added, here: $s = (0.25 + 0.12)h + (-1.75 + 2.1)$, $s = (0.25 - 0.12)h + (-1.75 + 2.1)$, etc. If there are m parameters, 2^m dotted lines are drawn, so you can have an idea of what could happen with this kind of imprecision in the data.

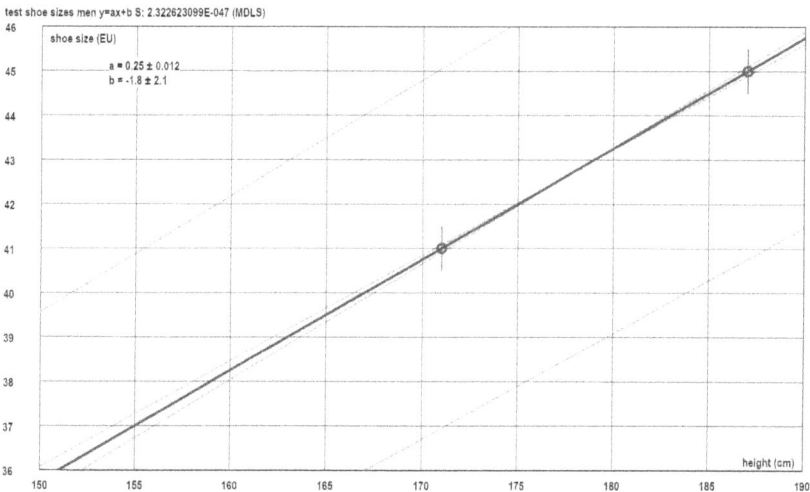

Fig. 3.35. The dotted lines reflect how the parameters could change with the given data imprecisions.

Remarks:
- If the given errors in the data are not standard deviations, but more like 95% confidence intervals, the estimated deviations of the parameters will also be bigger than the standard deviations, of course.
- If you repeat the estimation procedure, you might get slightly different results of course, but the order of magnitude should stay the same.
- Beware, this estimation is **only valid for these particular data and this model!** If we had picked two other men's data, the results would differ. Especially if the sizes were close to each other! Moral of the story is if you can gather only few data, pick them nicely spread out, and don't be afraid to include extreme cases! For more certainty about the relationship between height and shoe size, you will definitely have to collect more data! (See case study p. 287.)

Note that parameter b seems very uncertain, which suggests that it could just as well be zero. And that is very plausible: someone with length 0 probably has shoe size 0 too...

If you change the model to $s = ah$ (b = fixed to 0), then the uncertainty in a drops significantly: $a = 0.2402 \pm 0.0019$! You see that the distance between the dotted lines and the central line is now approximately the size of the error flags, as could be expected intuitively.

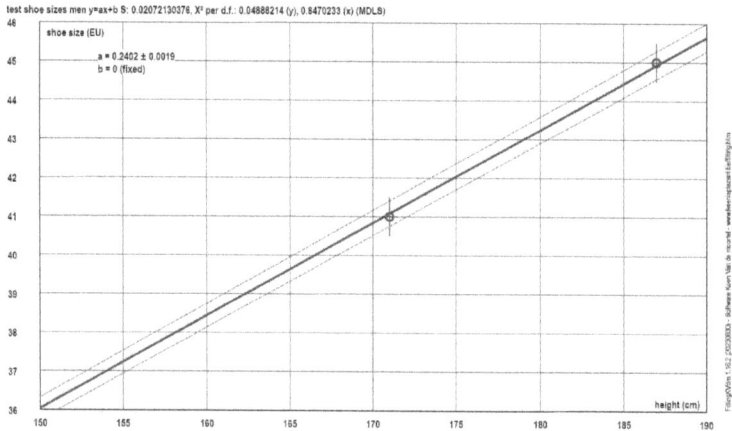

test shoe sizes men y=ax+b S: 0.02072130376, X² per d.f.: 0.04888214 (y), 0.8470233 (x) (MDLS)

Fig. 3.36. Possible variations in the best fitting line if b = 0.

This method takes into account the uncertainties on the data *and* the error propagation through the model function. It gives you an idea of the possible uncertainties in the parameters. It can't say without doubt if the model itself is good or bad though. We come back to that later.

3.3. Going Toward the Best Fit

3.3.1. Iteration, how does it work?

To find the best fitting curve $y = f(x)$ through a set of data points, we have to adjust the parameters in the formula of f so that the sum of squares (S, see above) becomes minimal.

How can that be done?
First, we need to choose starting values for all the parameters, and a range in which the parameters are allowed to move.
Then, a possible tactic would be to vary all the parameter values one by one in small steps, say 1% of the range, and calculate S each time. The combination of parameters that produces the smallest S, is taken as the new set of starting values, and then, for the next iteration, you would narrow the ranges and repeat this scan.
This will usually work, but slowly. If you have five parameters, each step requires $100^5 = 10^{10}$ calculations of S. If there are many data points and the model function is complicated, that might take a significant time.

A quicker method is similar to the Newton–Raphson algorithm to find the zero of a function: calculate the tangent line in the starting point, and then use the zero of that tangent line as the new approximation. Now, we don't want to find a zero but a minimum. So, we should apply the algorithm to the derivative of S (partial derivatives with respect to the different parameters). What we actually do then – and I find this a nicer way to visualize the algorithm – is, we find the *minimum of the tangent parabola* in each S vs. parameter graph and take this as the next approximation for that parameter. Do this for all the parameters and repeat until the values stay approximately the same. Fig. 3.37 shows this (p is one of the parameters).

What can go wrong? If the starting value is not chosen well, the tangent parabola might be "upside down" and its top might be further away from the minimum. In some cases (e.g., if f is periodic), the iteration can lead to a relative maximum, or even totally diverge toward infinity. In the shown example, if the starting value p_0 is somewhere between 7 and 11, the algorithm will converge to a maximum.

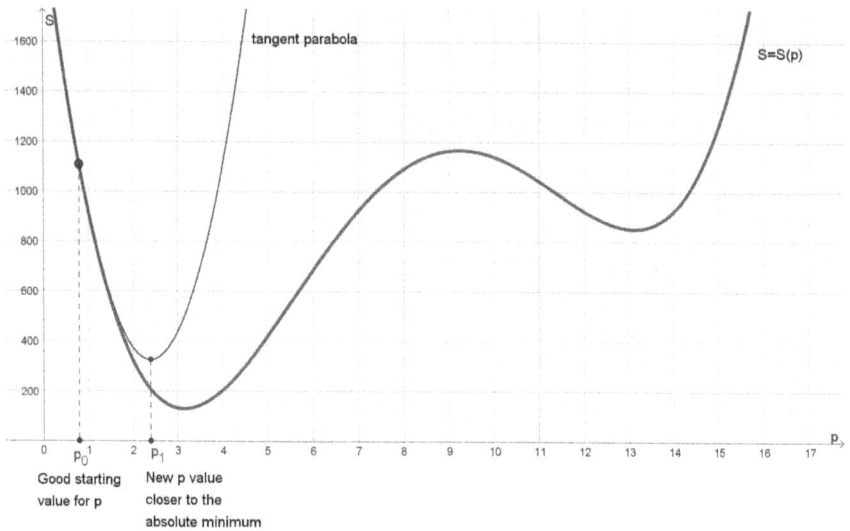

Fig. 3.37. Approximating the minimal "sum of squares" by calculation the minima of tangent parabolas.

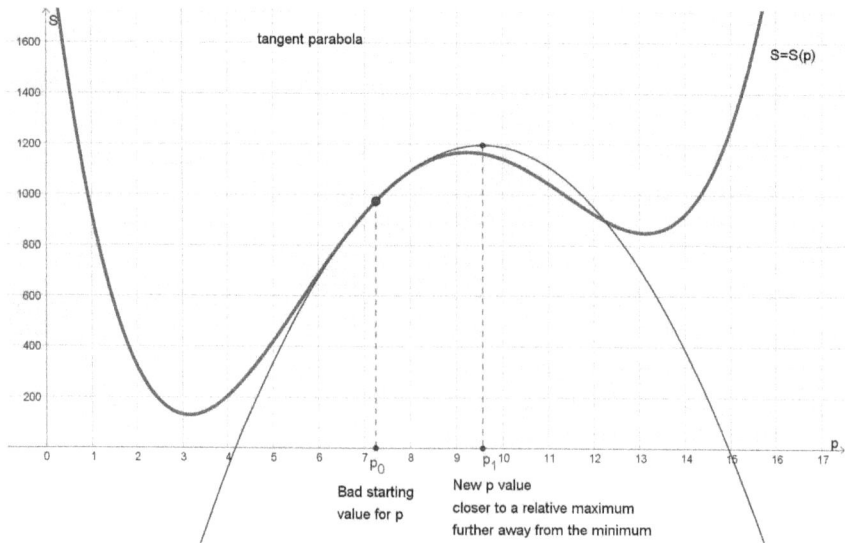

Fig. 3.38. Bad convergence due to a bad starting value for a parameter *p*.

Now, this is actually easy to avoid: the equation of the tangent parabola will be of the form $T(p) = ap^2 + bp + c$. If $a \leq 0$, we know that we are going the wrong way, so we should make an adjustment to p in the other direction, maybe repeatedly, until $a > 0$ again.

How do we find a, b and c? By demanding that S, S' and S'' be the same as T, T' and T'' for $p=p_0$ (S' is short for $\partial S/\partial p$):

$$\begin{cases} S(p_0) = ap_0^2 + bp_0 + c \\ S'(p_0) = 2ap_0 + b \\ S''(p_0) = 2a \end{cases}$$

So,

$$a = \frac{S''(p_0)}{2}, \quad b = S'(p_0) - 2ap_0, \quad c = S(p_0) - ap_0^2 - bp_0$$

Now, S' has to be approximated numerically. The simplest way to do that is

$$S' \approx \frac{S(p_0 + \Delta p) - S(p_0)}{\Delta p}$$

with Δp a "small" value, but the following is usually better:

$$S'(p_0) \approx \frac{S(p_0 + \Delta p) - S(p_0 - \Delta p)}{2\Delta p}$$

Now, here appears a problem: how small is "small"? A fraction like 1/1000 or 1/1000000 of the allowed p range will usually do for a few iterations, but once we get closer to the optimal p, Δp should be made smaller. Okay..., you might think, but no, not okay: computers have a limited precision! From a certain Δp value, the computer will not be able anymore to see the difference in the numerator! Try for example with your calculator: $(1 + 10^{50}) - 10^{50}$. The result will probably be zero. Starting from Windows version 10, Microsoft even decided to store real variables *less* precise in memory, can you imagine!

The consequence is that you can't keep iterating to any desired precision. Software modules for calculations with higher precision exist, but they make the iteration slower. Fortunately, most people are happy enough with about 10 significant digits.

In the example shown in the graphs, a starting value p_0 above 12 may lead to a convergence around 13. S has a minimum there, but not the absolute minimum. In FittingKVdm you can check the S vs. p graph, so you can manually choose a better starting value.

The algorithm should also check after every iteration if the new approximation of p is still within the given limits! If not, it has to be corrected, for example, to a value between the previous and the limit that was crossed.

In the example, any p_0 between 0 and 5 should produce a good result. In the next pages, we discuss some tricks to find good starting values.

3.3.2. Estimating initial parameters for the iteration

3.3.2.1 The general idea

Every iteration process starts with some initial values, and it usually helps to speed up things if they are more or less "realistic". So, how do we get an intelligent guess?

In principle, this is simple: if the formula of the model function f contains m parameters, we need m pieces of information, for example m of the n available data points ($n \geq m$) with different x values: (x_1, y_1) ... (x_m, y_m). If we put the data in the formula: $y_1 = f(x_1)$, $y_2 = f(x_2)$,... $y_m = f(x_m)$ we get a system of m equations and m unknowns, which usually has one unique solution. The obtained curve will fit exactly through those m points but not through the others; that's why the iteration has to be done.

Does it matter which points you select? Yes, the more evenly spread over the data point cloud, the better!

The parameter boundaries should be set a bit below and above the first estimations, leaving enough space for the iterations, but there might be restrictions (maybe they have to stay positive or below 1 or so).

Important: Always check during the iteration if a parameter comes too close to the limits, and adjust the boundaries if necessary.

3.3.2.2 Examples of easy cases

● If your model function is **linear through the origin** ($f(x) = ax$), you pick one point (x_1, y_1): the one that is the furthest away from the origin. Putting this in the equation, we get: $a = y_1/x_1$. You need to leave enough room for the iteration, so you could set the upper limit for a to, e.g., $a_+ = 100\,\sigma_y/\sigma_x$ (with σ_y the standard deviation of all the y values, etc.) and the lower limit a_- could then be set to 0 or even $-a_+$.

- For a **general linear model** ($f(x) = ax + b$), pick two points and solve the equations as shown on p. 12.

The two points should be chosen as far apart as possible to get the most probable estimation! Fig. 3.39 shows why.

Limits for b that are most probably safe: $b \pm 10(|ax_1| + \sigma_y)$.

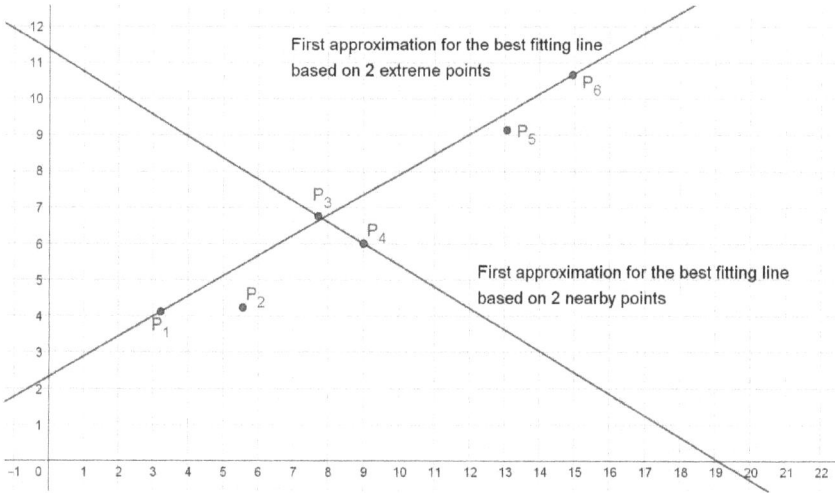

Fig. 3.39. First approximations of the best fitting linear function.

- For a **quadratic function** ($f(x) = ax^2 + bx + c$), take the leftmost, middle, and rightmost point and solve the three equation system. You should get

$$a = \frac{\dfrac{y_2 - y_1}{x_2 - x_1} - \dfrac{y_3 - y_2}{x_3 - x_2}}{x_1 - x_3} \qquad b = \frac{y_2 - y_1}{x_2 - x_1} - (x_1 + x_2) \cdot a$$

$$c = y_1 - a\sqrt{x_1} - bx_1$$

- For a **general polynomial model of degree** n: pick $n + 1$ independent points (i.e., having different x values) and solve the $n + 1$ equation system, at least, in principle this should work...

A problem with higher-order polynomials is that a small deviation in one of the data points can cause a huge difference in parameters. So, it's usually just as good to start with all zero coefficients. But the limits have to be set tight enough to prevent the coefficients to go crazy. For the linear part of the polynomial, you might set the limits like for a linear model, and then for each subsequent term, reduce the allowed parameter interval by at least an order of magnitude.

- **Power functions** ($f(x) = ax^b$):
Putting the values of two points in the formula, and dividing the two equations, leads to the following estimation:

$$b \approx \frac{\ln\left(\dfrac{y_2}{y_1}\right)}{\ln\left(\dfrac{x_2}{x_1}\right)} \qquad a \approx \frac{y_2}{x_2^b}$$

Limits like these will usually be fine, unless the data are extremely loaded with noise:
$a_+ = 10a,\ a_- = a/10 \quad b_+ = 10b,\ b_- = b/10$
(To be switched if $b < 0$.)

- **Homographic functions** (in the parameterization $f(x) = a - b/(x + c)$):
The parameters can still be calculated from three points, albeit with a bit more lengthy algebra. Start with three equations:

$$y_i = a - \frac{b}{x_i + c} \quad i = 1...3$$

Subtract the second and the first, the third and the second, and then divide the resulting two equations. After some manipulations (a nice algebra exercise for students), you should obtain

$$c = \frac{m_{23}x_3 - m_{12}x_1}{m_{23} - m_{12}} \qquad \text{with } m_{12} = \frac{y_2 - y_1}{x_2 - x_1} \quad m_{23} = \frac{y_3 - y_2}{x_3 - x_2}$$

$$b = m_{23}(x_2 + c)(x_3 + c)$$

$$a = y_1 + \frac{b}{x_1 + c}$$

Remark: There is a problem when $m_{23} = m_{12}$, but that's normal, since it would mean that the three points are collinear.

For many other functions, the equation system can't (easily) be solved, and we have to find some tricks to make at least a rough approximation, hoping it will work!

3.3.2.3 Some heuristic tricks for exponential and logarithmic models

- Exponential functions in the form $f(x) = ab^x + c$:

The parameters a and b are easy to obtain from two points *if* the "baseline" parameter c (asymptotic value) is zero or otherwise known (I leave the calculation to you as a simple exercise):

$$a = \left(\frac{y_1 - c}{y_2 - c}\right)^{\frac{1}{x_1 - x_2}} \qquad b = \frac{y_2 - c}{a^{x_2}}$$

For the boundaries, be careful: if the estimate of $a > 1$, set a_- to 1 and a_+, for example, to a^2 or a^3. If $a < 1$, switch them. The boundaries for b can be set to values 10 times bigger and smaller (not + or − some value, since they have to have the same sign as b).

If c is not known, you might estimate it as y_{min} or a bit lower (if Kendall $\tau > 0$) or y_{max} (if $\tau < 0$), but if the data points are still far from the asymptote, that might be a bad estimate, which can cause a bad estimate for a and b too.

There is another approach though, based upon a rough approximation of the derivative:

$$\frac{y_2 - y_1}{x_2 - x_1} \approx f'\left(\frac{x_1 + x_2}{2}\right) = b \cdot \ln(a) \cdot a^{\frac{x_1 + x_2}{2}}$$

Now repeat this with another combination of two points, and then you get a system of two equations that can be solved. Tip: divide both equations to get a, then get b from the first, and then c from entering one point in the function equation. When doing MDLS: move c a bit away from the center, to make sure it is outside the interval $[y_{min}, y_{max}]$, so the inverse can be calculated for each x value.

Remark: This trick can be used any time when there is a constant term in the model function, since it disappears in the derivative.

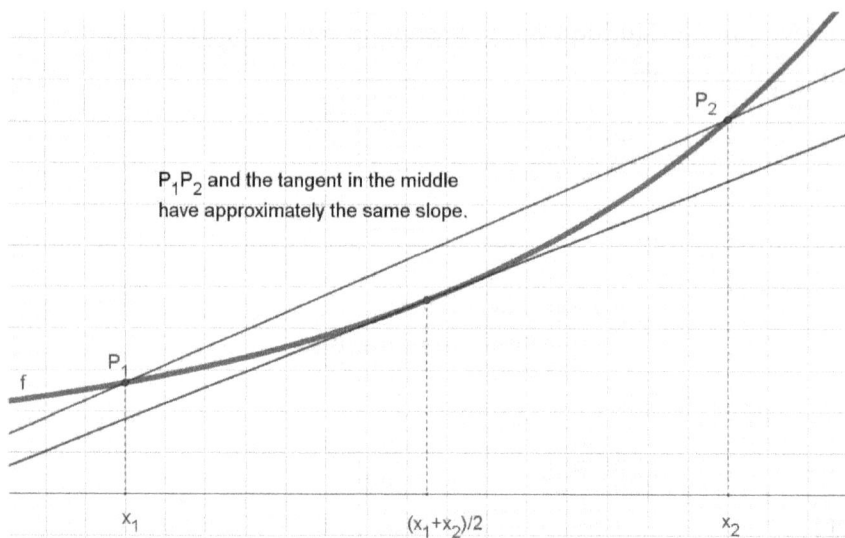

Fig. 3.40. Using the average slope to estimate the slope in the middle.

- $f(x) = \dfrac{a}{b} \cdot \ln(1 + bx)$

Take the leftmost point (x_1, y_1) and the rightmost (x_2, y_2). Then $a \approx y_1/x_1$. This can be a rough approximation, but it's close enough to get the iteration starting.

As with the exponential function, parameter b can be estimated from

$$\frac{y_2 - y_1}{x_2 - x_1} \approx f'\left(\frac{x_1 + x_2}{2}\right) = \frac{a}{1 + b\dfrac{x_1 + x_2}{2}}$$

- The **hyperbolic cosine** is an average of two exponential functions, see p. 56. If the top is shifted to (c, b), its position can easily be estimated from the minimal or maximal y value (if the points form a U shape or upside down). But for the stretch parameter, a little trick comes in handy: shift all the points by $(-c, -b)$, and then use the first terms of the Taylor series approximation for cosh:

$$\cosh(x) \approx 1 + \frac{x^2}{2} + \cdots$$

$$\Rightarrow a \cdot \left(\cosh\left(\frac{x}{a}\right) - 1 \right) \approx a \cdot \frac{x^2}{2a^2} = \frac{x^2}{2a}$$

Choose one measurement (x_i, y_i) far from the top and get the value of a from the approximation:

$$y_i \approx \frac{x_i^2}{2a} \Rightarrow a \approx \frac{x_i^2}{2y_i}$$

3.3.2.4 Logistic and similar models

In sigmoid models, there is usually one parameter for the upper limit and one for the lower limit like in the logistic (p. 28, 34) or the Gompertz (p. 30), or one parameter for the central y value (q) and one for the "amplitude" (r) like in the "transition" model (see p. 32, 32). The limits are then $q \pm r$.

If there are enough data points available to see the typical shape, the limits can be estimated as the lowest and the highest y values measured $(y_{min}$ and $y_{max})$, or better, a bit more away from the center. If q is needed, it can usually be approximated by the average of all y values, and then $r \approx (y_{max} - y_{min})/2$.

Again, if the points are good, the inflection point can be estimated as the average of the x values, and the slope at that point as the slope of the line between the two extreme points, or a few times that value. To find a better estimation of that maximal slope, you might calculate the slope of the line between points 2 and $n - 1$, then 3 and $n - 2$ and so on, until you get a maximal value. Or, you could find the maximal slope between two consecutive points, and then take the average of these points as the estimate for the inflection point. If there is a lot of noise in the data, that might be risky though.

Important: If the typical S shape of the sigmoid is not visible in the scatterplot, it will also be impossible to do a stable regression!

The function I described as "transition" function (p. 32, 32) poses an additional problem: the estimation of the parameter k that determines the "straightness" of the slope is not so easy to estimate, especially is the number of measurements is small. Here I would propose to do "**step-wise regression**": first, set k to a fixed value (e.g., 2) and estimate the other parameters and do some iterations until the graph starts to look stable. Then loosen k and iterate again.

3.3.2.5 Detecting a peak

How does one find a peak in a dataset?
First, we need to estimate the "baseline". If the model function is a distribution (meaning, the y values are counts), that is normally the x axis ($y = 0$), but Lorentzian peaks (see p. 34), for example, are usually on top of a background, or if the y values are a temporal evolution of some quantity (a rise and fall of temperature, for example), there is an "equilibrium value" of y. If some measurements were done far enough from the peak, this base value can be estimated as y_{min}. If not, we have a problem.

To find the top, we can just scan all points and take the point with the highest y value (y_{max}) as the peak position. We can then estimate the parameter connected with the peak position (e.g., μ if f is the Gauss function, describing the normal distribution, see p. 35) as the x value of the peak point. If the function is not "normalized" (to have a fixed integral), its height parameter is approximately $y_{max} - y_{min}$.

For the Gauss function, the parameter N is also proportional to the height, but it actually represents the *integral* (surface under the entire curve), so the best way to estimate it is

$$N = \sum y_i \cdot \Delta x$$

with Δx = class (bin) width, normally $(x_{max} - x_{min})/n_{bins}$. If there are some points missing, this is not correct, but the iteration will fix this.

In every peak describing function there is also a parameter describing the width. For example, in the Gauss function, $f(\mu \pm \sigma)$ is approximately the half of the height. So, to estimate σ, we can scan the points left and right of the peak to find out how far from μ the y values were approximately $y_{max}/2$.

In the Lorentz function, the width parameter r is exactly the distance from x_{top} where the function is at half the peak height. If there are points missing, linear interpolation might be needed to find this position. If the function describes an asymmetrical peak (like the "skewed peak" functions), you can start with the same values for the parameters connected to the left and right width; the iteration will do the rest.

Remark: For the Gauss function, we can also find μ as the weighted average of the x values (each x_i value has to be counted y_i times!):

$$\mu = \frac{\sum x_i y_i}{\sum y_i}$$

If the distribution is symmetric, this value should be approximately the same as the scanned peak position, but the average can be more influenced by outliers (e.g., one abnormal y value (frequency) far away from the peak). We can take the arithmetic average of both estimates to get a more stable estimate.

The width parameter can also be guessed as the usual standard deviation:

$$\sigma = \sqrt{\frac{\sum (x_i - \mu)^2 \cdot y_i}{\sum y_i}}$$

Again, this estimate can be more sensitive to outliers far from the peak. For a more stable estimation, we can combine this with the value obtained from the scan: take the *geometric* mean of both.

Things become more difficult if *two distributions* are mixed in your sample, for example, the heights of men and women. They might overlap, and it might not be easy to distinguish the two peaks. If we use a model that is just the sum of two peaks with centers μ_1 and μ_2, the iteration might not work well: it may lead to the same values or switch values. The solution for this problem is to use a different parameterization: replace μ_1 by $\mu - \delta/2$ and μ_2 by $\mu + \delta/2$, with $d > 0$. The parameter μ is now the average x value of both peaks, and δ the difference. The start value of μ can be set easily to the same as above.

Analogously, instead of using parameters N_1 and N_2, it's better to use N like before, and writing N_1 (# in group 1) as $N_1 = p \cdot N$ and $N_2 = (1-p) \cdot N$ with $0 < p < 1$. A start value of $p = 0.5$ will usually work. If we also assume, to start, that $\sigma_1 \approx \sigma_2 \ll \sigma$ (the value estimated as if it were one peak), then it can easily be calculated that

$$\delta = \sqrt{\sigma^2 - \sigma_1^2} \approx \sigma.$$

We might start with values like $\sigma_1 = \sigma_2 = \sigma/10$, which will often be too small, but for the convergence it's better to start with too small than too big values. Then, it's a good idea to shift μ a bit left and right to find a better start value because the S vs. μ curve can have a strange shape (not like a parabola and thus difficult for the iteration method).

3.3.2.6 Periodic functions

How can we estimate the parameters of a periodic function? Let's start with the simplest: the linear transformation of the sine function (see p. 45).

Since there are four parameters in the formula, you might think entering four points in the $y = f(x)$ equation and solving the system will work fine. Nope, unfortunately there is not just one unique solution; it was called the "aliasing problem", remember (p. 46)?

If you know the period (T) exactly or even approximately (like: one day, or one year), use it! Make it fixed in the formula and even if you start with the other parameters set to zero, the iteration will probably work.
If you don't have a clue, the first thing to estimate, is the parameter describing the average (m). If your data contain evenly spread points over an integer number of periods, that's pretty straightforward: just take the average of your y values. If not, you might have a serious problem; try to estimate it from studying the scatterplot or any information you have about your data.

So, suppose you have an estimation of m, how do you find T?
One way is to scan the data and see how many times the y values cross the $y = m$ line, and then dividing $y_{max} - y_{min}$ by two times that number. This works often if the data are well spread, but it is quite vulnerable with noisy data.

You might consider subtracting the average from the y values and then do a so-called "fast Fourier transform" (FFT). That is indeed an often used technique to find all the frequencies (hence periods) in data. But it works only in time series that have fixed intervals between the measurements! That shouldn't be a prerequisite for regression!

A tool that does work though, is called "**cross-correlation**". Basically, it works like this: first subtract the estimated average (m) from all the y values. Then imagine you lay a sine wave with a period T over the scatterplot; amplitude and phase don't really matter, so you can just take 1 and 0 (so $f(x) = \sin(2\pi x/T)$). Then multiply each $y_i - m$ value with the corresponding $f(x_i)$; if both have the same sign, you get a positive product, if not, it's negative. Now, add up all those products; that sum is the cross-correlation c:

$$c = \sum_{i=1}^{n} (y_i - m) \cdot \sin\left(\frac{2\pi x_i}{T}\right)$$

If there is a positive correlation between the measurements and that function, which means the data are going up and down at the same rhythm, c will be relatively high. With a strong negative correlation, c will be a big negative value. When there is no correlation, the positive and negative terms will more or less cancel out each other. Now, repeat this with a whole range of T values until you get the highest $|c|$, and that will most probably be the real period in the data.

Note: The cross-correlation can be "normalized" by dividing by the standard deviation of the $y_i - m$ values and that of the sine function ($1/\sqrt{2}$) and the number of points, so the result is between -1 and 1, but for this purpose, it's not necessary.

Which test range for T should you use? You need at least two points per period to see an up-down rhythm, so the shortest T should be in the order of two times the average distance between the x values. The longest T should be about $x_{max} - x_{min}$.
A good thing is that the T for which c is maximal is not critically dependent on the precise value of the estimated m.

Once you get T, you can also change the phase shift a bit to maximize c (not $|c|$ anymore, since you want the correct phase now, not the antiphase), and the amplitude can be estimated as $(y_{max} - y_{min})/2$. Start the iteration with these values, and all the parameters will be fine-tuned quickly.

In the example graphs below, the vertical green lines show where $y_i - m$

and $\sin(2\pi x/T)$ have the same sign. If they are in red, they have a different sign. In the left case (Fig. 3.41) they are all green, meaning there is a good correlation. In the right graph (Fig. 3.42) you see alternating green and red lines, which means a weak correlation.

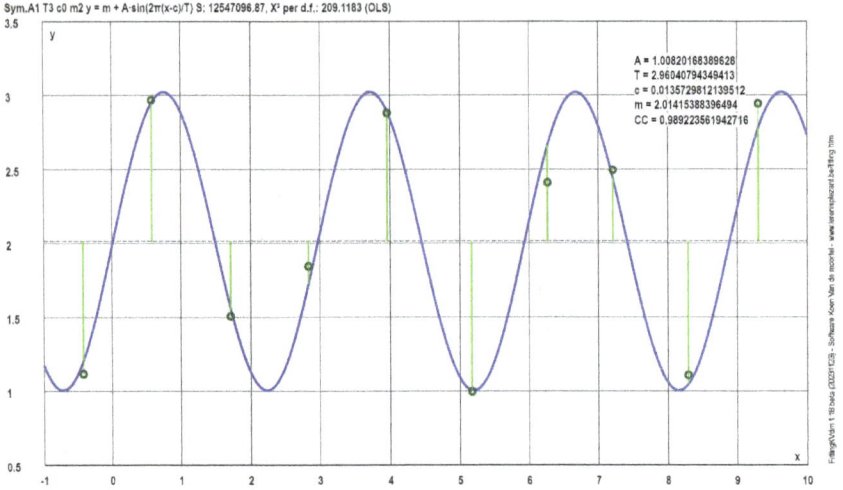

Fig. 3.41. A match: normalized cross-correlation ≈ 0.989.

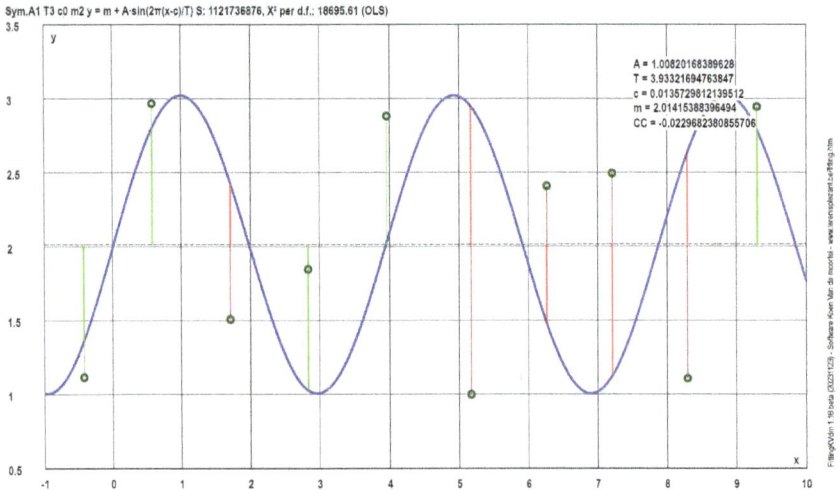

Fig. 3.42. No match: normalized cross-correlation ≈ -0.0230.

For periodic functions with higher harmonics ($T/2$, $T/3$, etc.) of the same base period, this will work too. Just start with zero values for the higher harmonics.

Also for periodic peak functions, you can use this technique. Just start with the sharpness parameter set to a value that produces something that looks like a wave, not like sharp peaks.

For skewed periodic functions, you can just start with the skewness parameter set to the symmetrical case, and the iteration will do the rest. Sums of unrelated periodic functions are much worse. Use any theoretical knowledge you have about the phenomenon you are studying!

3.4. Regression Analysis – Use it Wisely

3.4.1. Choosing the model

If you have collected data and you want to use them in a model, the first question should be: "Which kind?". Will you try a linear, quadratic, exponential, a power or a logistic function, or one of the many other existing functions? Each kind has a typical shape that can be modified to a certain extent by changing parameters in the formula. Just assuming "linear regression will do the job", like many people do, is a very bad idea!

In this choice, the absolutely most important consideration to make is as follows: **is some existing theory supporting the expected relationship between your data?**
For example, if your data are positions from moving objects, there will be helpful theories of gravity, friction, aerodynamics, etc.; if they are optical measurements, you might use Snell's law or interference theories, etc.; electronic circuits and chemical reactions also follow natural laws. Things get more fuzzy when dealing with living organisms and especially when they have a "free will"! "Laws" in biology, psychology, economy, etc. are much less rigorous, and that can make things very complicated.

Often, these "laws" or models originate from **differential equations**. A few basic examples are as follows:

• If you can assume that the *rate of growth* of a variable y is (approximately) proportional to that of a variable *x*, like: "the bigger the house, the more it will cost", then the (approximate) solution is a linear relationship:

$$dy = a \cdot dx \implies y = ax + b$$

• If you assume a constant force is pulling down a falling object, according to Newton's law, it has a constant acceleration ($-g$). Since acceleration is the second derivative of the position, the position above the surface (h) can be found by integrating twice and using the initial velocity (v_0) and position (h_0) as integration constants:

$$\frac{d^2h}{dt^2} = -g \implies v = \frac{dh}{dt} = v_0 - gt \implies h = h_0 + v_0 t - \frac{1}{2} gt^2$$

The rightmost equation is what you can observe; the leftmost is what you could call the "platonic" idea behind it.
See the case study "Falling pear" (p. 178).

• If you assume that a population can grow without limiting factors, in other words it has unlimited space and resources, the growth will be proportional (with factor *a*) to the already existing population. The solution for this kind of differential equation is an exponential function:

$$\frac{dP}{dt} = aP \implies P = P_0 \cdot e^{at}$$

Google "growth models differential equations" for more examples.
See case study "Population of Nigeria" (p. 284).

• If an object is brought out of an equilibrium position, and some force F, proportional to the deviation y, is trying to pull it back ($F = -ky$), you will find that a sinusoidal motion is the solution to Newton's equation:

$$m\frac{d^2y}{dt^2} = -ky \implies y = -A\omega^2 \cdot \sin(\omega t + \varphi), \ \omega = \sqrt{\frac{k}{m}}$$

(A and φ are determined by initial conditions.)

This is an idealized situation. In reality, also a friction force, proportional to the speed, will come into the equation, and then a damped sine function is the solution. See the case study about measuring gravitation with a pendulum (p. 194).

In more complicated situations, especially in biology, many such differential equations will join in systems that have chaotic solutions that only be solved approximately by numeric integration. This is done in simulation software like "Comsol" or artificial intelligence software.

If you want to see some fascinating examples, have a look at this old article by E.G. Briggs and R.N. Robertson, *Diffusion and absorption in disks of plant tissue*; it can be found at
https://nph.onlinelibrary.wiley.com/doi/pdf/10.1111/j.1469-8137.1948.tb05104.x.

Sometimes, we don't really know what is "behind" a pattern we observe, and we can only try to describe it as good as possible, with a function that has the right characteristics. In the chapter about functions, I have proposed some of those, not originating from differential equations, but most probably useful too.

So, if there is no theory available because the phenomenon you are investigating is too complex, or you are exploring a new domain, you start with a handicap, but that doesn't mean you can't have **some kind of a hypothesis**! Otherwise, you are just fooling around!

Try to choose your model wisely, using plain common sense. The first and easiest distinction you can try to make is as follows: **does the x–y scatterplot of your data have...**

(1) a periodicity?
You can try various periodic functions: some are simple like a sine wave, some have a more complex pattern, sharp or asymmetrical shapes, etc. The most difficult are those that seem to have variable cycles.

(2) one or a few peaks?
Are the y values counts of x values, like how many times a shoe size of x is observed? There might well be the peak of a Gauss function showing up, depicting a random distribution of values around a certain average.

If x varies, should there be some "optimal" y value, like maximal or minimal energy absorption versus an incoming frequency? You probably have encountered a resonance phenomenon that might be described by a Lorentz function, or more complicated peak functions like those describing RLC filters.

If you are studying the change of some variable in time, a sudden growth followed by a decay, you might want to try one of the other peak functions I described. And, ask yourself if there should be symmetry. That would mean the causes of growth and decay are the same.

(3) another shape?
Now, you have to consider many options:

● **Can x or y be negative or should they absolutely be positive?**
A length, a volume, a mass, a light or sound intensity, a frequency, a population, an age, the number of words in a book, a shoe size, a concentration, hardness, crop yields, etc., in fact most quantities, must be positive. An income, a voltage, a force, etc. can be positive or negative; even a moment in time, according to the choice of a starting moment. A temperature can only be negative if it is not expressed in Kelvin. These days, even an energy price can be negative. Some phenomena can be quantified in different ways, like "happiness", for example, in a scale from -5 to $+5$ (very unhappy to very happy) or from 0 to 10, according to your preferences.

Note that a linear function, and any polynomial with an uneven degree, always goes from $-\infty$ to $+\infty$ or vice versa, so if both your variables have to be positive, a power function might be more appropriate than a linear one. If only y has to be positive, an exponential function might be a choice.

● **Is there a theoretical limit for your values?**
A shoe size of 100 or an age of 300 years might never be seen, but it is imaginable. A voltage in a static electrical circuit, above the battery voltage, is not imaginable. Your model should show at least one horizontal asymptote in the latter case. Some quantities can only have values between 0 and 1 (100%), like survival chances or anything else that can be expressed as a percentage. In those cases, if that is the dependent variable, your model function should have two horizontal asymptotes, like a sigmoid shape! The simplest are symmetrical, but do

your data need to follow a symmetrical pattern? Cases where x is a fraction (percentage), are the most "annoying" since most common model functions don't stop at $x = 1$, except if they have a vertical asymptote, like the one I called "power-Möbius" (see p. 57).

● **Does a zero x value automatically imply a zero y value?**
Like zero volume must have zero mass; for zero distance, zero time is needed; zero force causes zero effect; etc.
This limits the possible model functions: it can be linear, but with zero intercept, for example. It can be exponential or logarithmic, but shifted so it goes through the origin.

If so, does y have to start going up very slowly with a flat slope in the origin, like power functions with an exponent higher than 1, or should it start very suddenly with an infinite vertical slope like power functions with an exponent between 0 than 1? Or should the slope in the origin have a controllable value? In that case, there is plenty of choice, see Fig. 3.43.

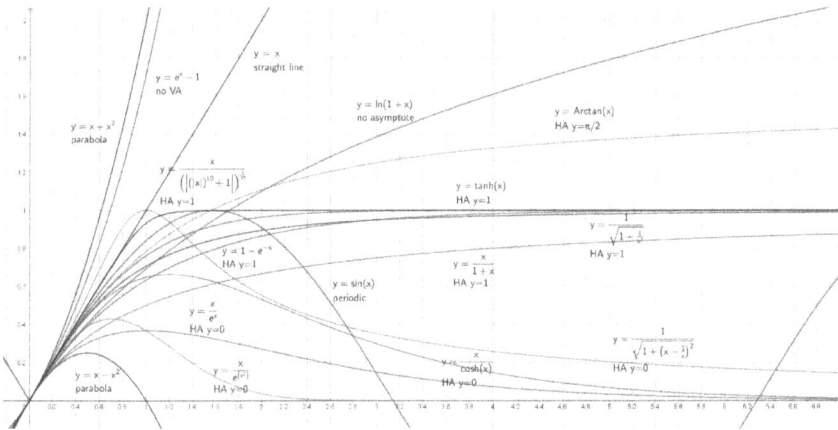

Fig. 3.43. Functions that go through the origin with a slope of 1 (adjustable of course).

● **Or, does y start at a certain fixed value when $x = 0$?**
For example, an exponentially growing population, or a radioactive decay cannot start from zero.

● **Is y more or less proportional or inversely proportional to x?**
For example, if the pressure on a gas increases, the volume decreases. In

such cases, you might need a power function with a negative exponent, for example.

We make all those considerations in the case studies.

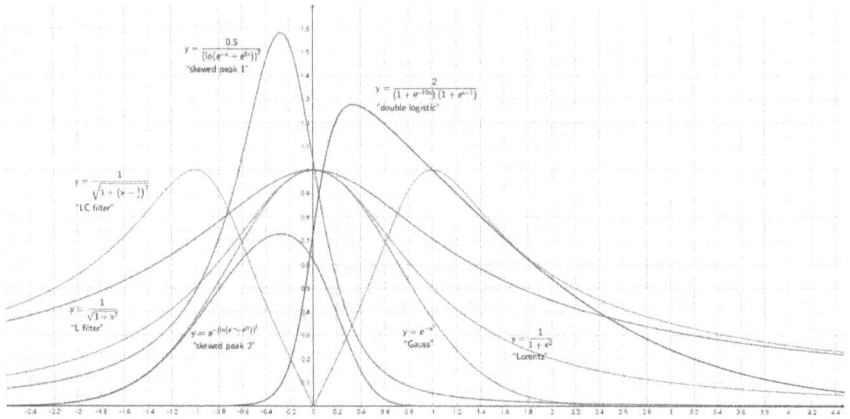

Fig. 3.44. Some peak functions. For the "LC filter", normally only the positive side is used.

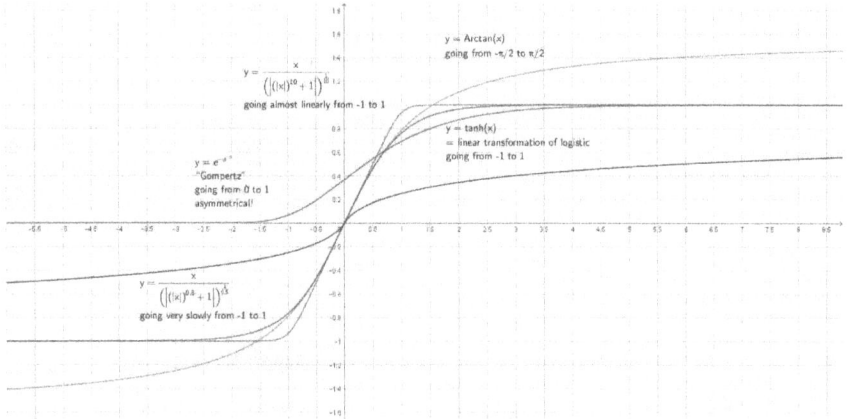

Fig. 3.45. Some functions that can depict a transition from one to another situation (sigmoids, etc.).

3.4.2. Checketh the extrapolation before thou choosest!

This might be the most important commandment for modelers! In fact, most of the questions to think about when choosing a model function come down to the question "Does extrapolation of the model make sense?".

In some situations, models can't be expected to be correct until infinity, but if they are extrapolable to a certain extent, it gives them credit. If they give absurd results when looking beyond the left and the right of the data point cloud, they are probably not trustworthy.

To illustrate what I mean, I'll give you some *bad* examples from the world wide web. They are all over the place, really not hard to find.

(1) This gentleman, an American teacher, is very convinced to do a *linear* regression on the **fuel consumption of cars versus their mass**, and showing that on YouTube. Now, you have to know that Americans tend to express this consumption in "miles per gallon" (MPG), not in "liters per (100) kilometer" like the rest of the world does. If it were in l/km, you might expect a relationship that looks more or less linear, although certainly not perfectly.

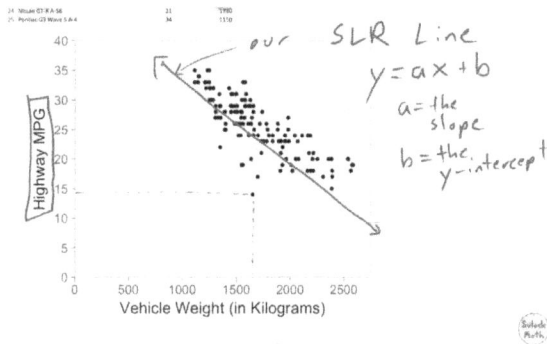

Simple Linear Regression | Least-Squares | Weight vs MPG | Residual | Slope Y-Intercept |TI-83 TI-84

Fig. 3.46. Screenshot from a presentation about regression on car fuel consumption data.

That means, if it is in "MPG", it should be approximately *inversely* proportional! A car with a very huge mass will run only a very short distance on a gallon; there must be a horizontal asymptote with value 0. Extrapolation of the model shown in the

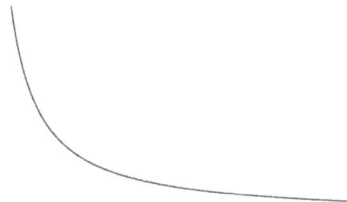

Fig. 3.47. A better model?

graph produces absurdities on the left and the right: cars with "negative masses" run the most miles per gallon, and very heavy cars would have a "negative consumption".
See the case study "car fuel consumption" (p. 207).

(2)   The next example is from a group of ladies who did a study about **speech recognition in noise (among normal hearing people) versus age**. This was published in a scientific psychology magazine.

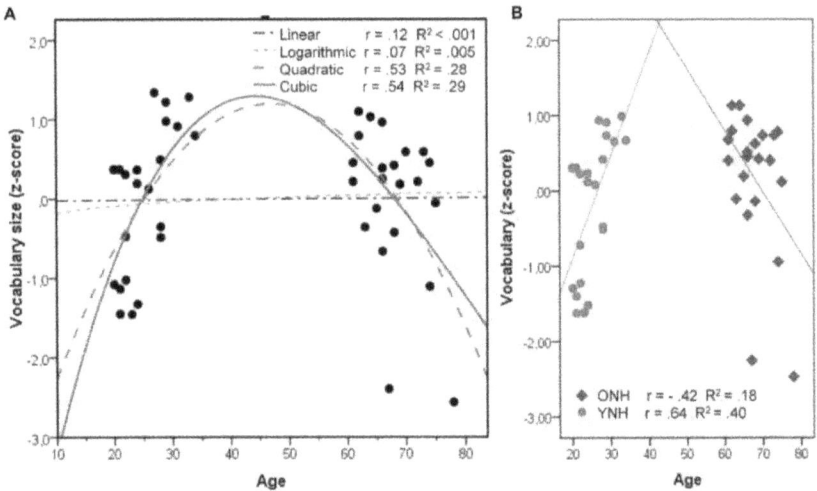

Fig. 3.48. Speech recognition vs age and some absurd attempts to fit curves through the data.

First of all, they split up their test persons in a "young" and an "old" group; why? If you want to recognize the function that connects x and y, it's better if you have an evenly spread group of x values.

Then they just tried a bunch of functions: linear, logarithmic, quadratic, and cubic, and they even tried a composite of two linear functions. I have no idea where they got their inspiration.

A function connecting these dots should start in the origin to begin with (newborns don't recognize anything), have a peak in the best years, and then slowly go down to a low value, as shown in the graph here. Is the limit at $+\infty$ zero or not? That is difficult to

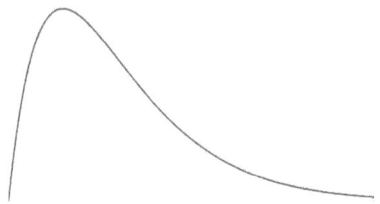

Fig. 3.49. A better model?

decide since we can't test people of age 1000, 10000, etc. and there is probably no hard theory about this.

(3) How does the **sign reading distance vary with the age of the driver?**

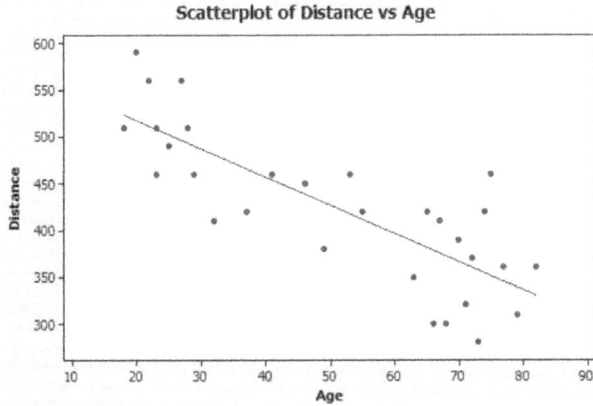

If you believe the lessons of this American college, it's linear... Once you are at a certain age, that distance becomes negative, doesn't it?

Fig. 3.50. Reading distance vs. age, linear?

How hard can it be to see that it should slowly go down, never reaching zero and certainly not negative values? Maybe exponentially, if the average eyesight diminishes a certain percentage year after year?

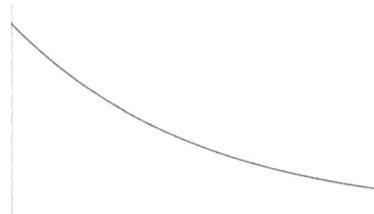

Fig. 3.51. A better model? Maybe the left part should be flatter?

(4) The next gentleman has a Facebook page about statistics, which is of course a noble use of this platform, but... as an example for *linear* regression, he presents a dataset of **weights (actually masses) vs. heights of teenagers**.

Measuring and Modeling by Example

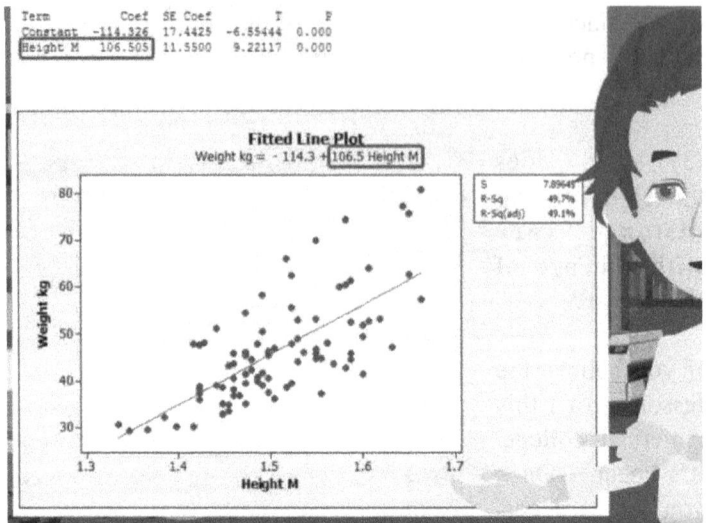

Term	Coef	SE Coef	T	P
Constant	-114.326	17.4425	-6.55444	0.000
Height M	106.505	11.5500	9.22117	0.000

Fitted Line Plot
Weight kg = - 114.3 + 106.5 Height M

S	7.89645
R-Sq	49.7%
R-Sq(adj)	49.1%

Fig. 3.52. Mass vs. height of people and a linear regression.

Linear, seriously? Even with a basic knowledge of geometry, you know that the volume and hence the mass of a three-dimensional body is proportional to the third power of its dimensions (if the shape stays the same, which is only approximately so with living beings). To define the so-called "body mass index" (BMI), the second power is used, which is still wrong but better, we come back to that later in a case study (p. 293).

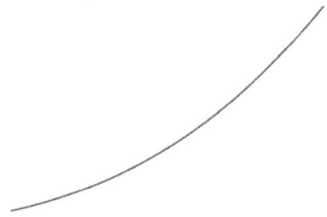

Fig. 3.53. A better model?

(5) What is the **best temperature in your office, for optimal productivity**?

Interesting question, and in 2006 some researchers published an answer: 22°C, and they gave a graph that showed their model function to predict

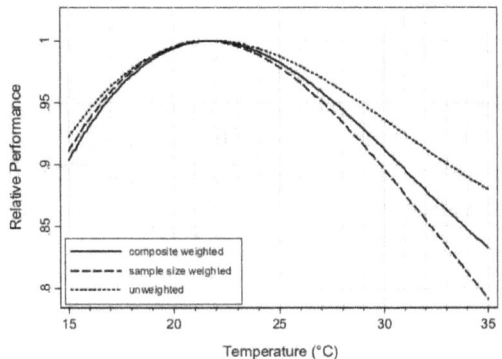

Fig. 3.54. Productivity vs office temperature – a polynomial model.

the "relative productivity" (P, 100% = optimal) versus temperature (T):
$P = 0.1647524T - 0.0058274T^2 + 0.0000623T^3 - 0.4685328$.

Besides the fact that their data were extremely noisy (how do you measure productivity precisely?), way too noisy to justify seven significant digits in these parameters, I don't have to tell you anymore what will happen if you expand the scale of the graph a little bit on both sides... a catastrophe of course.

The function to model this relationship should have one peak (most probably not symmetrical), and it should go down to zero on both sides, no possible doubt about that: you'll freeze or you'll boil eventually for sure. The skewed peaks I proposed (p. 38) might be useful here.

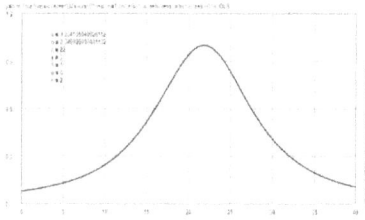

Fig. 3.55. A better model?

Exercise: Find more examples!

3.4.3. "Cleaning" the data?

Do your data need to be "cleaned" or "pre-processed" before feeding them into a regression algorithm?

● **Smoothing or other filtering?**
Your data look too noisy? You want to remove the ripples by replacing time series data by moving averages, for example, or using some other filter? Don't do it! You might remove interesting details in the data that might confirm or contradict your model. In fact, the regression you will do is a form of smoothing, but a structured one, that enhances the specific patterns that you are searching for.

Smoothing is like noise filtering in an electronic circuit: you can do it when you want to keep a clean audio signal that is nice to listen to, but you can't do it when you want to scientifically analyze an unknown signal to find out what is hidden in it! Applying low-pass filters on an image is good if you want to remove wrinkles, and high-pass filters are good to accentuate them, and deconvolution of a seismic time series helps find the position of reflection peaks,... but choose what you want to use: filtering or regression, not both. That's like swiping the floor and then trying to find footprints.

Other forms of pre-processing, like transformations, might cause harm too, see further.

• Outliers?

"Outliers" are data points that seem to be "out" of the pattern you have in mind. So, whether a measurement is an "outlier" depends totally on the model you use. A point may be far a way from the "best fitting line" through your data, but it may be close to an *exponential* function through your data.

Some solid reasons to remove outliers might be
• really wrong measurements that can't be true (because of a typographic error, disturbance of data transfer, bad observation, etc.),
• data that have gotten in the wrong set accidentally, see the case studies "Mammal heartbeat" (p. 303), "Car fuel consumption" (p. 207), and "Temperature vs. latitude" (p. 273).

A visual inspection of a scatterplot may suggest which points could be bad, although such points can also be hidden in the middle of your data cloud of course.

I'd like to quote Jochen Wilhelm (Justus-Liebig-Universität Gießen (D), Institute for Lung Health) here: "Statistical methods can help you to identify points that are 'far off from the rest'. If these values should or should not be excluded is not a statistical question but a subject-matter question. More often than not in research such 'outliers' are desperately trying to tell you that your assumptions are bad. Excluding such values just makes your wrong assumptions look better but your conclusions more wrong. Generally, outliers should be rare. If they are, then they are usually not a problem in your analysis. If they are not rare, your whole experiment/assay is in doubt.". (communication on a science forum, 25 Aug. 2023, see, www.researchgate.net/post/Can_someone_suggest_statistical_methods_t o_exclude_outliers/1).

Or, Adrian Olszewski (biostatistician at 2KMM, Katowice, PL): "Tests of outliers are nothing but numerical procedures telling you ONLY if your observed data match some assumed, theoretical pattern. Nothing more.", see, www.linkedin.com/feed/update/urn:li:activity:7125102353958993920/.

Actually, "outliers" that are decent measurements are a blessing! They can make your model a lot more stable. Watch this example: in Fig. 3.56 you see five points and a fitted line through them. The dotted lines show how uncertain that line is! In Fig. 3.57, one "far away" point is added. Suddenly, the fit and the precision of the parameters become much better. What a waste it would have been if you had mistaken it for an outlier! For another example, see p. 162.

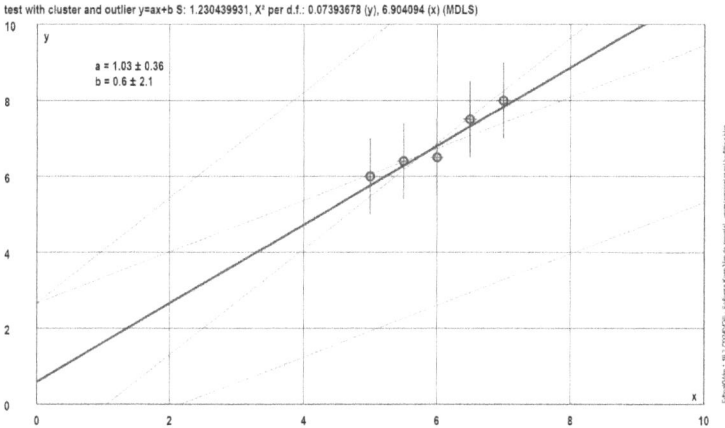

test with cluster and outlier y=ax+b S: 1.230439931, X² per d.f.: 0.07393678 (y), 6.904094 (x) (MDLS)

a = 1.03 ± 0.36
b = 0.6 ± 2.1

Fig. 3.56. Not a very stable fitting.

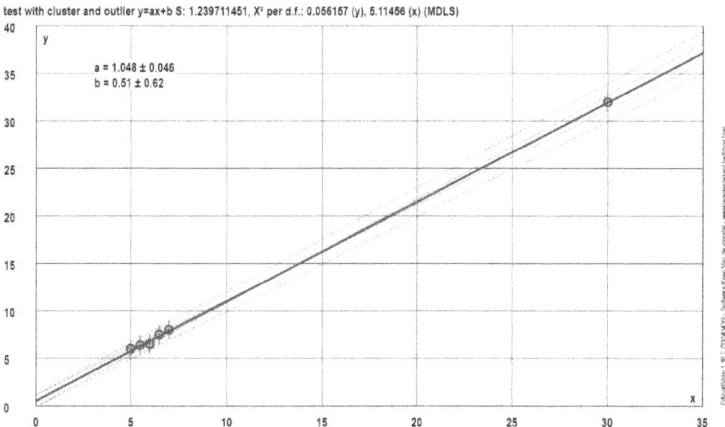

test with cluster and outlier y=ax+b S: 1.239711451, X² per d.f.: 0.056157 (y), 5.11456 (x) (MDLS)

a = 1.048 ± 0.046
b = 0.51 ± 0.62

Fig. 3.57. Much more stable because of one far away point.

A search for "abnormal" points *after* the regression is done can of course be very useful to provide insight into the data and the validity of the model! We come back to that later.

- **Is the distribution of the data points okay?**

Sometimes, your x and y values are normally distributed around their average values, for example, if you collect the heights and shoe sizes from randomly chosen people (see case study p. 287). That is interesting to see if you want to know how the population distribution might look like, but if you want to study the relationship between x and y, it's actually much better to have a more or less *uniform* distribution of x values. If they are clustered too much, instead of spread out, a curve through the points can vary a lot without changing the sum of squares much. In fact, the previous example illustrated this.

- One question that should be checked before you start modeling is as follows: **are x and y really independently measured?**

If you want to find a relationship between two variables, this is a prerequisite! I mean, for example, you can't
- prove Hooke's law using a classical dynamometer (a spring in a tube),
- prove that expansion of alcohol vs. temperature is linear, using a classical thermometer,
- prove Ohm's law by measuring U and I on a resistor with a classical volt and ampere meter (which is a galvanometer + serial and parallel resistors *based upon Ohm's law*; for Ohm's real experiment, see the case study "internal resistance of a battery", p. 222) (You can only use this to check if some unknown component also obeys Ohm's law.),
- prove that happiness is related to wealth if income is already included in the happiness score (see case study p. 347).

You may think this is so obvious that it is needless to say, but believe me, I have seen this being done in many a classroom or school book! If you notice such a contamination of your data, just throw them away; find better measurements.

- **Are some points "useless"?**

If a theory tells you that your data will follow a power function model with a positive exponent, it makes no sense to add the point (0,0) to your dataset; it's redundant and therefore useless. No matter what the parameters in the function are, it will always go through the origin, so it doesn't contribute to the sum of squares.

This is valid for any model that goes through the origin of course.

If, on the other hand, you would see a point like (0,5), or a negative y value in this dataset, that would be a contradiction and highly suspicious. Either such points are wrong or your model assumption is wrong.

FittingKVdm has a procedure to detect such points (see Fig. 3.58).

Points that occur multiple times ("**duplicates**") are *not* useless if they come from really independent measurements, e.g., you have three people with the same height and shoe size in your dataset. Such points will just increase the weight of this measurement; don't remove any of them!

"**Missing values**" are missing values, period. Don't fill them up with interpolations. If you need to do that to make a pattern more visible, don't use them in a regression because they contaminate it.

Fig. 3.58. Detection of useless points in FittingKVdm.

3.4.4. Preparing distribution data

I explained earlier that classifying data in bins causes a loss of data (see p. 82), but in one kind of situation, it's hard to avoid. If you want to study a distribution, i.e., a count of something, like the number of occurrences of a shoe size or the ages in a certain group, or the number of specimens from a plant or animal species at different altitudes, etc., you'll have to classify them in bins to get a visible pattern in your histogram.

Now, a good question is as follows: **how many bins?** What you will see, will depend a lot on that choice! I've done a simulation for you to make that clear: 1000 random values normally distributed with a mean value of 100 and a standard deviation of 10 were generated using: www.random.org/gaussian-distributions. (They produce truly random numbers based on noise.)

Then I split them in 1, 2, 3, 5, 10, 30, 100, 500, and 2000 bins and made the histograms. Starting from 3, a Gauss curve could be fitted through the data (it has three parameters). With too few bins, the pattern is very rough, and with too many, it gets buried in noise. The rule of thumb for the optimal number is the square root of the number of observations, in this case $\sqrt{1000} \approx 32$. The histogram with 30 bins seems the nicest indeed. Note that that one has the χ_y^2 the closest to 1!

# bins	3	5	10	30	100	500	2000
χ_y^2	5.7	0.1	0.22	0.86	0.55	0.62	0.38

So, if you need to divide n data points in bins, use approximately $\sqrt{n}$ of them.

Remark: This problem can be avoided by fitting to *cumulative* distributions.

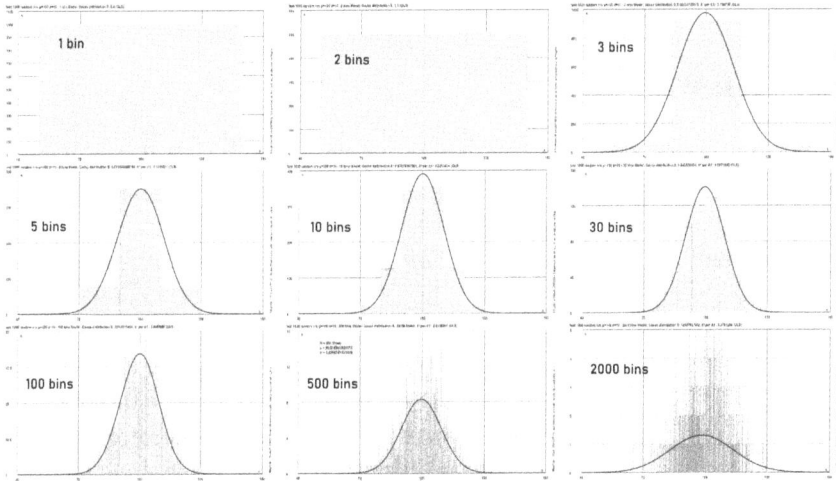

Fig. 3.59. 1000 random values classified in different numbers of bins.

3.4.5. Why non-linear transformations should be avoided

Among the people who work with regression, it is a very common practice to transform nonlinear models into linear ones by doing some transformation of the variables, like **taking the logarithms** or **inverting them**. For example, the following is a nonlinear relationship:

$$y = a \cdot x^b$$

But if we take the logarithms on both sides of the equation, and define $X := \log(x)$ and $Y := \log(y)$, this becomes

$$\log(y) = \log(ax^b) = \log(a) + b \cdot \log(x) \implies Y = \log(a) + bX$$

At first sight, this looks perfectly okay: there is a linear relationship between X and Y, definitely yes. And for linear regression, a simple straightforward algorithm exists to find the best fitting parameters, while exponential regression requires "time consuming" iteration. Textbooks recommend to use it, see, for example, *Numerical Methods* by Rao V. Dukkipati (New Age International Publishers, 2010), p. 207. And, most popular regression software programs, like the TI-84 calculator, Graphmatica, GeoGebra, and even Wolfram, known for its sublime math software, do it. See,
https://mathworld.wolfram.com/LeastSquaresFittingPowerLaw.html.

So what's the problem?

Problem 1: Weights

It's very simple: suppose the precision is the same for all the measurements, this will not be the case anymore for their logarithms! Each measurement obtains a different weight.
(The problem is mentioned in *The Engineering Statistics Handbook*, see, www.itl.nist.gov/div898/handbook/pmd/section1/pmd143.htm.)

Example: If $y_1 = 1000 \pm 10$ and $y_2 = 100 \pm 10$, then $\log(y_1) = 3 \pm 0.0043$ and $\log(y_2) = 2 \pm 0.044$, since $(\log(1010) - \log(990))/2 \approx 0.0043$ and $(\log(110) - \log(90))/2 \approx 0.044$. Smaller y values will therefore have a much larger leverage! (Strictly speaking, the errors are now slightly asymmetrical, but not much.)

Let's invent some data and feed them to the algorithms: point A (5,26), B (10,95) and C (14,141).

The TI-84 calculator, GeoGebra, etc. do the regression by taking the logarithms and then doing linear regression with – of course – the traditional OLS method. They produce $a \approx 1.8106$ and $b \approx 1.6760$.

If we put the same data in FittingKVdm, giving the same weights to the three points by entering 1 as "error", we obtain, also with OLS $a \approx 3.0896$ and $b \approx 1.4553$.

The difference is quite dramatic, and clearly visible in Fig. 3.60: with the first parameter values (f_1), the point A has more leverage; A is very close to the curve, but B and C are relatively far; $\chi^2_y \approx 13.509$. With the second parameter set (f_2), the curve moves nicely between all points, and $\chi^2_y \approx 9.632$, so this is objectively a better fit.

By the way, using MDLS, we get $a \approx 2.7202$, $b \approx 1.5111$.

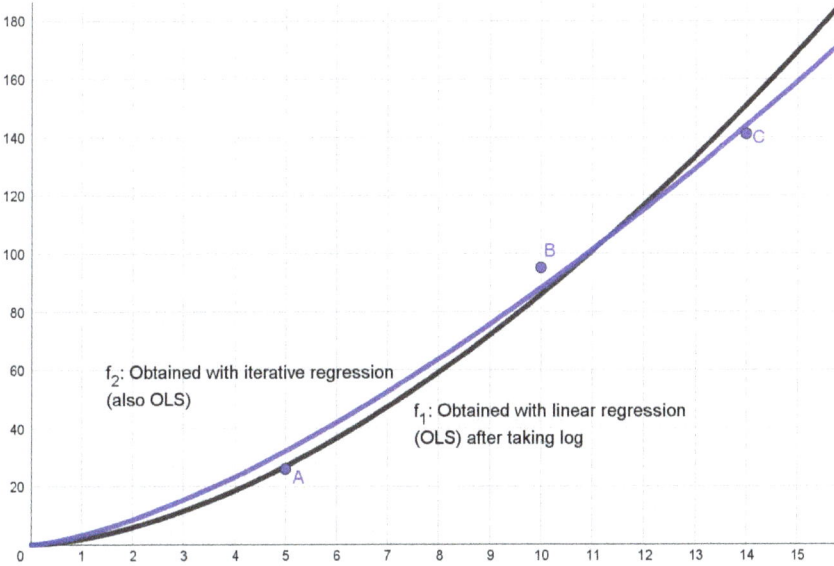

Fig. 3.60. The "best fitting" power function traditionally (f_1, black, by taking the logarithms of x and y) versus using direct iterative regression (f_2, blue).

Problem 2: Zeroes

Suppose you want to test the thermal insulation of a recipient. You put some hot water in it and measure the temperature in excess of the environment (let's call this $y = T - T_{env}$, in °C) at different times (t in minutes). According to Newton's cooling law, y will decrease exponentially to zero, see also the case study "Cooling down" (p. 217).

Suppose we have measured these data, using a thermometer with a resolution of 1°C:

t	0	1	3	10
y	16	9	4	0

Now, let's try and fit an exponential curve through these data: $y = b \cdot a^t$. FittingKVdm finds (using OLS) $a \approx 0.60487$ and $b \approx 15.781$.

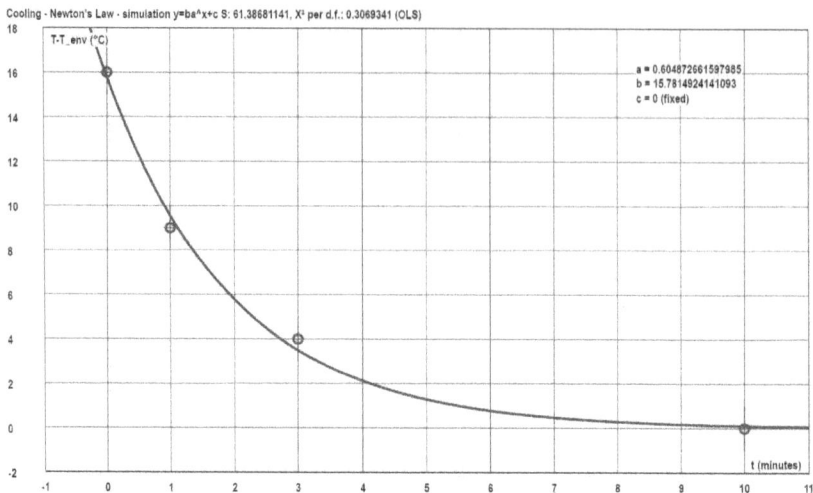

Cooling - Newton's Law - simulation y=ba^x+c S: 61.38681141, X² per d.f.: 0.3069341 (OLS)

$a = 0.604872661597985$
$b = 15.781492414109$3
$c = 0$ (fixed)

Fig. 3.61. Simulated data & best fitting exponential function.

If you enter the same data in a TI-84 calculator, Graphmatica, GeoGebra, Wolfram, etc., they will just all *refuse*! Why?

Theoretically, y can never be zero, so they *assume* it is safe to take log(y) to reduce the regression to a linear one:

$$\log(y) = \log(b) + \log(a) \cdot t$$

Now, practically, zero's *do* occur if the instrument can't distinguish between 0 and "a little bit above 0" anymore.

So, what should we do in this case? This zero is very significant; it would be a waste to leave out one of four measurements! What I saw people doing is as follows: "Let's replace the zero by a '*very small number*' and then it will work". That seems to make sense, but actually, it doesn't! Should we replace the 0 by 0.1? 0.01? 0.001? Does it matter? Or should we just leave it out?

Let's see what happens:

action	param.	FittingKVdm	GeoGebra
0 replaced by 0.1	*a*	0.60529	0.60299
	b	15.778	16.181
0 replaced by 0.01	*a*	0.60491	0.47179
	b	15.781	21.477
0 replaced by 0.001	*a*	0.60488	0.37002
	b	15.781	26.309
last measurement omitted	*a*	0.60531	0.63508
	b	15.778	15.242

Wow! Apparently, the choice of that "small number" doesn't have much influence on the parameter estimation using the iterative regression, but it has a dramatic impact in the traditional software: it can actually make the regression completely worthless, as you can see in Fig. 3.62.

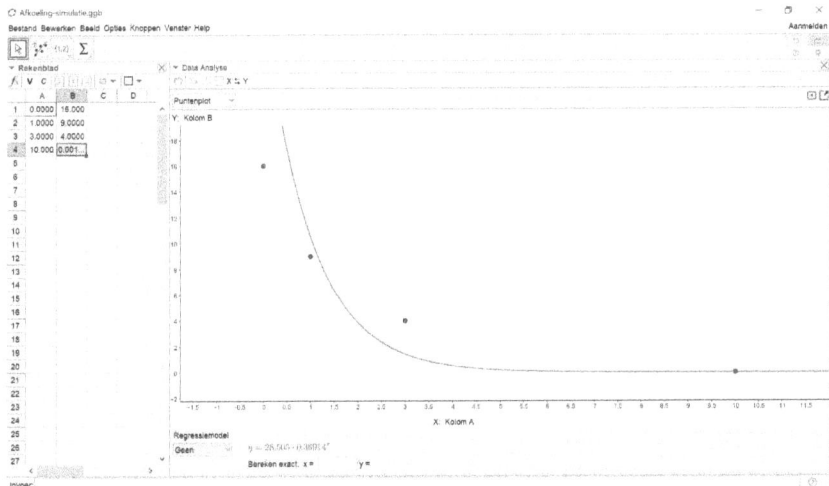

Fig. 3.62. Nonsensical "best fitting" exponential function using GeoGebra, the value $y = 0$ being replaced by 0.001.

As in the first example, the fourth point gets way too much weight because the logarithms were taken, and it pulls the whole curve in the wrong direction.

After all, it seems even better to just throw away the last measurement than to "invent a small number"! But much better is to use the iterative regression, clearly! Longer calculation times used to be a problem in the early days of computers, but nowadays, that's no longer an excuse for not using it!

Another example:

Suppose we have some data that are believed to be related homographically: $y = ax/(x + b)$ (called the "rational 1" model in FittingKVdm): $\{(0.5, 0.3), (1, 0.5), (2, 0.7), (3, 0.8), (7, 0.9)\}$.
(This dataset was actually calculated with $a = b = 1$, and the y values were rounded to the nearest tenth.)
Now, we want to find out the "best fitting" values of a and b. Doing this with the straightforward approach (i.e., iteratively) yields
using MDLS: $a \approx 1.01273$, $b \approx 1.01430$
using OLS: $a \approx 1.02839$, $b \approx 1.05314$

Now, if we define $X := 1/x$ and $Y := 1/y$, we can write the relationship also as

$$Y = \frac{b}{a} \cdot X + \frac{1}{a} = A \cdot X + B$$

This is a linear relationship between X and Y. So, you might now perform a linear regression, thinking it will not make any difference, but it does! We get
using MDLS: $A \approx 1.17301$, $B \approx 0.916733$
using OLS: $A \approx 1.18755$, $B \approx 0.902264$

So, the recalculated parameter values ($a = 1/B$ and $b=A/B$) become:
using MDLS: $a \approx 1.09083$, $b \approx 1.27955$
using OLS: $a \approx 1.10832$, $b \approx 1.31619$

Why are the results different (comparing both MDLS and OLS results)? Again, it's the weights!
Almost always, regression is done without taking the precision of the measurements into account, which means that all the points are given the

same weights. Now, let's assume all our original measurements have the same precisions, for example, $\sigma_x = \sigma_y = 0.1$, which could be realistic. What happens when we invert the values?

$$\sigma_X \approx \frac{\partial X}{\partial x} \cdot \sigma_x = -\frac{1}{x^2} \cdot \sigma_x$$

And the same for y, obviously.

So, in our dataset, this means that if the first y value is 0.3 ± 0.1, after transformation, we get $Y \approx 3.33333 \pm 1.111$. But the last y value 0.9 ± 0.1 becomes: $Y \approx 1.111111 \pm 0.1235$.

Since the weight of a measurement is proportional to $1/\sigma^2$, the weight of the last point should be much more than that of the first, more specifically: $(1.111/0.1235)^2 \approx 79.6$ times! That is, if we only bring the σ_y values into account (OLS)... With MDLS the weight difference becomes even more important. If one puts the X and Y values straight into a linear regression program, the obvious mistake made is that the first points are given the same weights as the last!

In Fig. 3.63, we see that the differences between OLS and MDLS are not very dramatic in this case, while the differences between the straight-forward (iterative) and the linearized fitting are very obvious.

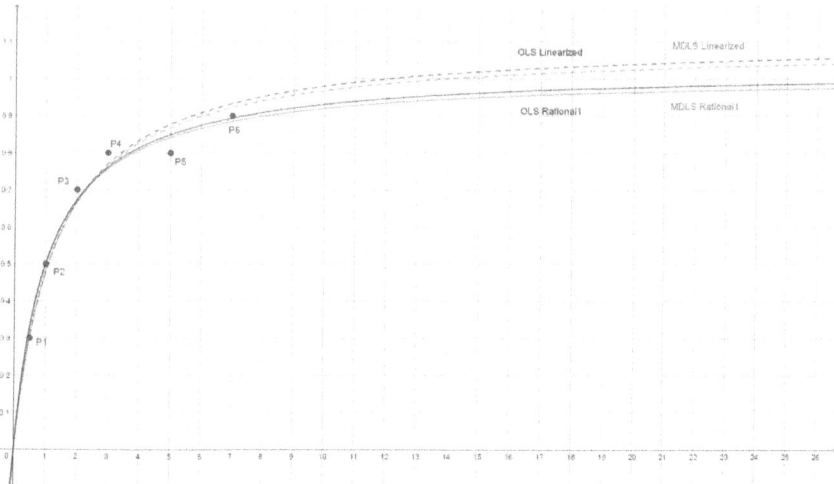

Fig. 3.63. Some simulated data and different ways to find the "best" fitting curve.

These were just a few common nonlinear data transformations. If you encounter other types, always be suspicious! Ask yourself how the measurement errors are transformed and what happens with zero x or y values?

More interesting thoughts about this, by Adrian Olszewski (10 Oct. 2022), *Why is transforming the response in regression analysis and hypothesis testing so dangerous?*. See, www.linkedin.com/pulse/why-transforming-response-regression-analysis -testing-olszewski/.

Remark: A *linear* data transformation will never do harm, and might sometimes be useful, but for a reason you might never think of.

For example, you may want to find an exponential growth pattern ($y = ba^x$) in a population, and feed some data in the software like (1950, 3452), (1970, 5683), (2000, 7451), etc.

Now, the algorithm will produce a small b value, which would be the theoretical population in the year zero, and parameter a will be 1.000... plus a very small number.

To do a prediction for the year 2020, you would have to calculate 1.0000 something to the power of 2020. Yes, the computer will do this, but even the smallest rounding error in b will cause a dramatic difference in the result. So, what you could do to minimize this risk, is to enter the x values as "years since 1950", for example, (0, 3452), (20, 5683), (50, 7451), etc. Now, you only have an exponent of 80 for your prediction.

The same can be done with periodic models.

3.5. Judging the Model

After performing a regression, there are many things you can do to rate the quality and the trustworthiness of your work.

3.5.1. Chi squared

If you look at Fig. 3.64, what can you say about the "goodness-of-fit"? You see a line that more or less follows the dots, but is it "good"?

test y=ax^b S: 1513.634289, X² per d.f.: 17.5059 (y), 44.32796 (x) (MDLS)

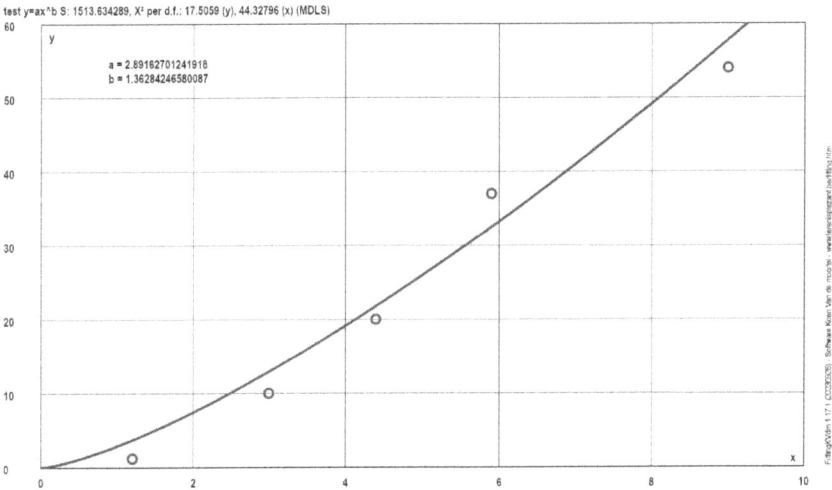

a = 2.89162701241918
b = 1.36284246580087

Fig. 3.64. A good or a bad fit?

Traditional number crunchers will attack the data with Pearson r, and square it, to obtain a number between 0 and 1 that is supposed to give "the" answer: near 1 is "good", near 0 is "bad".

In this case: $r = 0.9803$, so $r^2 = 0.9610$, so they would be very happy.

Now this r^2 is rather "sterile" because
• it tests only for linearity, so you can only use it for linear or linearized models. It can't say anything about the nonlinear curve fitted here, well, just that a straight line might also fit well through the dots,
• it totally ignores measurement errors, which is quite "unworldy".

Now, look at the next graphs (Fig. 3.65 and 3.66), with the same data, the same curve. The only difference is the error flags.

test y=ax^b S: 1513.634289, X² per d.f.: 17.5059 (y), 44.32796 (x) (MDLS)

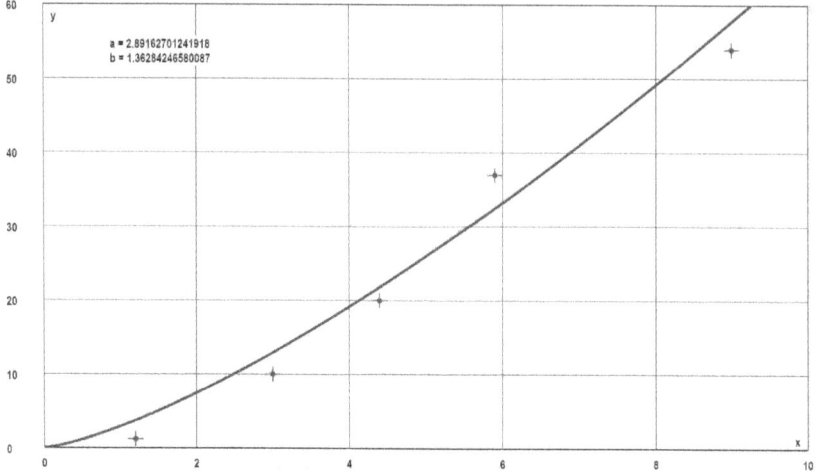

a = 2.89162701241918
b = 1.36284246580087

Fig. 3.65. Small error flags.

test y=ax^b S: 2.421814862, X² per d.f.: 0.7002356 (y), 1.773118 (x) (MDLS)

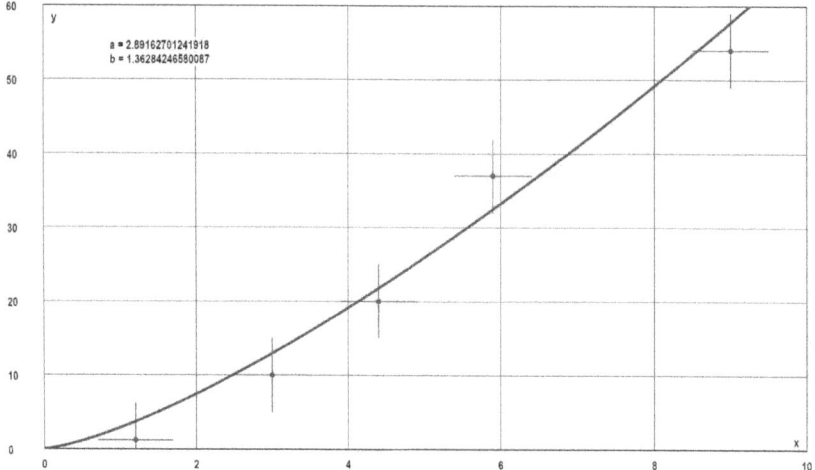

a = 2.89162701241918
b = 1.36284246580087

Fig. 3.66. Big error flags.

Can we say anything more now? Yes we can! If the measurement errors were really that small as shown in the left graph (Fig. 3.65), the model seems to "miss" something. For example: the leftmost point lies seriously below the model function curve. Its y value is 1.2 ± 1 while the model predicts 3.7, that's more than three times as much!

We might conclude here that either
- the model is not so appropriate,
- or there are other hidden variables influencing y,
- or the errors are bigger than we thought.

In the right graph (Fig. 3.66), the curve goes quite well through the error flags, which means that the data are not very precise, but the model might be quite good, on condition that the measurement errors were estimated realistically. To obtain more certainty about that, we might try to do more precise measurements, if possible, or just more.

Based upon this observation alone, we can't draw definitive conclusions, but it gives us a direction to explore further. This is one reason why using measurement errors is so important!

Can we summarize the observations in some "goodness-of-fit" number, better and more general than r^2?
Well, in the left graph (Fig. 3.65), the ratios of the residuals to the measurement errors have much bigger absolute values, as well in the vertical as in the horizontal direction. So, we might calculate something like the weighted sum of squares (S), but separately for the vertical and the horizontal direction, and then divide it by the number of points (n), to make the comparison independent of that number.

Okay, but we need to make a small adjustment. You might know that if you want to estimate the variance of some variable for a population, and you only have data from a sample of size n, there is an $n-1$ in the variance formula, not n, since you "used up" one data point to calculate the average first. So, if you want to calculate residuals, after fitting a curve that has m parameters, you have to realize that you have only $n-m$ independent points (or "degrees of freedom") left, so you have to divide the sums by $n - m$.

This statistic is called the "**reduced chi squared**" or "**chi squared per degree of freedom**", usually denoted as χ_v^2 (Greek letters "chi" and "nu"), but I use χ_y^2 here since we might do the same calculation with the horizontal residuals and errors if the model function is invertible (and then we'll call it χ_x^2):

$$\chi_y^2 = \frac{1}{n-m} \sum_{i=1}^{n} \frac{\left(y_i - f(x_i)\right)^2}{\sigma_{y,i}^2}$$

$$\chi_x^2 = \frac{1}{n-m} \sum_{i=1}^{n} \frac{\left(x_i - f^{-1}(y_i)\right)^2}{\sigma_{x,i}^2}$$

With the data and errors in Fig. 3.65, we get
$\chi_y^2 = 17.5059$ $\chi_x^2 = 44.3280$
and with those from Fig. 3.66
$\chi_y^2 = 0.7002$ $\chi_x^2 = 1.7731$

If both values were zero, the fit would be perfect, and that rarely happens in reality. In fact, we can be happy when χ_y^2 and χ_x^2 are about 1 since that means that the residuals and the measurement errors are in the same order of magnitude. We don't like a big chi^2, but sometimes there is a good explanation for it; see, for example, the "Running records" case study (p. 289).

Remark: *If* the model function is linear, and *if* that is really the appropriate function to describe the data, and *if* the regression is done with OLS, and *if* the residuals are normally distributed (in other words, the data are "homoskedastic"), an estimation for the parameter uncertainty can be calculated, and from that a "*p* **value**", which is the probability that the found parameters are not zero. These are very popular, and produced by most software programs, because they give the illusion of certainty about the following question: "Is there really a *significant* linear relationship between my variables?". Usually, if $p < 5\%$, the answer "yes" is taken for granted.
Of course, this 5% is quite arbitrary, and very often a linear model is used when it shouldn't be, and worse, this estimation doesn't take measurement errors into account! So this p value becomes really quite fuzzy since the measurements also have some fuzziness, and the other assumptions might not be fulfilled! It's really difficult to produce an absolute probability value here; it might be better to use common sense. At least, we have a method to estimate the parameter imprecisions that works for any kind of model (see p. 111). See,
https://en.wikipedia.org/wiki/Simple_linear_regression#Normality_assumption.

3.5.2. Best fit = best model?

So, if this χ^2 tells us something about the "goodness-of-fit", should we conclude that the best model for a given dataset is the one that fits best, meaning the one with the lowest χ^2 or another statistic like Pearson r^2? The author of the online course from which you see a screenshot here (Fig. 3.67), seems to think so. The "curved" function fits better than the straight line, so the hypothesis that there is a "curvature effect" can be accepted, he says. Really? He seems to forget that a function that has more parameters can automatically adapt itself better to the data and hence follow the curvature!

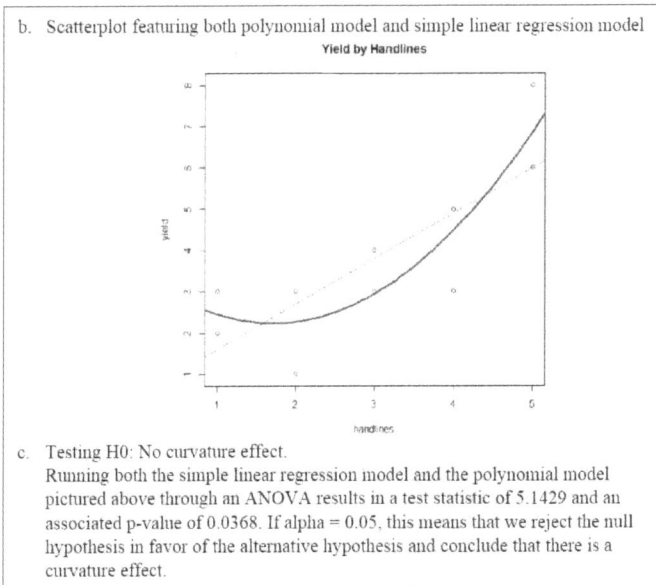

b. Scatterplot featuring both polynomial model and simple linear regression model

Yield by Handlines

c. Testing H0: No curvature effect.
   Running both the simple linear regression model and the polynomial model pictured above through an ANOVA results in a test statistic of 5.1429 and an associated p-value of 0.0368. If alpha = 0.05, this means that we reject the null hypothesis in favor of the alternative hypothesis and conclude that there is a curvature effect.

Fig. 3.67. NOT the good way to judge a model! (Screenshot of an online course).

Some software programs even offer the possibility to choose "automatically" the "best" model according to Pearson r^2. Users like that of course! That allows them to blame the software instead of their own (lack of) thinking.

Let's have a look at an example. In the following, you see the same five data points and three different functions that seem to fit nicely. OLS was used in the three cases to allow comparison. All of the χ^2_y values are quite high, which probably means that the errors on the data were higher than assumed. Just by looking at the numbers, one would be tempted to choose the function in the center. Now, that also happens to be the one with the most parameters: four! That means the risk for "overfitting" is very real here! With five points, there is only one degree of freedom left. The χ_y^2 calculation takes that into account, but still... Fig. 3.68 shows a quadratic function (three parameters) and Fig. 3.69 a cubic function, which *always* fits better than a quadratic one, since the third degree term corrects the other terms. A fourth degree polynomial would fit perfectly. (Don't worry about the small font in the graphs; just look at the dots and the curves.)

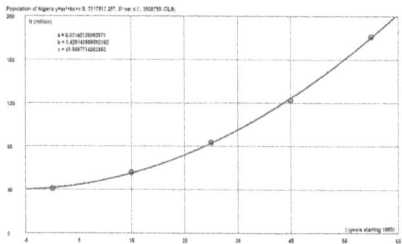

Fig. 3.68. $\chi_y^2 \approx 3.5$ million.   **Fig. 3.69.** $\chi_y^2 \approx 1.6$ million.

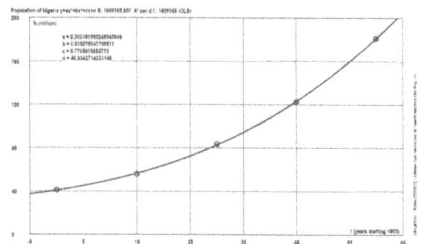

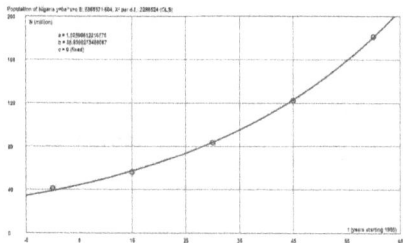

Fig. 3.70. $\chi_y^2 \approx 2.3$ million.

The truth becomes much more exposed when we widen the x range of the graphs, to see how the extrapolations look!

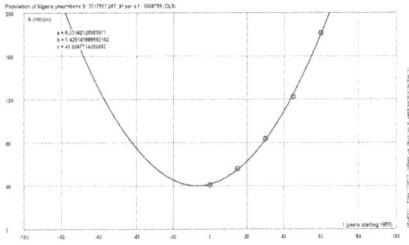

Fig. 3.71. Quadratic model extrapolated

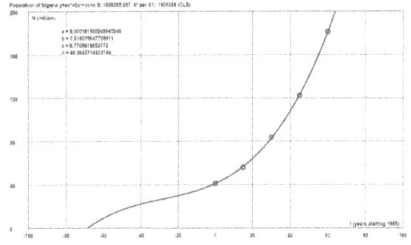

Fig. 3.72. Cubic model extrapolated

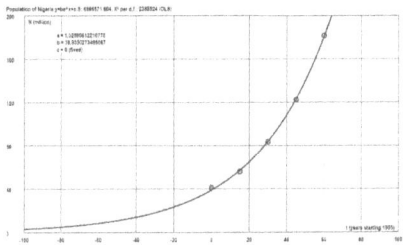

Fig. 3.73. Exponential model extrapolated

Now, if I tell you that the graphs show population data, you might realize how ridiculous the first and the second model are! The third model (Fig. 3.70) is an exponential function, which assumes that there is a constant percentage growth going on, and that seems to be plausible, in a good approximation. The first model would "predict" the same growth backwards, and the second would "predict" a negative population some time ago. See also the case study "Population of Nigeria" (p. 284).

We could have known this *before* doing the regression... I told you so!
But even if you "chose wisely", it's still a good idea to **check the extra-polation, again**, since there is always a risk for **overfitting**, meaning that accidental fluctuations (the "noise") can be seen as "reality" and hence fitted into the model... See for another example, the case study "Temperature in one day" (p. 269).

So, the lesson also to be remembered from this observation is that you should **try to use as few parameters as possible**! If a model with two parameters describes the data as nicely as one with four, the one with two is probably better. And that sounds a lot like the principle of **"Occam's razor"**! (Google that if you have never heard about it!) If you have to choose between two models with the same number of parameters, the one producing the smallest χ^2 might be the best, but it's no proof.

3.5.3. Watching the iteration process

In most software programs, you write some lines or you click a button, and you get "the" results. And you have to thrust them as they are. In some cases, you get an "exact" result using a nonlinear transformation, and you know what can go wrong with that. In some iterative programs, you get some intermediate results after 1, 2, 3, etc. iterations, which is better than nothing.

In few programs, like FittingKVdm, you can follow the whole iteration process visually. You might think that's a waste of time, but no, it isn't, especially if you are exploring unfamiliar data.

First of all, it's fun and exciting to watch the parameters come, especially if you can follow it visually, like in FittingKVdm. Normally, you'll see the numbers and the graph converging and stabilizing quickly. So, what could go wrong?

(1) The iteration converges quickly, but to a not expected value. For example, this can easily happen when looking for periodicities if there are not enough data available.

Example: If you try to fit a sine wave through the points in Fig. 3.74, and you set the initial T value to 1, it will nicely converge to $T = 1.071$, but you see with your naked eye it looks suspicious. If you start with $T = 3$ though, it will converge to 3.008, and it seems to look better.

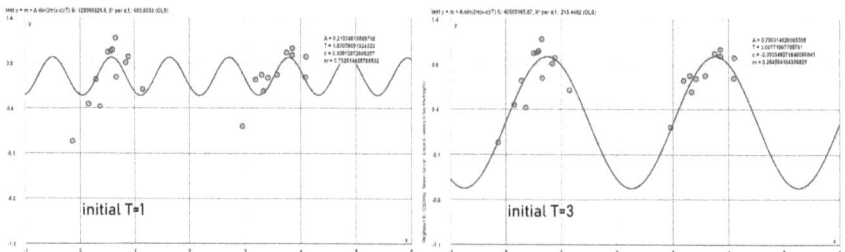

Fig. 3.74. Different initial values can cause different convergences.

Is it objectively better? Well, if you look at the *S* vs. *T* graph (Fig. 3.75, done with FittingKVdm), you see indeed that the deepest minimum is around $T = 3$.

This kind of situation can sometimes be solved if you can provide better initial values, or even fix some parameters (keep the value unchangeable). If you know that there is a 24 h or one year cycle in your data, put it in your model, so the software can hunt more efficiently for the other parameters.

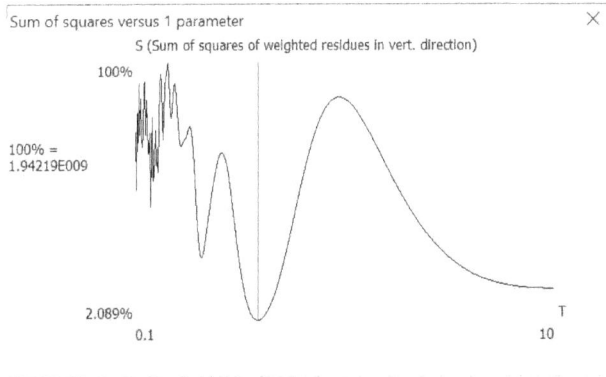

Fig. 3.75. S vs. T for the above data.

(2) Worse is the situation when the parameters don't want to converge, although the graph doesn't change visibly anymore.

This happens when the algorithm searches for a typical feature like a peak, an inflection point, a horizontal asymptote, an exponential behavior, etc. that it can't find in the data. This can only be solved by

• collecting more data, but at least some points in an interesting area, to make the special features visible,
• fixing some parameters according to a theoretical assumption, for example, if you know the period, the average, or the limit, set it to a fixed value,
• or by choosing a more appropriate model.

Example: Try to fit a logistic function through 10 data points that lie more or less on a line; see Fig. 3.76. You need a lot of phantasy to see that the *y* values go to some limit if *x* goes to ±∞. So, the algorithm has difficulties seeing the pattern as well. After 1000 iterations, it thinks the final limit is 1.93. After 10000 iterations, it went up to 3.12, but the graph, and the *S* value, almost didn't change (*S* went from 1839051.5 to 1824837.7).

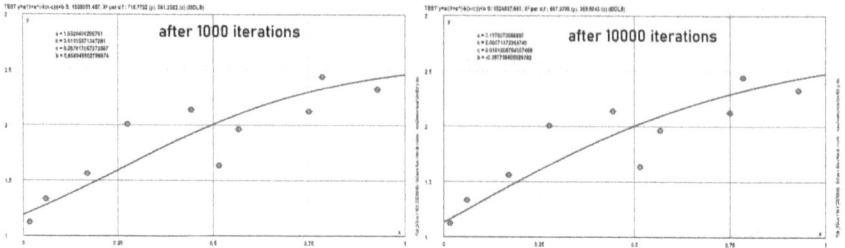

Fig. 3.76. A suspicious fitting: almost the same curves, very different parameters!

Now, if we add one point on the left and one on the right, we get a rock solid convergence: the parameters practically don't change anymore after 50 iterations.

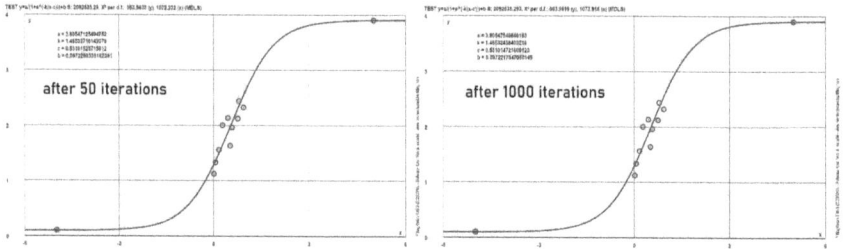

Fig. 3.77. A solid fitting!

So, studying the iteration process can give you a lot of insight and help you to see which additional data might be required to get a stable model.

See also the "Concrete strength" (p. 209) and the "wind speed" (p. 201) case studies.

3.5.4. Watching the worst case deviations

Using a Monte Carlo technique, adding noise to the data, and refitting, we can estimate the precision of the obtained parameters, as I explained before (see p. 111).

Studying the graphs of the model functions with the most deviating parameters can give us some more insight about the stability of the model. If a slight change in the data, within the range of the measurement errors, can have a disastrous influence on the model, that's not a good sign. We would like to see the "worst" curves still going through the error flags of the data points.

For example, some data, forming a pattern like a peak, were fitted with four different peak shaped functions, and then this Monte Carlo technique was applied (using FittingKVdm v. 1.6), as you can see in the next graphs (Fig. 3.78). (Don't try to read the small letters, just look at the dots and the lines.)

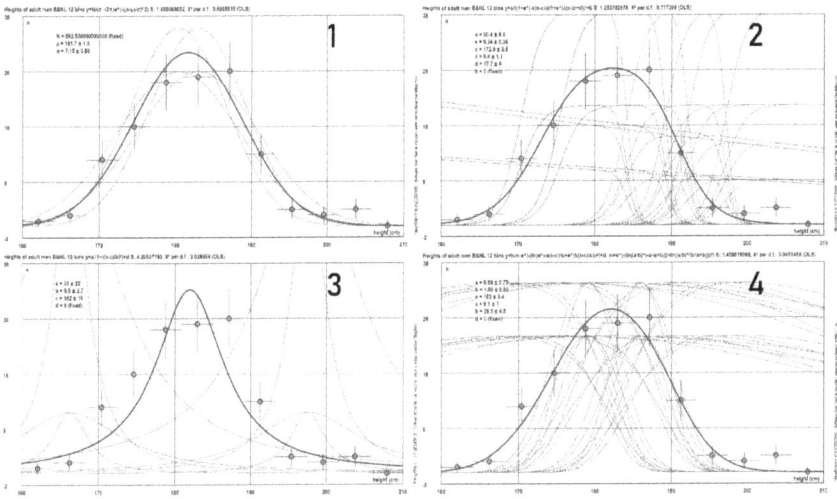

Fig. 3.78. Four fitted models through the same data points (solid lines) and worst case scenarios if some noise (with the same amplitude as the measurement imprecisions) is added to the data (dashed lines).

Only the first model seems to be stable enough to withstand the "shaking" of the data!
Very chaotic things happen with models 2, 3, and 4; they seem to be very sensitive to small data changes. So, if you are not sure which model is best, this might give you an indication. It's not a proof but a hint. Anyway, it's always the best if the choice of model is inspired by a theory. By the way, this happens to be the case since the data were the height distribution of random adult men, and the first graph shows the Gauss function, depicting a normal distribution.

3.5.5. How do the residuals look?

• Let's have a look at Fig. 3.79. In the leftmost graph you see some data points and a best fitting curve through them. You see that some points are a bit below the curve and others are above, and overall it looks like a good fit. The rightmost graph shows the vertical residuals: the observed minus the calculated values by the model function, so that's like a "microscopic view" of the measurement dots with the curve pulled flat in the middle. They look randomly scattered around the center with a higher concentration around the line in the middle. Out of the 20 points, 15 are between the dotted lines of $\pm\,\sigma$ and only one is outside the $\pm\,2\sigma$ band, so the residuals are more or less normally distributed around zero. This is **what we expect when the model function is appropriately describing the data**.

Remark: The χ_x^2 and χ_y^2 values are kind of high (88 and 1665), but that might mean the $\sigma_{x,i}$ and $\sigma_{y,i}$ values are higher in reality than assumed, or another independent variable has a very small influence.

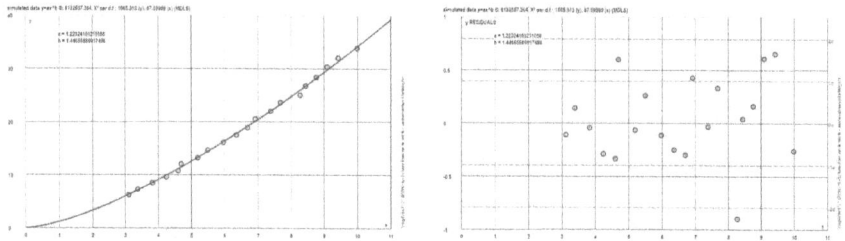

Fig. 3.79. Some data, a fitted curve and more or less normally distributed vertical residuals.

• In the next graph (Fig. 3.80) you see the same points but fitted with a different function. Although the points are still reasonably close to the curve, you see a *systematic* (not random) deviation, which is magnified in the residuals plot on the right: in central part of the graph, the points are above the curve (positive residuals), while the leftmost and the rightmost points are below the curve (negative residuals). There is clearly a pattern.

What does this observation teach us? With a "correction factor" depending on x (more than 1 in the center, less than 1 in the left and right parts), we might push the curve better between the dots. That means that the **model function needs to be adjusted** to properly describe the data.

There might be a hidden variable that is correlated with x too, like the temperature in the "diode" case study (p. 236).

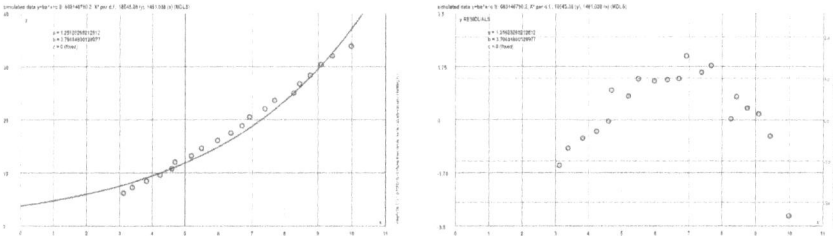

Fig. 3.80. A pattern in the residuals: the model function might be inadequate.

● In Fig. 3.81, we see something strange: one point obviously stands out, and all the residuals except one are negative. This is what looks like a real "**outlier**". Apparently, this one data point has had the force to pull up the whole curve. Most probably, this point was a faulty measurement, or a typo, and for sure, you should investigate it. If it was a mistake, it should be removed or corrected, since it messes up the whole fitting. If it was not a mistake for sure, you might have to reconsider the choice of your model function. Could there be a peak somehow? Anyway, you have to undertake something. See also the case study "Planet orbits" (p. 192).

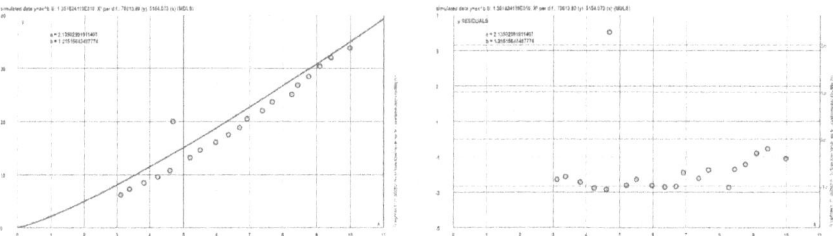

Fig. 3.81. An "outlier" disrupting the fit. Graph on the right: residuals.

● Fig. 3.82 shows a different pattern: on the left side, the measurement points are close to the fitted curve, but then gradually the residuals become bigger and bigger, positive and negative. In other words, the variance of the residuals seems to be somewhat proportional to x. This phenomenon is called "**heteroskedasticity**".

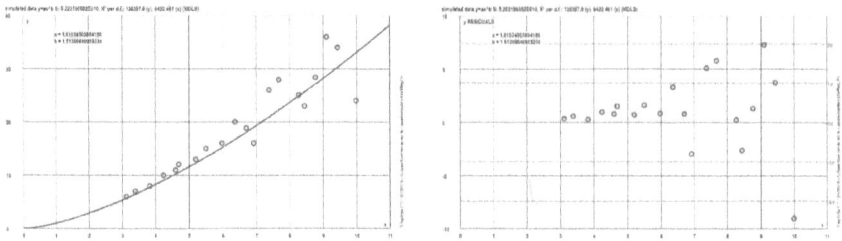

Fig. 3.82. "Heteroskedasticity" with respect to the fitted curve. Graph on the right: residuals.

What kind of a situation can we imagine producing such a plot? A few examples are as follows:

(1)   Suppose x = the age of a person and y = his/her body mass. During childhood, most people's masses will stay more or less near a "standard curve", of course with deviations of, say ± 30% or so very well possible. But then, in adulthood, some people, who have a healthy lifestyle, stay slim while others, who eat too much sugar, don't sport, etc., might grow morbidly obese. The character traits that lead to good or bad habits, and genetic differences, are probably not related to the age, but... the *cumulative effects* of these habits or genes grows with age, so that's why we see a different path for healthy people on the downside of the graph and another extreme path on the upper side.
See also the case study "Food expenditure" (p. 367).

(2)   Suppose x = the number of words in a book or a text, and y = the number of different words. Make a dataset from various writers. For short texts, the proportion y/x will not differ very much among writers, but if the books become thicker, the cumulative effects of the authors's vocabulary richness or lack thereof will become more and more visible. Again, the property of having a large vocabulary or not, is not related to x, but its *cumulative effect* does become more visible when x is bigger.
See also the case study "Vocabulary of a writer" (p. 354).

(3)   Suppose x = the length of a river and y = the flow rate at its mouth. We can expect the relationship to be more or less quadratic, since the water comes from the area near the river (the basin). The longer the river, the bigger the probability for an extremely low or high flow rate, since the river might cross a wide arid or hot area, or on the other hand, a humid or cold area. Again, the climate in the river basin has nothing to do with the length of the river, but the *cumulative effect* of that climate

shows up more if the length is bigger. See also the case study "Rivers" (p. 280).

Remark 1: Some people, especially econometrists, seem to make a huge problem out of this heteroskedasticity, like it is some kind of "data disease that has to be eradicated". Why? It is *not* true that the residuals have to be normally distributed for a regression to be valid, as you can read often. No, the fitted parameters will still be optimal, but... the **p values** for the parameters (see p. 156) will not be correct anymore, because they assume a normal distribution of the residuals. For more details, google "**Gauss–Markov theorem**".
Since econometrists (most of them, is my impression, sorry if you are one of the exceptions, I don't want to offend you) seem to have a dogmatic belief in the "slope, intercept and *p*-value" narrative, they desperately need that "*p*" to decide if a hypothesis can be accepted or rejected. But, on the other hand, they ignore measurement errors, they apply linear regression to almost everything, even without asking themselves if the intercept shouldn't be zero. If they see "heteroskedasticity", they panic, and they find all kinds of weird tricks to "clean it up", like dividing each *x* and *y* value by the square root of *x* (*assuming* another linearity, that of the residual variance vs. *x*), or taking the logs of variables, just because "it looks better". That's like taking a painkiller without trying to understand the causes of the symptoms. Instead of searching for a more appropriate model function, or for other ways to judge the validity of a hypothesis, they believe the *p* values produced by their linear regression software are sacred numbers, even after molesting the data and therefore changing the weights of the measurements. Wishful thinking, in my humble opinion! So, don't mess with your precious data. They should be studied as they are, not botoxed! Or, as Nobel Prize winning economist Ronald Coase (1910–2013) said it: "If you torture the data long enough, they will confess."
If you want to know more, have a look at this video by Justin Zeltzer ("Zedstatistics" channel): www.youtube.com/watch?v=2xcUup_-K6c.

Remark 2: If all the data points in a set are nicely near some curve, it doesn't really matter if you leave out some of the points or if some are accidentally missing.
But, there is a danger with heteroskedastic data (and in fact with all kinds of "cloudy" data): if you select more of the lower or the higher points, be it accidentally or deliberately (if you want to "prove" something for political purposes, for example), the resulting parameters will

turn out to be very different, and you would be totally unaware of this bias if you just look at the scatterplot! So, the points have to be selected very randomly, and all the available data have to be used!

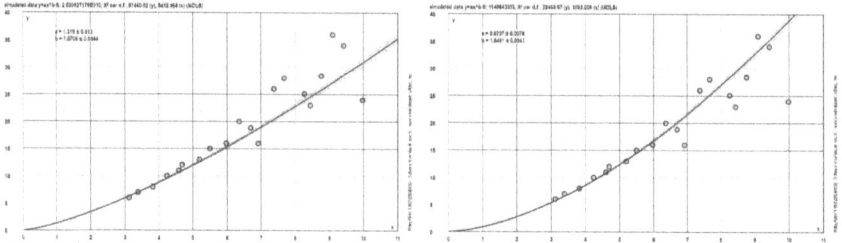

Fig. 3.83. Heteroskedastic data: the model is very sensitive to biased sampling! Left: three higher points (in red) were left out of the fitting; right: three lower points were left out.

4

Case Studies

Learn from examples.

What is the best model in a given situation, and why?

Recognize patterns!

Learn to ask yourself always if your experiment or data gathering could be improved.

Real life data reveal secrets that simulations never will.

Explore with common sense!

Look at the important details!

All the shown data files are included as ready-to-use files in the FittingKVdm software package!

4.1. Physics, Chemistry, Engineering...

4.1.1. How much does a coin weigh?

Suppose you want to know how much a 2€ coin weighs.
Not difficult, you think... We just put a coin like that on the letter scale and we know it? Okay, "7 grams" says our kitchen scale. Hmm, not very accurate, huh? That should mean the real mass is somewhere between 6.5 and 7.5 g. That's a margin of error of plus or minus 7%. Unfortunately, we don't have a more precise scale. What should we do if we still want to know the mass more accurately?

The smart reader now comes up with the obvious solution: just weigh several pieces and divide the total mass by the number! The smarter reader will now say that the masses of the pieces might differ a little bit. Okay, so we won't know the precise mass of one particular coin, but at least we'll know the average mass.

OK, let's add one.

Fig. 4.1. Coin on a low-precision scale.   Fig. 4.2. Oops! Magic?

Oops, what happened now? One weighs 7 g, two together weigh 16 g? Was the second coin perhaps heavier? We put them on again, but in reverse order... Same result, so that was not the explanation.

And with three pieces it gets even crazier: 25 g! With each subsequent piece, 8–9 g was added, so we can assume that the real mass is somewhere in between. With the 34 pieces we collected, we came to 287 g, so 8.441 g per coin. In the best case scenario, *if* the scale was properly calibrated, the accuracy should now be 34 times better, so ± 0.015 g instead of ± 0.5 g.

But what's the problem with the scale now? We had pressed the "tare" button in the beginning so that it was nicely showing zero. Well, maybe it needs a bit of "starting weight" to start working properly (linearly)? After all, such a scale is often used like this: you put a bowl on it, press the tare button until the screen shows 0, and then you pour the desired amount of flour, sugar, or whatever in the bowl.

So, we repeated the measurements with a bowl (of 116 g). One coin now weighed 8 g, two 16, three 25, etc., which looks more "normal". Strangely enough, the 34 coins now weighed only 284 g together, which means that one coin weighed an average of 8.353 g.

Hmm... that 3 g difference on the total cannot be explained with that 1 g too little on the first coin on the empty scale. Is there anything else going on? The scale must have some bias or maybe even nonlinearity.

To find out, we can do a linear regression analysis with x = the number of coins and y = the total mass. You can do this with many programs, e.g., Graphmatica, GeoGebra, and the TI-84 calculator, but we want to do it as precisely as possible. In FittingKVdm, we *have* to enter a "measurement error" next to each number. Now, for the y values, we can assume ± 0.5 g but that is only realistic if the scale is perfectly linear and calibrated, which doesn't seem to be the case. So, ± 1 g might be safer. Note: As long as the errors are the same for each y value, it won't change the fitted parameters, only their confidence intervals.

But what do we do with the x values? Aren't they integer numbers with zero "error"? If we count seven coins, it's seven coins precisely, not 7 ± 0.1 or so, isn't it? Well, actually it's not that simple: "seven coins" means: the weight of seven coins divided by the weight of seven average coins, so it *could* be 7 ± 0.1, but we don't know the variability of the coin masses. If the masses of coins have a standard deviation σ, then the standard deviation of a group of x randomly chosen coins is $\sigma \cdot \sqrt{x}$. I have no idea how much tolerance vending machines have, but I guess it won't be more than 1%. So, we can safely assume one coin has a mass of about 1 ± 0.01 average mass, two coins weigh 2 ± 0.014 average masses, etc. Again, the precise value of σ doesn't matter for the parameter fitting, but the fact that the weight is different for each y value will have some influence.

MDLS regression with the "linear" model produces: $y = ax + b$, with
$a = 8.492 \pm 0.026$ (= the actual expected mass per coin)
$b = -0.74 \pm 0.52$ (= the bias of the scale)

The confidence intervals for the parameters have to be taken with a grain of salt, because our σ_x was also a rough estimation. But anyway the a value is obviously much more reliable than the b value.

The χ^2 per degree of freedom in the y direction of 0.56 suggests that the assumed σ_y value was realistic. In the x direction, $\chi^2 = 7.3$, which suggests that the variation in the coin masses might be a bit greater than 1%.

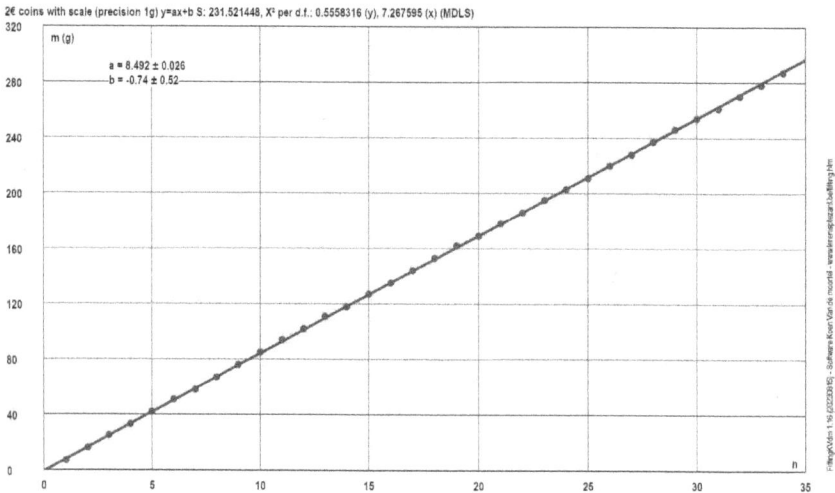

Fig. 4.3. The measurements and the best fitting line.

Using OLS here doesn't change much to parameter a (8.506 ± 0.030); b changes a bit more (0.95 ± 0.51) but still not significantly. The dots seem to be very nicely on the line, so if our scale had some "nonlinearity", it's certainly not much.

Okay, the really pedantic reader might want to check if a quadratic fitting ($y = ax^2 + bx + c$) performs better. This produces

$a = -0.0059 \pm 0.0026$
$b = 8.672 \pm 0.095$
$c = -1.46 \pm 0.69$

The graph looks just the same because the quadratic term is indeed very small. Model functions with more parameters (three instead of two) will always perform better, but let's also compare the residuals.

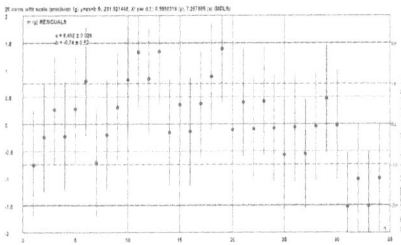

Fig. 4.4. Linear fit residuals.

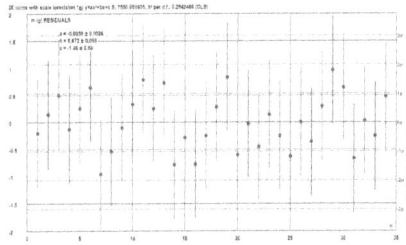

Fig. 4.5. Quadratic fit residuals.

Here we do see a small difference: the residuals from the linear fit (left) seem to form kind of a "hill", while those from the quadratic fit seem more normally distributed around the zero line. This means the scale seems to have a slightly quadratic characteristic. I wouldn't worry about that too much when using it though.

Some explanation is as follows: the operation of such a scale is based on electrical resistance or capacitance changes due to strain, and strain is approximately, but not perfectly, directly proportional to the force that causes it. For more information about the possible mechanisms of scales, take a look at https://en.wikipedia.org/wiki/Load_cell.

Conclusion: The average mass of a 2€ coin obtained by linear regression (8.492 g) is a bit more reliable than the one obtained by just taking the average (8.441 g). Using a bowl makes the average a bit worse in this case (8.353 g). The confidence interval of $\pm$ 0.026 g from the regression is probably more realistic than the one estimated by just averaging and assuming a perfectly calibrated scale ($\pm$ 0.015 g).

FittingKVdm example data file: "Coin mass.dta1".

4.1.2. Friction between surfaces

When you try to push or pull an object that is standing on a surface (e.g., the floor), you experience a resistance, a friction force. That friction (F_f) is proportional to the weight of the object; $F_f = \mu mg$ (m = mass, g = gravitational field strength ≈ 9.81 m/s^2). The proportionality coefficient μ depends on the kind of material on the bottom of the object and the floor. In fact, there are two versions of μ: the so-called "static" μ_s and the "dynamic" μ_d; the first applies when the object is still standing, and the second when it is already shifting. From experience, we all know that $\mu_s > \mu_d$; once a chest is moving, it's easier. Example values are for steel on ice, $\mu_d \approx 0.03$, for rubber on dry asphalt, $\mu_d \approx 0.9$. See, www.engineeringtoolbox.com/friction-coefficients-d_778.html

How can we measure this μ_d and check if is really constant? Well, if you pull the object with a constant speed, the net force on the object is zero, which means the friction and the pulling force are equal in absolute value. You can show this with a simple home or classroom experiment: hook one chair to a dynamometer and drag it on the floor, while keeping the speed as constant as possible; the dynamometer should give a more or less stable reading then. Then put a second chair on top on the first, and repeat. Then a third, etc.

My test measurements were done on a ceramic floor, with chairs having some kind of plastic (?) caps on their legs. The dynamometer was one of those typical things people use to weigh their luggage before going to the airport. The instrument gave a reading in kg, with a resolution of about 0.1 kg, but the actual measuring error was bigger because of the fluctuations during the movement; I would estimate the accuracy to be about 0.5 kg. If you want the force in Newtons, just multiply by 9.81, but that is not needed since these multiplicators cancel out.

Only one chair was weighed (since it was impossible to hang more than one chair on the dynamometer), with a precision of ± 0.1 kg, so the error on the mass of n chairs was $\pm n{\cdot}0.1$ kg. With more than four chairs, the situation became unstable.

As expected, a linear relationship with $b = 0$ fitted through the data within the error margins. The value for the dynamic friction coefficient obtained was: $\mu_d \approx 0.233 \pm 0.023$, which seems a bit low, but probable, since the floor was quite smooth.

Pulling force vs weight (chairs) y=ax+b S: 13.57464125, X² per d.f.: 0.2855887 (y), 27.51645 (x) (MDLS)

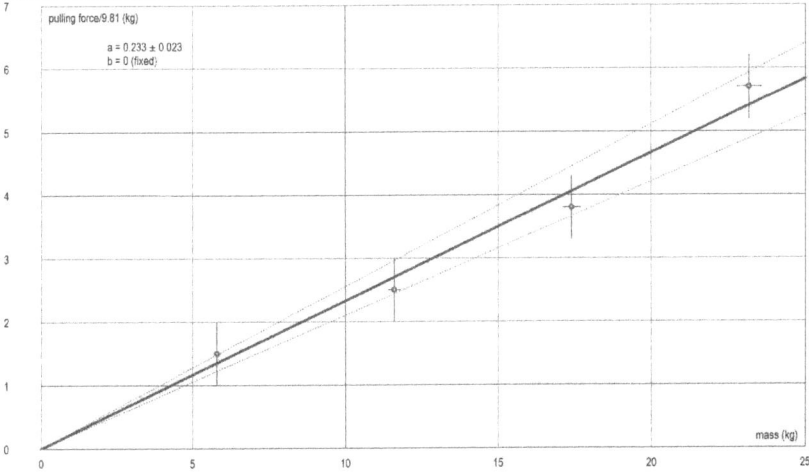

Fig. 4.6. Pulling force needed vs weight.

An easier version of the experiment might be done with a carton box in which you put 1, 2, 3, etc. books, and drag it on a table.

Is μ_d really constant and independent of the velocity? To answer that question, we would need more accurate measurements, for example, using a motorized winch to drag the object with a more constant speed, and an electronic strain gage with a data logger.

FittingKVdm example data file: "Friction of chairs on floor.dta1".

4.1.3. Does a sponge obey Hooke's law?

If you press pull a spring, the force (F) you need is proportional to the distance (x) you press or pull it away from the equilibrium situation: $F = kx$ (often written as $F = -kx$ to reflect the opposite senses of the force and the x direction). The proportionality parameter (k) is called the spring constant. This "law", named after the English physicist Robert Hooke (1635–1703) is the basis for a dynamometer and a lot of weighing devices. Would this pattern also show up with a sponge? Or might it be more like $F = ax^b$ with $b \neq 1$?

Let's define the question a bit more precise: suppose you want to push a vertically standing ruler in the middle of a dry sponge standing on its side; how much force would you need for a distance of x millimeter?

How can we measure this force easily? Placing it on a weighing scale is probably the simplest! That's what I tried to provide you with some data! To keep the ruler easier in the same position, a Lego construction was built (see Fig. 4.7), including a lamp to make it easier to read the ruler. The ruler was pushed in slowly step by step, and each time, the mass given by the scale (m, in grams) was recorded. So, the force F in Newtons was given by $m/1000 \cdot 9.81$. Although the resolution of the scale was 1 g, I estimated the error on the mass measurements to be about ± 2 g, increasing to at least ± 10 g, since it was very difficult to keep the ruler totally stable, and hence the reading fluctuated a bit. A more sophisticated mechanical construction would improve the precision here.

Fig. 4.7. The experiment set-up.

Problem: Once the force was about 1 kg, serious *plastic* (inelastic) effects started to disturb the measurements; after keeping the ruler a few seconds at the same position, the resisting force lowered, because the sponge stayed somewhat deformed, causing a kind of hysteresis; lifting the ruler up and back down didn't cause the same resisting force. So, the simple model will no longer be valid.

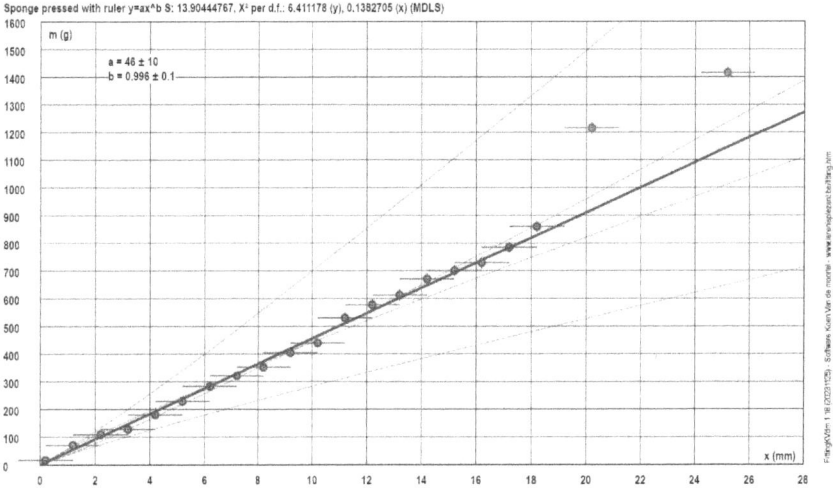

Fig. 4.8. The measurements and the best fitting power function.

Fitting the measurements with a power function, gave an exponent $b = 0.996 \pm 0.10$. That means that the assumption that Hooke's law is obeyed, is very probable.

To find k, we have to refit with a fixed $b = 1$. That produces $a = 45.610 \pm 1.7$ g/mm (1 g/mm = 1 kg/m), so $k = 9.81a = 447 \pm 17$ N/m.

FittingKVdm example data file: "sponge-ruler.dta1".

4.1.4. Observing a falling pear to find "g"

When we drop an object, it begins to accelerate downwards. If friction is negligible (if the object is small and heavy, unlike a leaf or a plume), the speed (v) increases linearly with time (t): $v = gt$, where g is the **gravitational acceleration** caused by the Earth's attraction force.

The distance (s) traveled is easy to calculate as the *average* speed times the time: $v_{av} t$. Since the speed changes linearly, you can take as an average: the average of the initial and final speed, $v_{av} = (0 + v_{end})/2 = v_{end}/2 = gt/2$, so after a time t elapses, we have $s = \frac{1}{2}gt^2$.

Now, how big is this g? If you paid attention in school, you remember that the *average* value on Earth is 9.81 m/s², but how can we measure the precise value of g, for example, on the spot where we are right now?

Transforming the above formula, we get: $g = 2s/t^2$, which means that we just need to measure how long it takes an object to fall from a given height. Now, that's easier said than done. Just try dropping an object from a few meters high and chronometer it! It moves so fast that it is virtually impossible to look and press the button accurately, and you don't know how much of the measured time is actually your own reaction time. If you could drop something off a very tall building, it might be a little easier, but still... Fortunately, today we have technical tools that Galileo Galilei did not have with his fall test from the Tower of Pisa!

First of all, acquire an accomplice who can drop an object from a measured height. Newton did it with an apple, but it works with a pear too! Take your smartphone, or better, a high resolution camera that can film, set it to the highest speed, and make sure there is a lot of light (to keep the shutter time small). Sixty frames per second is much better than thirty. Ask your assistant to drop the object and film it. Send the video to your PC via a direct connection (Bluetooth or a USB cable, for example), not via Facebook Messenger or something like that because that crushes all your files and makes them 30 frames/s! Open the video with a player that can show frame by frame, such as Windows Media Player (go to "options", "playback speed settings"). Find the exact start of the fall, and then use the "arrow-right" button to go to the next image a number of times, and don't forget to count, because unfortunately you cannot read the time at the bottom precisely enough.

That is what I tried. In my set-up, the pear fell 1.90 m, and I counted 38 at 60 frames/s, although in the first and the last it was not entirely clear whether the fall had already started or was still in progress... This gives us a value for g:

$$g = \frac{2 \cdot 1.90}{\left(\frac{38}{60}\right)^2} \approx 9.47368 \frac{m}{s^2}$$

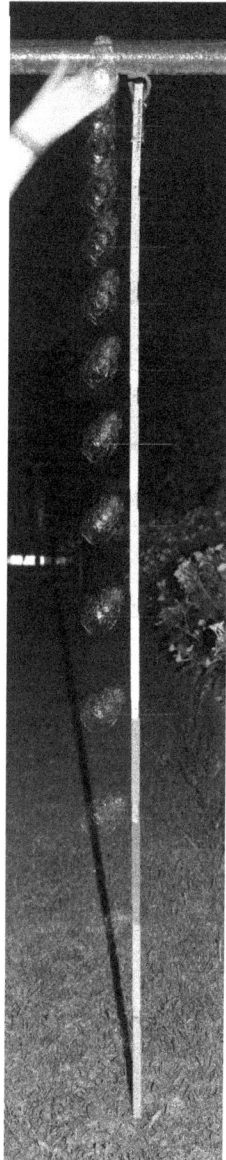

Hmm... that is "somewhere near the expected value", but we can't be really satisfied about this result.

How exactly *can* we determine g in this way? Like I said, it could just as well have been 37 or 39 statues, and that number is squared in the denominator, so it has a major influence. Indeed, with 37 frames we would get $g \approx 9.99270$, and with 39, $g \approx 8.99408$.

Wow! Such a small deviation in the time measurement makes a dramatic difference! So... if you want to do this properly, you need a camera that can film at a much higher speed, but that's expensive. Or an electronically controlled system to release the object and a sensor that measures the precise time when it hits the floor and ideally everything should be done in vacuum, but that's more expensive.

Is there no other simple way to properly determine that gravitational acceleration? Well yes of course, and regression will help us! We could use all the frames and look on each one how far the pear advanced, and find the best fitting quadratic function.

But, it's more fun with a *stroboscope*, you know, one of those things that flashes intensely many times per second. Some camera flashes can also be used as a stroboscope.

Fig. 4.9. Long exposure photo of the falling pear, lit with a stroboscope.

If you take a photo with a long shutter speed of the object falling next to an upright ruler, and flash a number of times during that time, you can read the distance fallen with each flash. So, I tried, with a 2 m folding meter hanging from a swing, a Canon reflex camera (EOS 5Ds with 40 mm lens, but it can be done with a cheaper one of course) on a tripod, and aimed at half height perpendicular to the meter, and a 600EX II-RT "speedlite".

How did I find the best possible settings? The drop height was 2 m, so we expected a fall time of no more than 2/3 s, so a 1 second exposure time would certainly be enough. The aperture and sensitivity had to be set so that the image without the flash was just dark enough. This worked with the aperture at f/4 and the ISO at 200. (Hence, f/5.6 and ISO400 would also work.) The flash was set at 1/128 of its full power (so as not to overload it) and at 25 flashes/s. Why not more? Simply because the images of the object then overlap and the positions can no longer be read. By using the self-timer, I was able to avoid the need for an assistant.

Experiments are always a bit more difficult in practice than in your mind: in the first photo, nothing of the pear was visible, while the rest of the setup was neatly lit. How did that happen? Well, this way the moving pear got one short flash at each position, while the background got 25! Using a stronger flash was not a solution, because then the measuring stick would be completely overexposed and therefore un-readable! Now what? The pear didn't reflect the flash enough, so I had to take care of that... A piece of aluminum foil can do wonders! Now it was just lit enough except during the last flash (see Fig. 4.9).

Using MS-Paint, I drew horizontal red lines from the lightest zone of the pear to the measuring stick, to make it easier to read. The first pear images were a bit too unclear, so I left them out.

The measurement results are as follows:

t (s)	0	0	0.1	0.12	0.16	0.2	0.24	0.28	0.32	0.36	0.4	0.44
s (m)	0.1	0.15	0.22	0.32	0.42	0.55	0.69	0.85	?	1.21	1.41	1.63

So, apparently our clock started actually when the pear had already fallen a distance s_0 of 83 mm. At that time, it already had an "initial" velocity v_0.

The formula describing the fallen distance versus time (neglecting friction and the extremely small difference in gravitational field strength between those two meters) is

$$s(t) = \frac{1}{2}gt^2 + v_0 t + s_0$$

See, for example, https://en.wikipedia.org/wiki/Free_fall.

So, I put the measurements in a regression program (FittingKVdm, but it can also be done with a TI-84 calculator), choosing a quadratic model $y = ax^2 + bx + c$ (x = time; y = s; a = $g/2$; b = v_0; c = s_0). The measurement errors on the times could be assumed to be quite small (in the order of 0.005 s if the flash worked as it should). The positions were measured in millimeters, but since it was not always obvious to indicate the precise position of the pear's center of gravity, it seemed safer and more realistic to assume a precision not better than 5 mm.

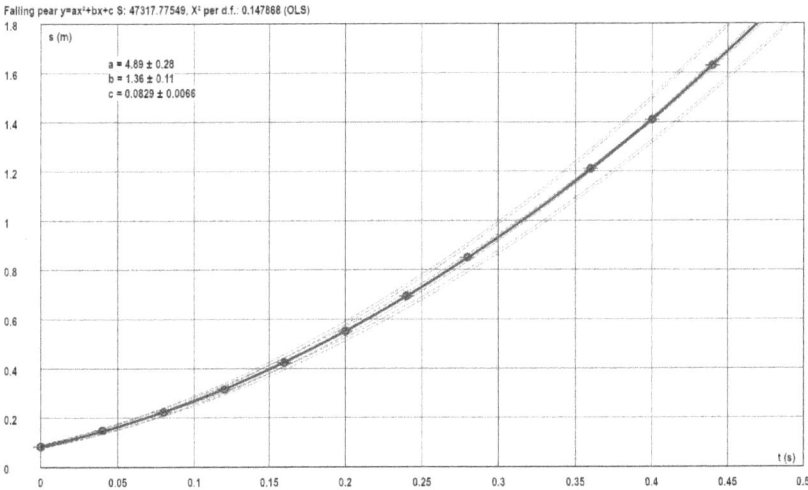

Fig. 4.10. The distance traveled by the pear vs time, and the best fitting quadratic function.

The obtained $a \approx 4.89479179$, gives us: $g = 2a \approx 9.78958358$ m/s^2. If the assumptions about the measurement precisions were correct, $\sigma_g = 2\sigma_a \approx 0.56$ m/s^2, so: **$g = 9.79 \pm 0.56$ m/s^2**.

The uncertainty seems bigger than reality, if you see how nicely the dots follow the curve. That is confirmed by the low χ^2 value of 0.15.
The c value of 0.0829 (= t_0) is pretty much as expected.

Of course, there is room for improvement here, besides the expensive solutions: use a smaller, shiny, round object, like a metal marble or so. That would make it a bit easier to indicate the center of gravity, and it might fall a tiny bit faster, resulting in a slightly higher g value. Doing the measurements in a vacuum would have the same effect.

Can we check how well we did this experiment?
Well, there is an approximate calculation method for g that takes into account the flattening of the Earth. You can find it, for example, at: www.sensorsone.com/local-gravity-calculator.
If you enter the geographical latitude (in our case, 45.2°N, near Novi Sad, Serbia) and the height above sea level (200 m), this gives g = 9.80576 m/s², indeed slightly smaller than the Earth's average. Be careful, this does not take geological irregularities into account!
For some countries, you can find maps with measured variations; google "gravity anomaly map", but I didn't find any from Serbia.

There is, however, an interesting object about 70 km from our measuring location, in the center of Belgrade: a marking pyramid on which you can read a precisely measured value:
g = 9.8060226 m/s².

Our location was about 100 m higher, which should reduce g a tiny bit.

Fig. 4.11. The gravitational field strength in the center of Belgrade, Serbia.

There are also smartphone apps that measure g, using the microscopic dynamometer sensors that are built in. For example, "Physics Toolbox" and "Accelerometer" for Android.

FittingKVdm example data file: "Falling pear.dta1".

4.1.5. Measuring the height of a building

How do you measure the height of a tree, a building, etc. without having to climb on it?

A very simple way is to measure the length of its shadow and compare it with the shadow of an object with a known height at the same time. The shadows are proportional to the height if both objects stand with the same angle relative to the surface. That's the basic geometry invented by Thales of Milete. (See p. 75 for a remark about the precision.)

But what if the weather is cloudy, or you want to know the height of a specific mark on a building? No problem, you could measure the observation angle, and the distance, and use basic trigonometry, see p. 75. You just need a precise sextant, which costs at least 300€...
And what if you can't come close to the point just below the one you try to measure, if the building is wide near the surface, or there is a fence, or a canal in between?

Also no problem, you just need to do two independent measurements from different distances, and from the difference in viewing angle (the "parallax"), you can derive both the minimum distance and the height, with some basic geometry knowledge. Our brains do this all the time when they process information from both our eyes to estimate distances.

You would still need a sextant, but that can easily be solved with a rolled up piece of paper, some tape, a weight on a rope, and a simple protractor, as you can see in Fig. 4.12.

Fig. 4.12. A home-made sextant.

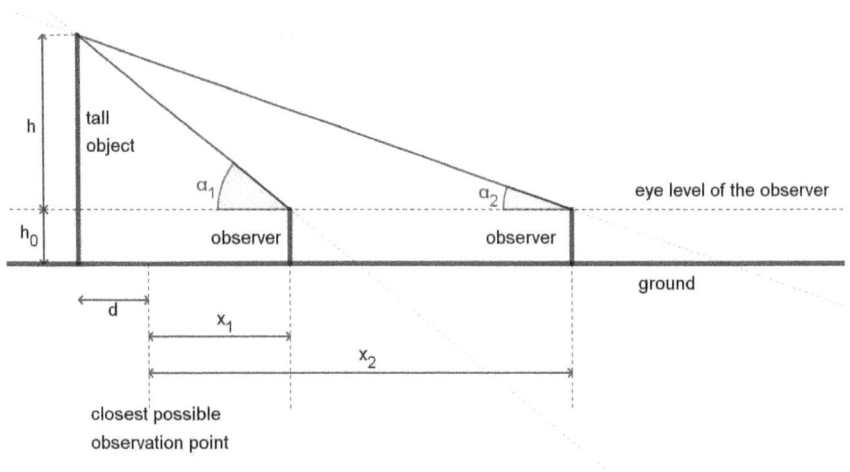

Fig. 4.13. The geometry of the measurements. d = ?, h = ?.

For all the observations, we have

$$\frac{h}{d + x_i} = \tan(\alpha_i)$$

Two of these equations would suffice to get a system we can solve for d and h, but we can get a better precision if we do more measurements and do regression with this model function (in FittingKVdm called "Parallax" model):

$$\alpha = \operatorname{Arc}\tan\left(\frac{h}{x + d}\right)$$

Note: Make sure the Arctan is in degrees. In programming languages, it's usually in radians; if so, the model function has to be multiplied with $180/\pi$.

I used this equipment to determine the height of the gutter of a house. The following are the measurements:

x (m)	0	2	4	6	8	10	12	14	16
α (°)	65	48	35	29	22	19	12	11	10

The error on x was estimated as 0.1 m since the observer's eye had to be exactly above the marked spots, which was not easily done. The angle's precision was probably not better than ± 1°. This is how the regression looks:

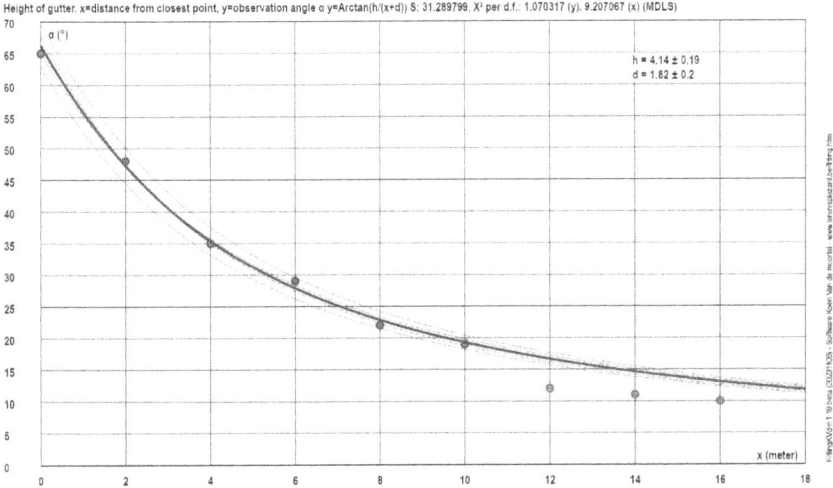

Height of gutter. x=distance from closest point, y=observation angle α y=Arctan(h/(x+d)) S: 31.289799, X² per d.f.: 1.070317 (y), 9.207067 (x) (MDLS)

h = 4.14 ± 0.19
d = 1.82 ± 0.2

α (°)

x (meter)

Fig. 4.14. Observation angles measured at different distances, and the theoretical curve. The last three points seem suspicious.

Since there is no doubt here about the choice of the model – it follows from the geometry – it seems some systematic error was made with the last three measurements, as you can see in the plot. So I fitted with those points deactivated.

The obtained height was 4.14 ± 0.19 m, so the real height of the building must have been 4.14 + 1.60 = 5.74 m since h_0 = 1.60 m was the eye level of the observer.

χ_y^2 ≈ 1.1, meaning that the error on the angles was realistically estimated. The error on x might still have been underestimated, it seems.

FittingKVdm example data file: "Height_of_gutter.dta1".

4.1.6. Determining the absolute zero temperature with a mayonnaise jar

If you heat a filled balloon, it will swell. The particles in the air fly faster and collide harder with the wall, pushing it outwards. If that air is trapped in a hard container that cannot expand, the pressure will increase. Now, there is something very remarkable that can be seen. If you then measure the pressure for a number of temperatures, and you put those measurements on a scatterplot, they will line up nicely, and if you extend that trend to the left, there should be a temperature at which the pressure becomes zero, and the particles should therefore have come to a standstill, at least according to classical theories. It looks like it couldn't get any colder than that... This had already been discovered by the French inventor and physicist Guillaume Amontons (1663–1705), but the measurements were much improved and published by the better-known French physicist and chemist Louis Gay-Lussac (1778–1850). The British physicist William Thomson (1824–1907), who would later receive the title of Lord Kelvin, then proposed using a more logical temperature scale that started from this so-called "**absolute zero**". With that scale, the relationship between pressure and temperature should be directly proportional. The results of these and some other gentlemen were summarized as the well-known "**ideal gas law**".

How can we make a good estimate of that zero point? Well, just by repeating the experiment. All we need is a container that we can seal airtight, a barometer, and a thermometer. A classic sterilization jar will do just fine! Ideally, the measuring instruments should both be in the jar; in any case, the barometer. They must of course be readable without opening the jar.

As a barometer I used a handy portable electronic device from the Chinese brand "Sunroad", which can also serve as an altimeter. There is also a thermometer in there, but unfortunately temperature and pressure cannot be displayed simultaneously. The thermometer is also anything but reliable because it shows more than 3 °C too much compared to other thermometers. The barometer seemed more reliable.
I did not have a thermometer available that also fit in the pot, so I used a multimeter from the well-known brand "Fluke" with an included thermocouple that gave a reading with a resolution of 0.1 °C, but the real precision was closer to 1–2 °C according to the manual. If you are handy

and you can drill a hole in the lid to insert the measuring tip of the thermocouple, and then close that hole very well, great! If not, it will probably suffice to hold the tip very close to the pot and wait long enough for both to reach the same temperature.

I placed the pot (mayonnaise jar) in places with the widest possible range of temperatures, for at least an hour at a time, to ensure that the temperature inside had reached equilibrium with that outside.

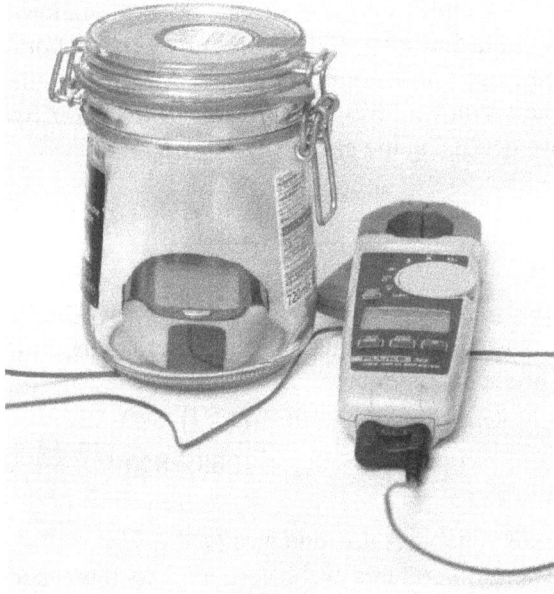

Fig. 4.15. The equipment for the experiment.

Examples are given in the following.

θ (temperature in °C)	p (hPa)	place
-21	840	freezer
5.3	928	fridge
10.5	953	outside
22.3	1001	living room
33.3	1025	0.5 m above the stove
39.5	1028	edge of the stove
52.2	1068	on the stove

The first value was measured with a different thermometer, as the Fluke only went down to -10 °C. That gave a reading to within 1 °C, the Fluke to 0.1 °C, but the accuracy of both was probably the same; the Fluke was often fluctuating up and down by several tenths of a degree and then I had to estimate the average.

In principle, two measurements are sufficient to find what you are looking for, after all, the pressure p is proportional to the absolute temperature T in Kelvin, and $T = \theta - \theta_0$, with θ the temperature in °C and θ_0 the unknown absolute zero in °C . So, for two measurements 1 and 2, we get: using the gas law ($pV = nRT$):

$$\frac{T_1}{p_1} = \frac{T_2}{p_2} \Rightarrow \frac{\theta_1 - \theta_0}{p_1} = \frac{\theta_2 - \theta_0}{p_2}$$

Converting this equation and entering the coldest and the hottest measurements produces:

$$\theta_0 = \frac{p_2\theta_1 - p_1\theta_2}{p_2 - p_1} = \frac{1068 \cdot (-21) \cdot 840 \cdot 52.2}{1068 - 840} \approx -290.6842...°C$$

How reliable is this outcome?
The temperatures were certainly no more accurate than 1 °C and the accuracy of the pressures was certainly no better than 1 hPa, so deviations like these are certainly conceivable:

$$\theta_0 = \frac{1067 \cdot (-20) - 841 \cdot 51.2}{1067 - 841} \approx -284.9522...°C$$

$$\theta_0 = \frac{1069 \cdot (-22) - 839 \cdot 53.2}{1069 - 839} \approx -296.3165...°C$$

This is relatively close to the value that is now known from very precise measurements, namely, -273.15 °C, but since this value does not fall within our margin of error, so we have to conclude that the measured temperatures and pressures were even more inaccurate than we thought.

Note that a systematic percentage error has no effect on the pressure; if all pressures were measured, for example, 50% too high, or in a different unit, this would not change the fraction. It would be more dangerous if there were a systematic *shift* in values. If the deviations are purely coincidental, it certainly makes sense to perform many more measurements. I did about 40 of them.

As you can see in the scatterplot (Fig. 4.16), they are fairly well aligned, but strange jumps seemed to occur, especially at higher temperatures. During some measurements on the stove, the pot was placed directly on it, and the thermocouple was between the pot and the metal of the stove. The temperature measurement was probably higher than the actual temperature in the pot. These measurements were therefore repeated with a piece of insulating cork under the pot, and even with the pot and the measuring wire in a more or less closed plastic box so that the temperatures just next to the pot and in the pot were as equal as possible.

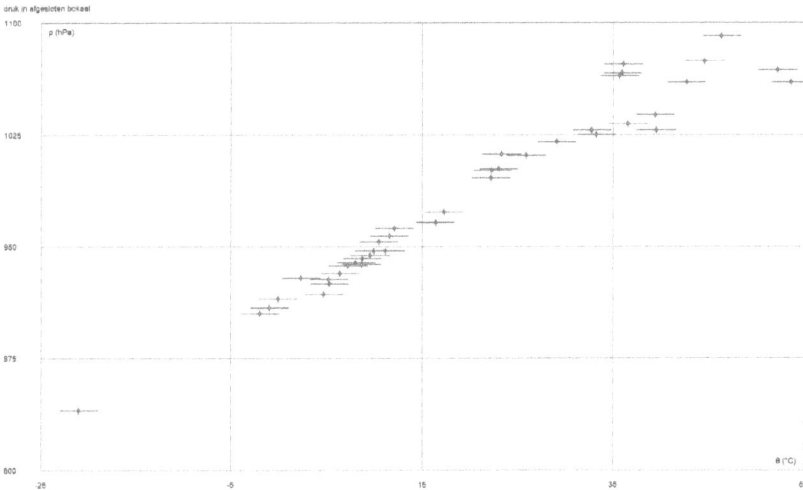

Fig. 4.16. The 40 temperature and pressure measurements.

Now, a linear regression (obviously MDLS since it shouldn't make a difference wether you choose temperature or pressure as the independent variable) gave us a much more reliable result:

The parameters for the best fitting line $p = a\theta + b$ were $a = 3.33 \pm 0.10$ and $b = 918.5 \pm 3.1$.
The zero crossing of this line should be the "absolute zero":
$\theta_0 = -b/a \approx -275.9441... \,°C$

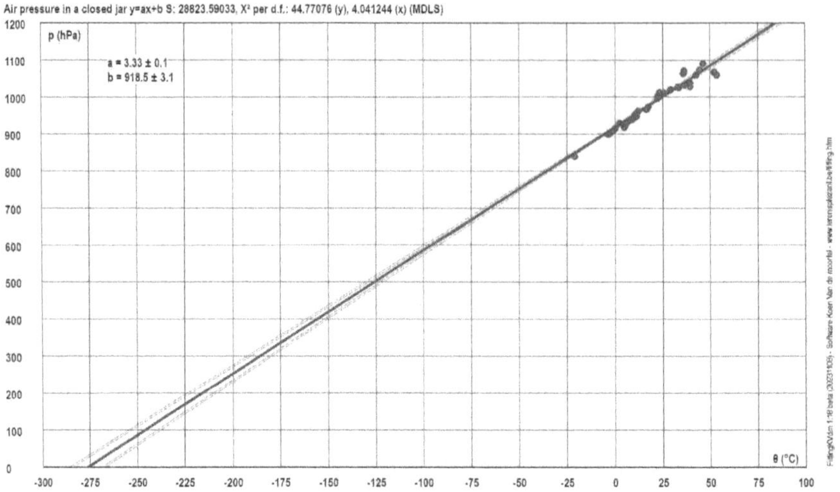

Air pressure in a closed jar y=ax+b S: 28823.59033, X² per d.f.: 44.77076 (y), 4.041244 (x) (MDLS)

Fig. 4.17. The measurements and the best fitting line, extrapolated to see where it crosses the horizontal axis.

The error on θ_0 can be estimated as:

$$s_{\theta_0} = \sqrt{\left(\frac{0.10}{3.33}\right)^2 + \left(\frac{3.1}{918.5}\right)^2} \cdot \left|-275.9441\right| \approx 8.6$$

So, our best estimate becomes: $\theta_0 = -275.9 \pm 8.6$ °C, which is pretty close, considering the simple equipment! By the way, OLS produces -265.2 ± 5.7 °C.

Notes for people who want to do this at home:

- Do not place the jar in the sun, because then the temperature inside and just outside can differ seriously! Not even when the thermometer is in the jar, because then the measured temperature can differ greatly depending on whether the tip of the thermometer is light or dark!

- If you measure with a thermocouple, keep the attached multimeter at room temperature if possible. After all, the thermocouple only produces voltage if there is a difference in temperature between the measuring point and the connection to the multimeter, where the reference temperature is measured with a thermistor.
  If the measuring tip and the jar are in the freezer or refrigerator, first

read the temperature, and then open the door to read the pressure, because the latter changes much more slowly.

- Of course, also pay attention to the maximum permitted working temperature of the instruments!

FittingKVdm example data file: "Pressure_vs_temperature_in_jar.dta1".

4.1.7. Volume and pressure in a gas

One of the scientists contributing to the "ideal gas law" was the Irish (al)chemist Robert Boyle (1627–1691). He proved that the pressure in a fixed amount of gas is inversely proportional to the volume you squeeze it in.

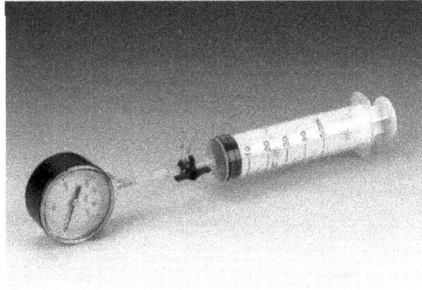

Fig. 4.18. Simplest equipment for this experiment: a syringe and a manometer (www.eurofysica.nl)

His original measurements from 1662 where published in London (GB) in the book *A defence of the doctrine touching the spring and weight of the air, propos'd by Mr. R. Boyle in his New Physico-Mechanical Experiment... by the author of those experiments.* They can be found at https://web.lemoyne.edu/~giunta/classicalcs/boyleverify.html.

The variables here are: x = volume (V) and y = pressure (p) in arbitrary units. No information about the precision was given, so I assumed the pressure was ± 1 and the volume ± 0.0125 since all measurements were multiples of this value.

Of course, the appropriate model describing the relationship here, is $y = ax^b$ (power function), and the regression should be done to check if the exponent b is indeed equal to -1.

The hypothesis is very nicely confirmed: with MDLS you get $b = -0.998 \pm 0.040$, and with OLS $b = -0.996 \pm 0.055$.

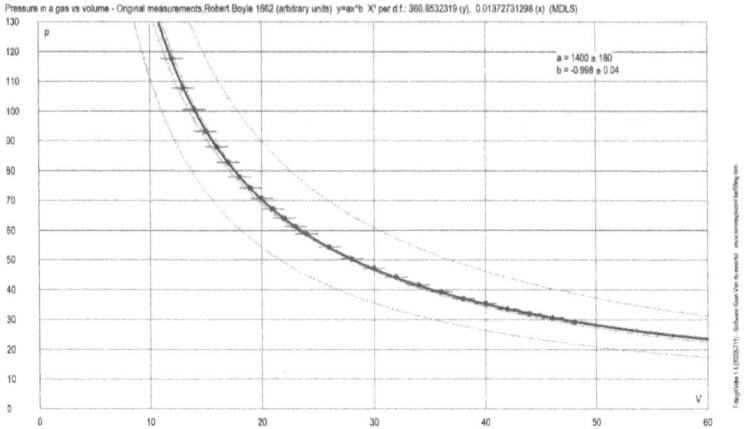

Fig. 4.19. The original measurements of Robert Boyle and the best fitting power function.

FittingKVdm example data file: "Boyle.dta1".

4.1.8. Planet orbits

Johannes Kepler (1571–1630) formulated three laws that describe the motions of planets around the Sun. Isaac Newton (1642–1726) proved later that they were equivalent to his gravitation law. The third law says that the square of a planet's orbital period (T) is proportional to the cube of the length of the semi-major axis (d) of its orbit. Or shorter, $T \sim d^{1.5}$. Strictly speaking, this is only valid if the planet's mass negligible in comparison to the Sun's mass and the other planet's masses.

You can easily check this law by fitting the best possible power function through the known (d, T) dataset from the Wolfram Alpha Knowledgebase 2018, published, for example, at https://en.wikipedia.org/wiki/Kepler%27s_laws_of_planetary_motion. The d values are given in Astronomical Units (AU), and the T values in days. The estimate for the precision used for the regression weights was the least significant digit.

This law is pretty precisely obeyed, as you can see.

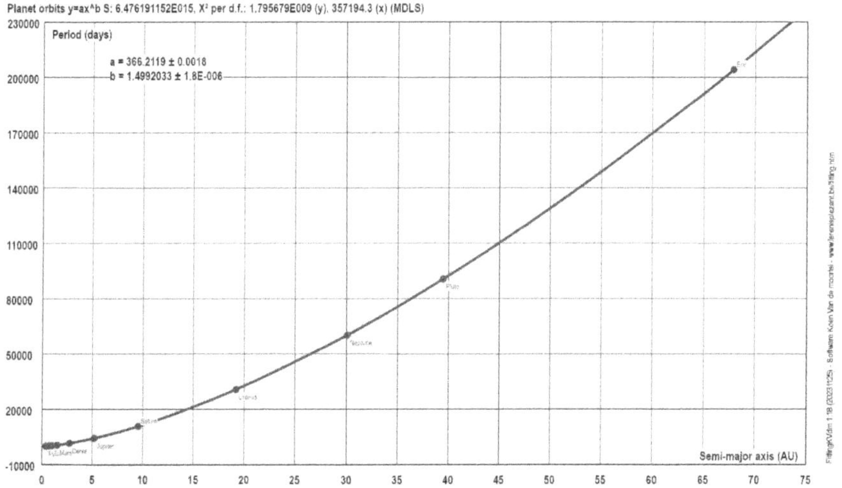

Fig. 4.20. Orbital times vs the semi-major axis of objects orbiting around the Sun, and the best fitting power function.

MDLS produces an exponent:
$b = 1.4992033 \pm 0.0000016$, which is 0.05% below 1.5,
and OLS:
1.5006013 ± 0.0000021, which is 0.04% above 1.5.

The proportionality constants differ more:
MDLS: $a = 366.2119 \pm 0.0018$, OLS: $a = 365.2097 \pm 0.0015$.

The furthest object, Eris, which seems to fit pretty well, is actually the most deviant with its $d = 67.864$ AU, $T = 204199$ days. That is normal since the relative attraction of the big outer planets becomes more important in comparison with the Sun's. MDLS predicts: $T \approx 204048$ days (151 days off), while OLS predicts $T = 204693$ days (494 days off).

FittingKVdm example data file: "Planet orbits.dta1".

4.1.9. Measuring the gravitation (*g*) with a pendulum

We have previously measured the strength of the Earth's gravity field (= the gravitational acceleration *g*) by observing a falling object. But we can also do this in another way, by measuring the period of an "ideal" pendulum. By this we mean a small heavy object that hangs quietly on a very thin, flexible but non-elastic cord without too much friction.

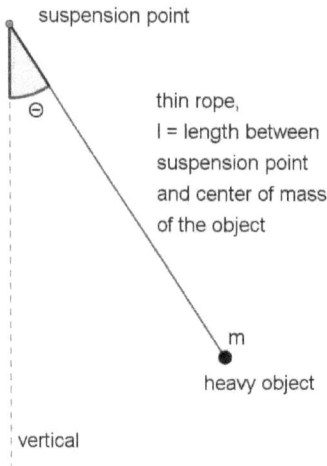

suspension point

thin rope,
l = length between
suspension point
and center of mass
of the object

m

heavy object

vertical

Fig. 4.22. Pendulum.

From the theory about this, which you can find in all textbooks, it follows that this period can be approximated as follows:

$$T \approx 2\pi \sqrt{\frac{l}{g}}$$

where *l* is the length from the suspension point to the center of gravity of the weight. The formula is valid for "small" deflection angles. As long as that is less than about 10 degrees, the deviation is less than 0.1%.

More details can be found, for example, at
https://en.wikipedia.org/wiki/Pendulum_(mechanics).

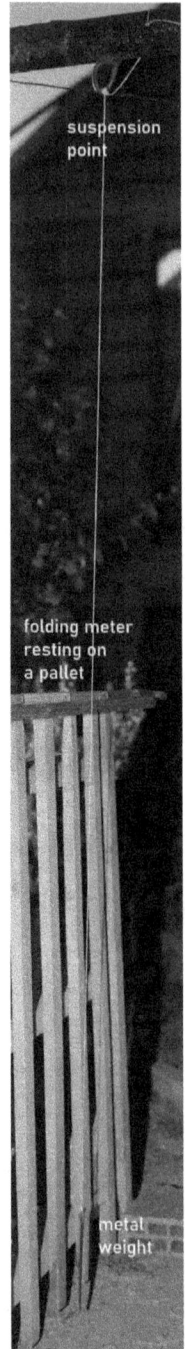

suspension
point

folding meter
resting on
a pallet

metal
weight

Fig. 4.21. The simple experiment set-up

If we can measure l and T precisely, we can transform the formula to cal-culate g. This is an often done classroom experiment. I'll tell you how I gathered some data for you, at the same location as previously.

I started working with material I had at my disposal. It is very important that the object is heavy enough to properly tension the cord, and so that it does not start to swing by itself at the slightest breath of wind, so do the experiment inside if you can. With a heavier weight, the relative in-fluence of friction is also less. It should also be as axially symmetrical as possible so that it would not affect the chord deflection reading if it were to rotate on its axis. Make sure the object itself doesn't start to wobble, as the movement can become chaotic!

Fig. 4.23. Finding the center of gravity of the object.

To determine the length l, you must find out as accurately as possible how far its center of gravity is below the suspension point. This is done simply by balancing it on a narrow object: if it remains balanced, the center of gravity is on the support line. I measured $l = 174.7$ cm (point of suspension to top of weight) + 6.2 cm (top of weight to cen-ter of gravity) = 1.809 m.

To determine the period, we must be able to read the position clearly somewhere. To this end, a folding meter was placed horizontally (level!) just behind the cord, 93 cm below the suspension point, with the "100 cm" position as close as possible behind the equilibrium position of the cord. You could now measure the time span of say, 100 passages through the equilibrium position with a chronometer, or you could film the movement and read a set of randomly spread elongation values from that, and fit a damped sine function through the data, which will give you a very precise time value.

For the latter, you need something to film with, preferably at 60 or more shots/s: a mobile phone or a camera on a tripod or a stable position neat-ly perpendicular to the measuring stick. I used a camera at the lowest resolution (to save disk space) but at the highest speed.

Getting the thing moving is not as easy as it seems. I therefore started with a fairly large deflection and allowed the pendulum to slowly wobble until its movement became nicely harmonic with a maximum deflection of less than 9°.

Fig. 4.24. Camtasia screenshot.

Then you need a software program that allows you to watch the video frame by frame! Camtasia 2020 (Techsmith Corporation) is well suited for this job.

The measured y values were converted to elongation angles:

$$\theta = \text{Arc}\tan\left(\frac{y - 100}{93}\right)$$

(h = 93 cm, see Fig. 4.25)

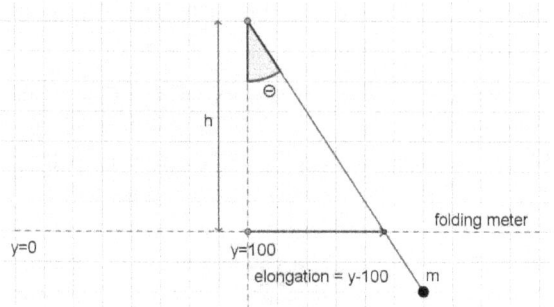

Fig. 4.25. The y values were measured.

The errors on θ were estimated assuming the errors on y were ± 1 cm (quite large, to be sure).

The regression was done with 51 (t, θ) values and this model function:

$$\theta = A \cdot \sin\left(\frac{2\pi}{T}(t - c)\right) \cdot e^{-\lambda t}$$

This resulted in
$T = 161.842 \pm 0.029$ frames = 2.69737 ± 0.00048 s, so:

$$g = l \cdot \left(\frac{2\pi}{T}\right)^2 \approx 1.809 \cdot \left(\frac{2\pi}{2.69737}\right)^2 \approx 9.8156325 \frac{m}{s^2}$$

That looks like a very decent value. What can we say about its precision? The length of the pendulum probably had a precision of about 2 mm, so we can estimate its influence on g as follows:

$$\sigma_{g,1} \approx \frac{\partial g}{\partial l} \cdot \sigma_l = \left(\frac{2\pi}{T}\right)^2 \cdot \sigma_l \approx \left(\frac{2\pi}{2.69737}\right)^2 \cdot 0.002 \approx 0.011$$

And this is the error on g caused by the error on T:

$$\sigma_{g,2} \approx \left|\frac{\partial g}{\partial T}\right| \cdot \sigma_T = \frac{2 \cdot (2\pi)^2 \cdot l}{T^3} \cdot s_T \approx \frac{8\pi^2 \cdot 1.809}{2.69737^3} \cdot 0.00048 \approx 0.0035$$

So, a good estimate for the error on g is

$$\sigma_g \approx \sqrt{0.011^2 + 0.0035^2} \approx 0.012 m/s^2$$

So, the result becomes $g = 9.816 \pm 0.012$ m/s². This is most probably more precise than the result from the previous experiment.

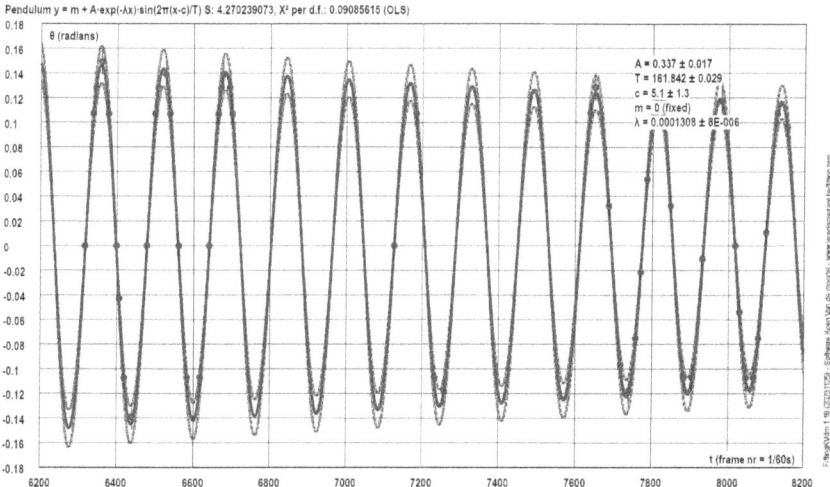

Fig. 4.26. The measurements and the best fitting damped sine function

FittingKVdm example data file: "Pendulum.dta1".

4.1.10. Measuring *g* with a pendulum – different approach

Instead of fitting a damped sine to the movement of a pendulum, we can also measure the period (*T*) with varying lengths (*l*) of the rope. A changeable LEGO construction can make this easy!

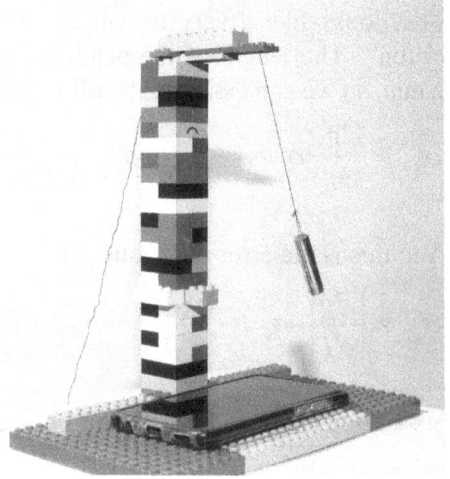

Fig. 4.27. The experiment set-up. Actually, putting the phone more towards the end of the plate near the biggest elongation, improved the peak shape (one per period instead of two).

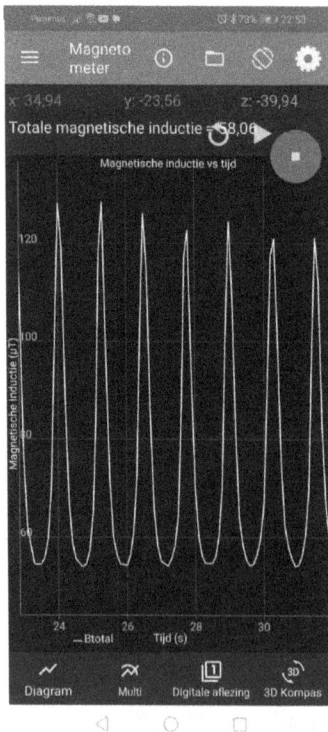

Fig. 4.28. The recorded magnetic field variations (uncorrected times). T = time between peaks.

To measure the period, you might film it and count the frames, but I found an easier way this time: use a magnet as the weight, and a smartphone app to register the magnetic intensity every time when it passes. I used the "magnetometer" in the "Physics Toolbox" app from www.vieyrasoftware.net.
Unfortunately, it has a bug that showed up on my Huawei phone: the time measurement is very wrong, so I had to calibrate that with another phone.

The measurements (with corrected times: measured "seconds" multiplied by 1.3724) are as follows:

l (mm)	σ_l (mm)	T (s)	σ_T (s)
144	1	0.776654884	0.0061758
213	1	0.9345385248	0.00658752
263	1	1.0293	0.0150964
281	1	1.0587084734	0.0130378
337	1	1.173402	0.013724
374	1	1.23516	0.0130378
413	1	1.291304884	0.01248884
448	1	1.3419025272	0.010293
488	1	1.411609468	0.0130378
537	1	1.47533	0.01139092
584	1	1.5413108748	0.01056748
641	1	1.6230125916	0.01193988
705	1	1.692622092	0.01193988

I measured 20–30 periods each time to have a 20–30 times better precision. Measuring more periods was not really possible since the pendulum started to move more and more in circles. That was most probably because the construction was not sturdy enough.

Since we know that

$$T \approx 2\pi\sqrt{\frac{l}{g}} = \frac{2\pi}{\sqrt{g}}\cdot\sqrt{l} = a\cdot l^b$$

I did a regression (MDLS of course since l and T can be switched) to find the best power function model. This yielded:
$a = 0.0644 \pm 0.0011$
$b = 0.4985 \pm 0.0029$, very close to the expected value of 0.5.

If we fix $b = 0.5$ according to the theory, we get:
$a = 0.06381 \pm 0.00018$.

Pendulum variable length y=ax^b S: 118.8566218, X² per d.f.: 0.4730006 (y), 12.70915 (x) (MDLS)

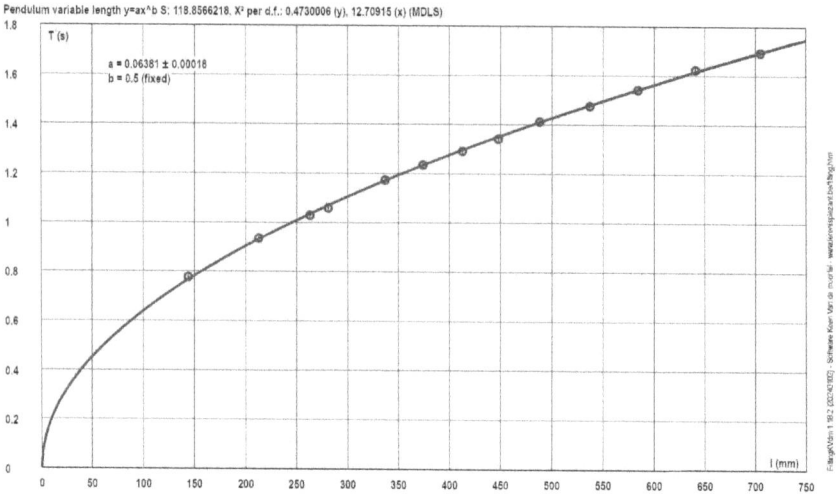

Fig. 4.29. Pendulum period vs length & best fitting square root function

So,

$$g = \left(\frac{2\pi}{a}\right)^2 \cdot \frac{1}{1000} \approx 9.69540\frac{m}{s^2} \qquad (/1000 \text{ because } l \text{ was in mm})$$

How precise is this?

$$\sigma_g = \left|\frac{\partial g}{\partial a}\right| \cdot \sigma_a = \frac{8\pi^2}{1000a^3} \cdot s_a \approx 0.055\frac{m}{s^2}$$

That is, if the errors on l and T were estimated well (probable) and if there are no hidden systematic errors...

The value for g is lower than expected (9.8116 m/s² according to the International Gravimetric Bureau). I suppose the main culprit is the flexibility of the LEGO tower, which probably caused the slightly chaotic movement that was observed. The previous experiment with the big pendulum attached to a sturdy structure didn't have this problem. By the way, precise pendulum clocks don't use cords, but a stiff metal rod. The formula changes then, because its mass can't be neglected.

So, you know what to do and what to avoid if you want to do this at home or in a classroom!

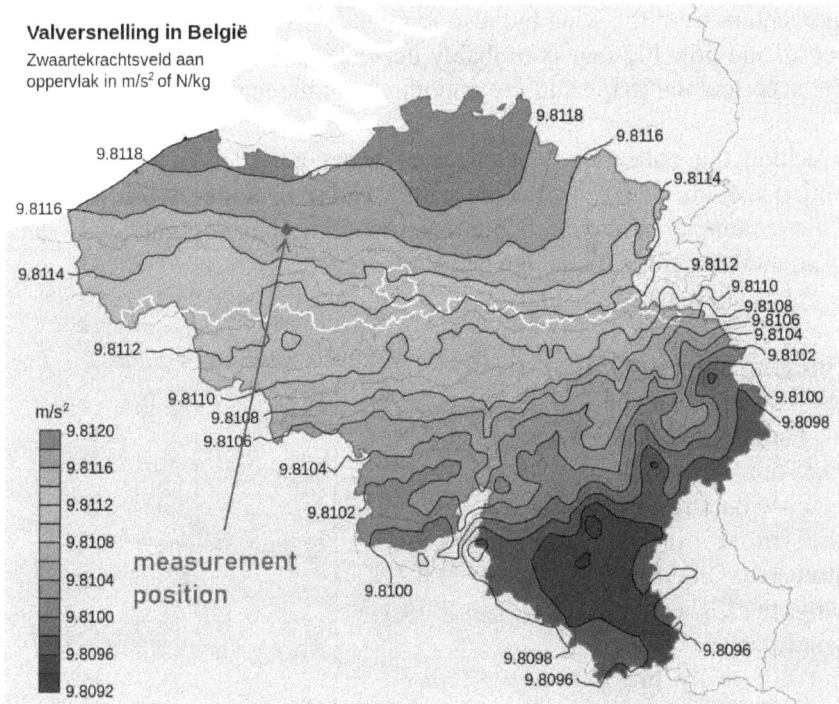

Fig. 4.30. Gravitational field strength in Belgium.
Source: Wikimedia Commons, based on the values from the International Gravimetric Bureau.

FittingKVdm example data file: "Pendulum variable length.dta1".

4.1.11. Measuring the speed of the wind

How strong is the wind? A simple question, but finding out the answer turns out to be a little less simple. There are all kinds of tricks conceivable:

Direct measurement:
You throw a very **light object into the air** (a feather, a balloon, dust, etc.) and you see how fast it flies away, for example by filming it or seeing how much time it takes to reach the distance between two positions. That seems doable, but... is it actually really the speed of the wind you measure this way? After all, that balloon receives not only

momentum from the wind but also resistance from the air in the opposite sense, and how big that is probably depends on a lot of factors, such as the mass and the shape and the smoothness of the material.

Tracking the collective movement of water particles in the air is an option though. That is possible with a **radar or sodar wind profiler**. This technique is beyond home experimenting, but you can study data from public organizations like NOAA: psl.noaa.gov/data/obs/datadisplay/.

Hanging object:
We can also watch a hanging object like the **windsock** you see on airports: the more it is stretched horizontally, the higher the windspeed. Okay, but not really precise, and anyway, how do you calibrate it? You need a *wind tunnel* with adjustable speeds to do that.

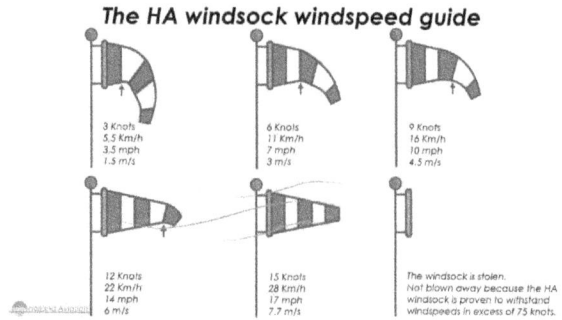

The HA windsock windspeed guide

Fig. 4.31. A calibrated windsock. Image: https://www.hollandaviation.nl/nl/windsock-windspeed-guide/

"Wet finger":
You can literally do it "with your wet finger": if you wet your finger, you can feel where the wind is coming from, and the harder the wind blows, the faster it will dry. You can make an electrical "finger" by heating a resistance wire with current and then measuring its temperature. The more wind, the more it cools down. Google "**hot wire ane-mometer**" to find out more. Again, the measurement is easy, but calibration is more complicated.

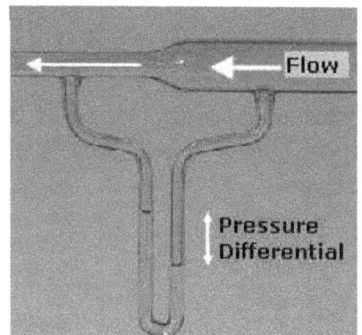

Venturi tube:
This device basically consists of a tube with a narrowing and a manometer. The

Fig. 4.32. Venturi tube.
Image: ComputerGeezer & Geof, Wikimedia Commons

higher the speed of the incoming air, the more pressure it causes on the fluid in the tube below it, because it is hindered by the narrowing. The difference in the fluid level near the exit is proportional to the kinetic energy of the flowing air, so the speed can be calculated from that. But the result will also depend on the temperature and humidity, and turbulences can influence the reading too. So, calibrating it is not that easy.

Classic "anemometer":
You can also make a simple version with a few cups and straws, as shown in many YouTube videos. The more wind, the faster it turns around, so you might think let's film that pinwheel and then count the number of frames between each revolution, and since the distance traveled by each cup is 2π times the

Fig. 4.33. Lots of funny anemometer designs explained on YouTube...

radius, we know the speed. Yes... but that cup will never move as fast as the wind. How fast exactly depends on the shape, the friction around the shaft, etc. The rotational speed is approximately proportional to the wind speed, so if you attach a dynamo to the rotational shaft, you could deduce the wind speed from the generated current. Again... the calibration?

Actually, you don't need to do all this construction work yourself. There is a solution for lazy people: just grab a cooling **fan** from a discarded laptop. If you don't have one, you can buy a new one for about three euros or dollars. Such a thing works in two directions: if you apply power, it turns and creates wind, but blow on it, it also generates electricity. The principle is the same as a wind turbine. All you have to do is mount it on something, so you can hold it in the wind and connect it to an amperemeter. Of course, the airflow must not be hindered, so use a thin stiff wire or something to attach it, see Fig. 4.34.

But again... how do we *calibrate* this set-up? Comparing with a calibrated anemometer is cheating; we want a procedure "from scratch"!

If you had a wind tunnel, you could adjust the wind speed exactly as desired and link the electrical measurements to it. Now, to provide you with some sample data, I didn't have anyone in my social environment

with a wind tunnel in his
backyard, so I had to im-
provise something else.
And that actually turned
out not to be too difficult:
you just stick the thing
through the window while
you're driving a car. You
do need a long straight
calm stretch of road
where you can drive at
various stable speeds, and
no police around prefer-
ably... And you need an
assistant who always
notes the measured cur-
rent while he/she holds
the device neatly perpen-
dicular to the air flow.
Naturally, you also have
do this on a windless day,

Fig. 4.34. "Anemometer", personal patent.

otherwise you will simultaneously measure the tailwind or headwind.

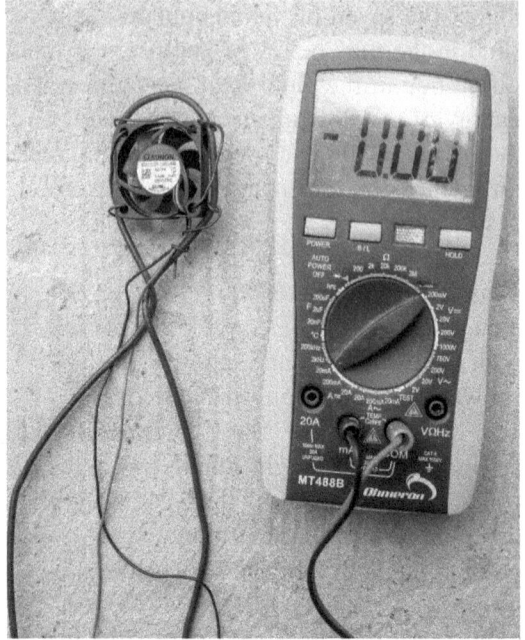

I managed to do it on a calm summer day in the plains of northern Serbia, kindly assisted by my wife. The following are the measurements,
using a small "Sunon" fan:

v (km/h)	30	35	40	40	45	45	50	55	60	65	70
I (mA)	6	7	12	8	17	17	18	22	30	33	40
v (km/h)	70	75	80	80	85	85	90	90	95	100	105
I (mA)	39	41	42	42	45	45	49	49	50	58	59

The error margin on the speed was assumed to be at least ± 3 km/h since
it is not that simple to maintain exactly the same speed for a sufficient
period of time for each measurement!

Although the amperemeter itself was more precise, I estimated the error
on each current measurement to be ± 2 mA, in the first even ± 3, since

the values always went up and down, probably due to vibrations and turbulence.

The faster we drove, the greater the current obviously, but what is actually the correct relationship between wind speed (v) and current (I) that we expect?

The kinetic energy contained in a certain amount of air is proportional to the square of the speed: $E = mv^2/2$. Now each air particle covers a distance $v \cdot t$ per unit time, so the mass that flows per unit time through a cylinder (imagined or not) with a cross-sectional area A is $m = Avt\rho$, with ρ the density. This means that the energy actually becomes proportional to v^3.
If the same percentage of energy is always converted into electricity, the generated power ($P = E/t$) is also proportional to v^3.
If R is the resistance of the coil in the fan, then $P = RI^2$, so $I^2 \sim v^3$, or: $I \sim \sqrt{v^3} = v^{1.5}$. There are more factors playing a role, but hopefully this will be sufficient for now.

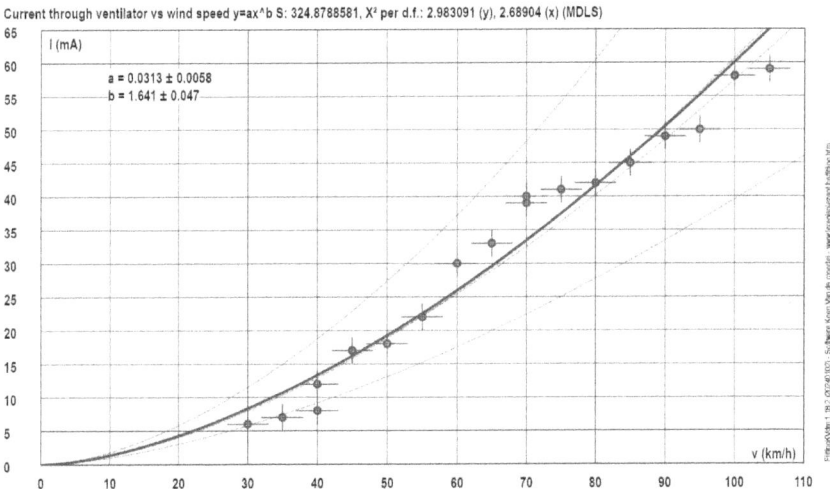

Current through ventilator vs wind speed y=ax^b S: 324.8788581, X² per d.f.: 2.983091 (y), 2.68904 (x) (MDLS)

a = 0.0313 ± 0.0058
b = 1.641 ± 0.047

Fig. 4.35. The measurements and best fitting power function with floating exponent.

Fitting the data in a power function model $I = av^b$ with a free floating exponent yielded
$b = 1.641 \pm 0.047$,
slightly deviating from the theory. But you can see in the scatter plot there is a lot of "noise" in the measurements...

If we force the theoretical exponent, we get:
$a = 0.05793 \pm 0.0011$, with which we can now estimate the wind speed
conversely based on a current measurement:

$$I = 0.0579 \cdot v^{\frac{3}{2}} \Rightarrow v = \left(\frac{I}{0.0579}\right)^{\frac{2}{3}}$$

Current through ventilator vs wind speed y=ax^b S: 438.5213952, X² per d.f.: 2.466563 (y), 3.185444 (x) (MDLS)

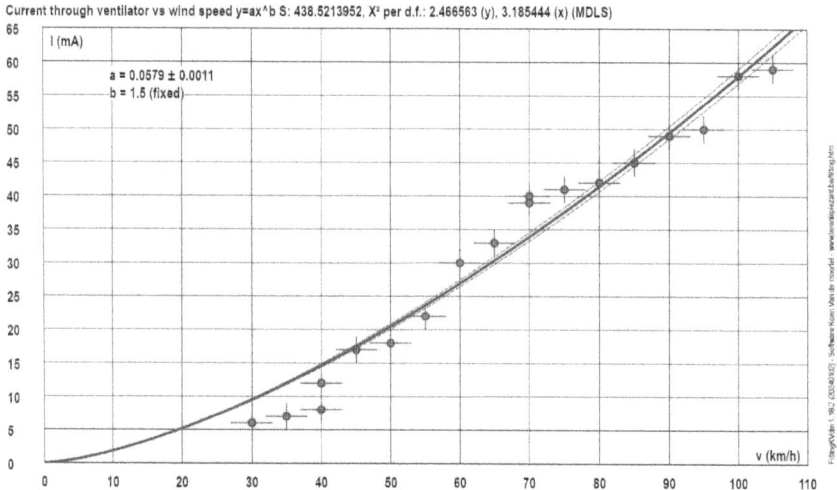

Fig. 4.36. The measurements and best fitting power function, exponent 1.5.

If we measure a current of, for example, 30 mA, we can calculate that
the wind speed must be 64.49 km/h, well, not that precise, because
32 mA already gives 67.32 km/h. So, the expected practical accuracy is
in the order of ± 3 km/h *if* the parameters are reliable...

Remark: The obtained parameter might be valid for very dry and warm
air (about 32 °C), but I can imagine if we would do the test on the
wintery North Sea coast, the density of the air (and therefore the force on
the propeller) could be somewhat greater, and high in the mountains it
will be smaller. For every situation, we need a recalibration.

Warning: The following are the many "mistakes" we have to learn from
here:
• No measurements could be taken at less than 30 km/h because the fan
  had a high static friction: it needed quite a high minimum force to
  start. Using a fan with larger blades and a lower working voltage
  would probably do better here!

- The fluctuations in the measured current may also be partly caused by the fact that we used a DC motor. This actually produces a rectified alternating current. An alternating current motor could be more stable, or a capacitor in parallel with the direct current fan would reduce the fluctuations too.

- The following are the possible systematic errors in our measurements:
(1) Most car odometers indicate a slightly too high speed. In fact, we should first calibrate it using a GPS!
(2) We also assumed that the resistance of the motor coil would remain constant but that could increase with increasing speed (induction). It would actually be better to measure current and voltage at the same time, so that we can calculate the real power generated, but then we need two devices. And ideally we should just log the GPS speed, current, and voltage automatically. The assistant could then focus on keeping the fan in the correct position and angle without having to take notes.
(3) Ideally, the fan should also be mounted in a very fixed position, far enough from the body of the car, since that might cause compressions and turbulences.

Conclusion: After all it might be easier to just buy an anemometer that is calibrated in a wind tunnel. But I hope you enjoyed the reasoning…

FittingKVdm example data file: "Electrical current from wind.dta1"

4.1.12. Car fuel consumption

A big, heavy car consumes more fuel for the same distance, that's quite logical since it has a higher kinetic energy at the same speed, it loses more energy when braking, it has more friction with the road and maybe also more resistance from the air, etc. How much exactly is difficult to calculate because too many variables play a role. We might expect some pattern though, if we take a sample of "similar" types of vehicles like typical passenger cars with the same kind of fuel, for example.

I found some useful data (benzine cars built between 2018 and 2021) at:
www.autoweek.nl/carbase/?bouwjaar=2018--2021&brandstof=benzine.

The variables here are x = empty mass of a car (in kg) and y = average gasoline (benzine) consumption (in l/100km) of the car when driving in urban areas.

The simplest model function to try as a model, besides the – too simple – linear one, is of course a power function. And indeed, that fits reasonably well with an exponent about 1.5.

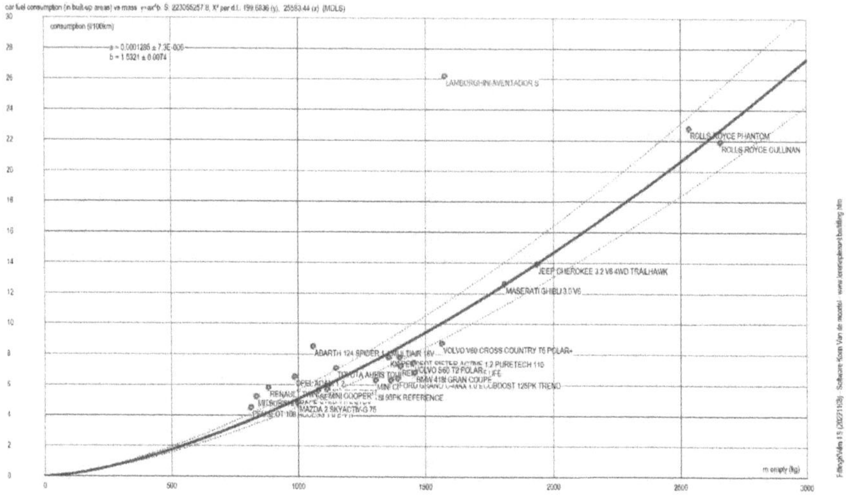

Fig. 4.37. Gasoline consumption vs. car mass, and the best fitting power function.

One type of car seemed to consume way above expected: the Lamborghini Aventador S, which – I hope you agree – can be considered to be a real outlier in this group of "normal" passenger cars, so I left that one out for the fitting.

A quadratic function with the constant parameter c = 0 also fits well, but I'm not sure if that makes more sense. It is less sensitive to the Lamborghini outlier though.

Some sources suggest an exponential relationship, but that would be weird, since a zero mass should have a zero consumption. I have even seen a guy on YouTube explaining a linear regression on the consumption expressed in the American way (miles per gallon) versus mass, but that is crazy of course!

FittingKVdm example data file: "Car fuel consumption vs mass.dta1".

4.1.13. Concrete strength

When you build a construction, you'll want to know the properties of the concrete you are using. A very important one, is the compressive strength. This depends on many variables, like the amount of cement, water, sand, pebbles, additives, etc. that are used, and the how long it was drying. It's really difficult to make a useful model with all these variables, and it is certainly not linear!

If you want to try, an interesting database of tests is publicly available; see https://data.world/uci/concrete-compre ssive-strength/workspace/file?filename =Concrete_Data.xls.

Fig. 4.38. Measuring compressive strength. *Image:* Wikimedia Commons.

To understand the influences playing, it's interesting to vary one independent variable at a time, while keeping the others (more or less) constant.

The easiest is probable the **age** (drying time). In the source above, we found age data for a mixture of 310 kg cement, 192 kg water, 970 kg coarse aggregate, 850 kg fine aggregate, and no additives.
So, the variables in this subset are: x = time (in days) after this specific concrete mixture was made; y = compressive strength after x days of hardening (in MPa). The uncertainty on x was unknown, hence set to 1; the original y values must have been derived from something else, not pascals, since they had way too many digits (maybe from kilograms?) so the uncertainties were not sure (my guess: 0.01). Anyway they were probably all the same, so the absolute values of σ_x and σ_y only matter for the parameter uncertainty estimation.

A fitting model should obviously start from the origin, since concrete starts from a muddy mixture, and a horizontal asymptote: the maximal strength after a long time, so a homographic relationship $y = ax/(x + b)$ is a good candidate, and it seems to fit quite well; see Fig. 4.39.

Another interesting variable is the **ratio of the amount (mass) of water used and the amount of cement**. We could calculate some values from the same database. Mixtures without additives were selected; the amount of water varied between 146–203 kg per 1000 kg, coarse aggregate: 838–1125 kg, and fine aggregate: 594–945 kg.

So, $x = m_{water}/m_{cement}$ and $y =$ compressive strength after 28 days of hardening (in MPa). The uncertainty on x was calculated assuming the uncertainty on the masses was 1.

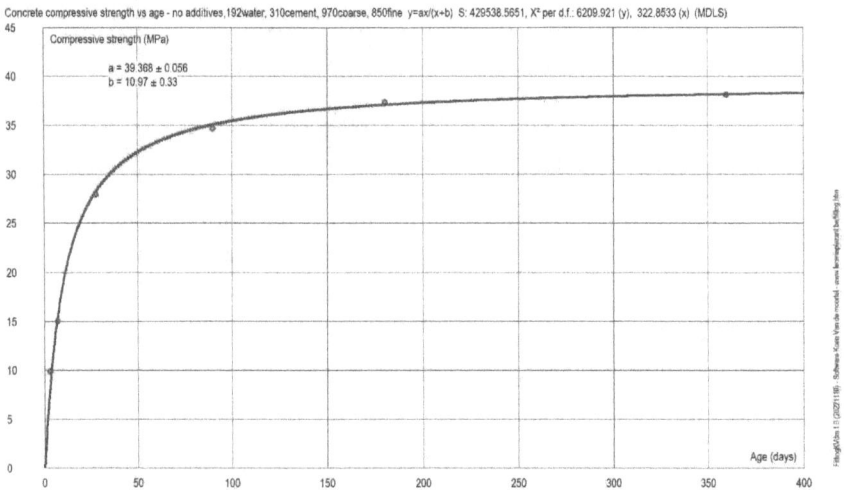

Fig. 4.39. A possible curve depicting the hardening of concrete.

Number crunchers unaware of the meaning of the data might try to fit several models through the data, see Fig. 4.40. But if you know what the data are, you see that none of those make sense: the linear model becomes negative once you use enough water, and the quadratic goes up again. And every one of these is not zero when no water is used. The power model even "predicts" infinitely strong concrete with no water. Now, even a child knows that you need some water to make mortar. Without it, you just have some non-cohesive powder. So the curve should start in the origin. If you have some basic knowledge of chemistry, you know that you need a certain amount of water to react with all the lime in the cement. So the strength should increase by adding water, until the reactants are exhausted. Adding more water can't make the concrete any harder (unless maybe if you add some other chemicals that react with it), and with too much water, it will inevitably become weaker and weaker. The product of a power function and an exponential function with a negative exponent ($y \sim x^n e^{-kx}$, see p. 42) seems a logical choice, see Fig. 4.41.

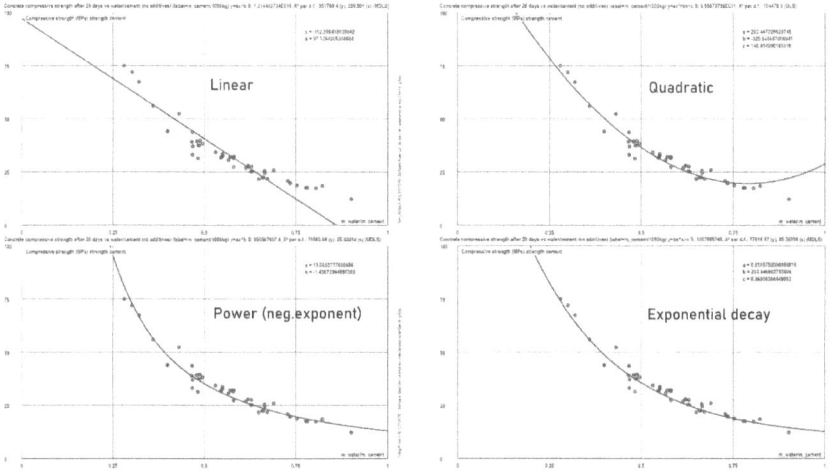

Fig. 4.40. Possible (bad) models describing concrete strength vs the water/cement ratio.

Unfortunately, the dataset doesn't have measurements with very little water; it starts from an x value of 0.28. So we can't really see which exponent for the power function is best. Fitted with a very small exponent, the maximum is much higher than the highest measurement, and with exponents bigger than 1, the predicted maximal strength is too low. So an exponent $n = 1$ (see Fig. 4.42) seems plausible since I suppose there would have been measurements with less water if that was known to produce a better concrete. It's not perfect, since the peak is probably too much to the left, but it's better than the other ones. Possibly, fine-tuning with an additional exponent in the exponent (like in the Weibull decay function) might improve the fit.

Concrete compressive strength after 28 days vs water/cement (no additives) (label=m_cement/1000kg) y=a(x/(bn))^n·e^(-x/b+n) S: 8.501858654E011, X² per d.f.: 127542.8 (OLS)

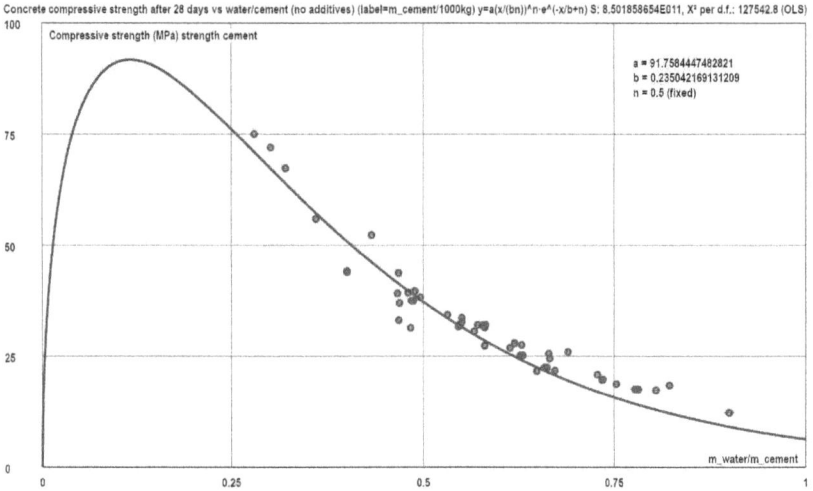

Fig. 4.41. Compressive strength vs. water fraction and model $y \sim x^n e^{-kx}$ with n = 0.5: predicted maximum is too high.

Concrete compressive strength after 28 days vs water/cement (no additives) (label=m_cement/1000kg) y=a(x/(bn))^n·e^(-x/b+n) S: 1.086888256E012, X² per d.f.: 175163.1 (OLS)

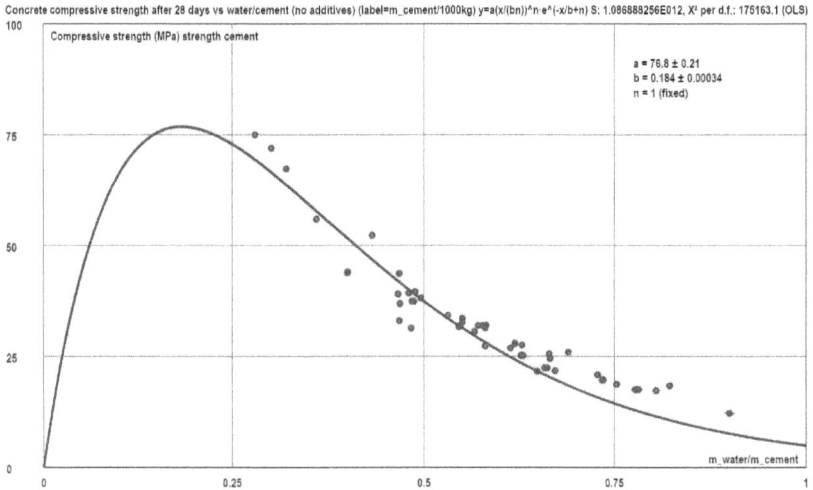

Fig. 4.42. A plausible model for the strength of concrete vs the water/cement ratio: $y \sim x^n e^{-kx}$ with n = 1.

FittingKVdm example data files: "Concrete compressive strength vs age - no additives - 192 water.dta1", "Concrete-compressive-strength vs water-cement ratio age28 no add.dta1".

4.1.14. A hanging chain

What happens when you hang a chain, a flexible cable, a washing line, etc. between two suspension points?

Fig. 4.43 shows a necklace chain I could borrow from my wife, attached with two pins to a piece of cardboard. The shape looks very much like a parabola, but it isn't one.

You can find that out easily; it's a fun classroom experiment. Make sure the bottom of the graph paper behind the chain is perfectly horizontal and there is no friction between the chain and the paper; it should really "hang" freely.

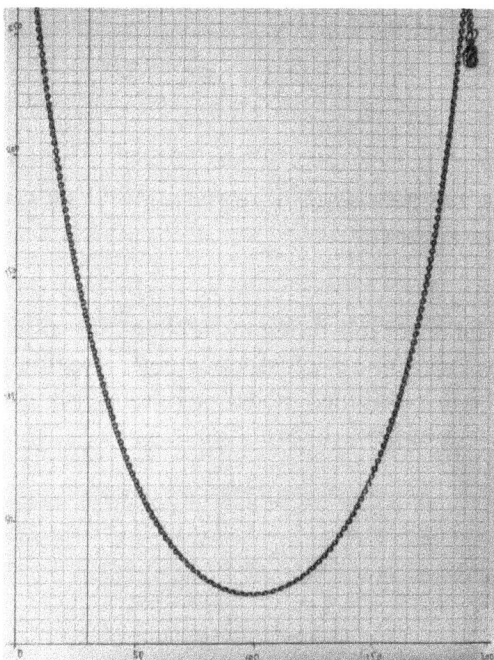

Fig. 4.43. A suspended jewelery chain.

Then read a number of coordinates (x, y) from points on the chain. It's easier if you take a photograph (from a position above the middle of the paper, camera parallel with it), and read the coordinates from that.

I collected some data for you (in mm, ± 1 mm). Some contrast enhancement and lens correction were done to ease the measurements:

x	21	31	41	51	65	82	100	125	135	145	155	163	170	178	184
y	175	125	90	65	40	25	20	31	43	59	81	105	135	175	210

By putting the points in a regression software program, and trying to fit a quadratic function, you will see something is wrong. Especially since the residuals show a non-random pattern!

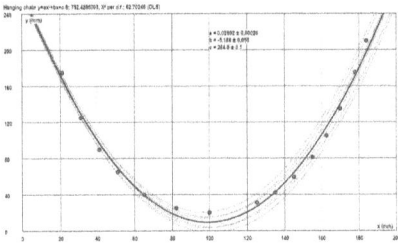

Fig. 4.44. The measurements and the best fitting parabola.

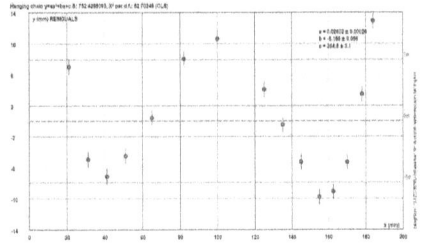

Fig. 4.45. Residuals from the fitted parabola.

The only well-fitting model here, is – of course – a "catenary", which is in fact a linear transformed cosh function (hyperbolic cosine, see p. 56), the "Chain line" model in FittingKVdm, since that is what mechanical theories predict. See, e.g., https://en.wikipedia.org/wiki/Catenary.

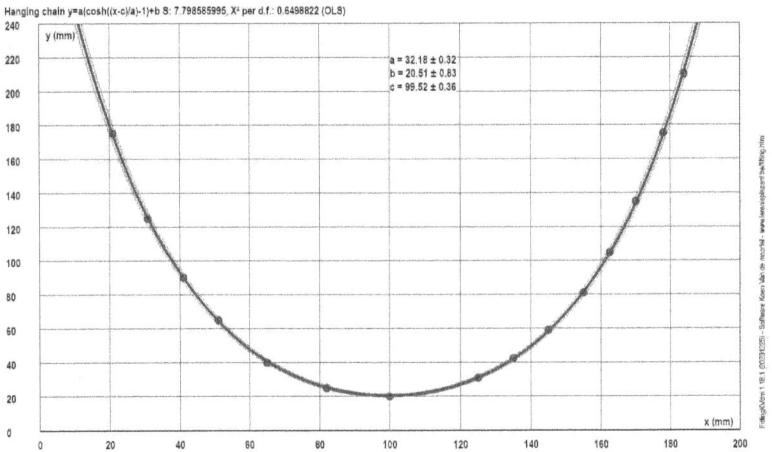

Fig. 4.46. The measurements and the best fitting chain line or catenary (cosh).

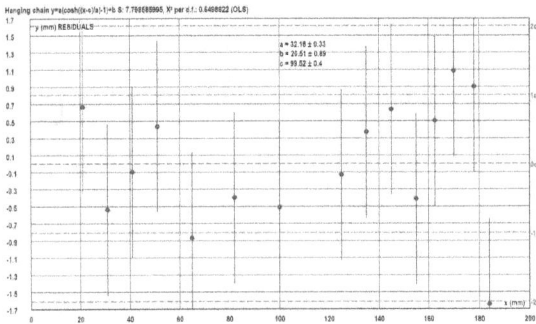

This time, the residuals seem to be randomly distributed without a clear pattern, as expected.

FittingKVdm example data file: "Chain.dta1".

Fig. 4.47. Residuals from the best fitting chain line.

4.1.15. Janka hardness versus wood density

Sometimes, we want to determine some quantity that is difficult (i.e., expensive, time consuming, etc.) to measure directly, while there is some other easy to measure quantity related to the first. Measuring concentrations indirectly with a spectrometer is a widely used example (see further).

But also hardness tests to determine the suitability of wood for flooring are a bit more time consuming than just measuring its density, and the two are well related, if you stay within a limited variety of wood types. See, e.g.,: www.engineeringclicks.com/density-of-wood.

The test named after the Austrian-American Gabriel Janka (1864–1932) requires a device that measures the force needed to press a steel ball with precise dimensions halfway into a piece of wood, see, https://en.wikipedia.org/wiki/Janka_hardness_test.

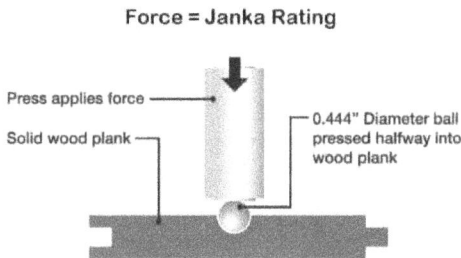

Force = Janka Rating

Press applies force ————

Solid wood plank —

— 0.444" Diameter ball pressed halfway into wood plank

Fig. 4.48. The Janka test.
Image: tinytimbers.com/resources/janka/

Some measurements to illustrate this are quoted from E.J. Williams (1959), as example nr. 334 in *A Handbook of Small Data Sets*, by D.J. Hand *et al.* (Springer 1994). The variables here are as follows: x = density (in the old Anglo-Saxon unit system: lb/ft^3) of 36 Australian Eucalyptus wood samples; y = their hardness on the scale of Janka.

The authors propose to fit a polynomial relationship through the data, but you know by now that this usually makes no sense, theoretically. Since it is logical that "zero density" would be equivalent to "zero hardness", and the data don't really seem to be on a straight line, a power function is the simplest model to try. MDLS should be used here since the conversion should work in two directions!

As you can see, it fits reasonably well with the data; as a matter of fact, the relationship seems to be quadratic, not perfectly but well enough to be useful in practice. The errors on the parameters seem a bit

Measuring and Modeling by Example

unrealistically small, so the precision of the measurements might have been worse than assumed.

Note that using OLS would produce a "best fitting exponent" of only 1.86 instead of 1.98.

Janka hardness vs density of eucalypt wood (Williams, E.J. 1959) y=ax^b S: 2.534136825E010, X² per d.f.: 112059.3 (y), 2338.879 (x) (MDLS)

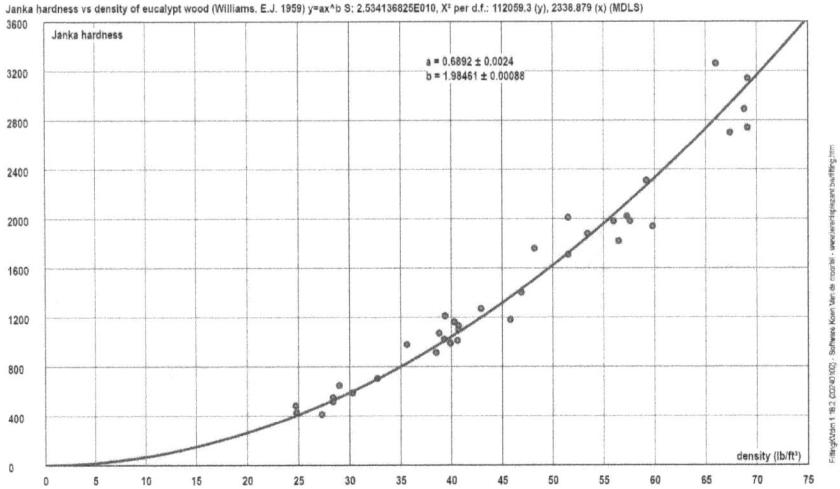

Fig. 4.49. Janka hardness vs. density measurements of eucalypt wood, and the best fitting power function.

Other similar procedures to quantify hardness exist, but they are not so easy to compare. An interesting attempt to do this, using regression, can be found at
Koczan, G., Karwat, Z. & Kozakiewicz, P. (2021), "An attempt to unify the Brinell, Janka and Monnin hardness of wood on the basis of Meyer law.", *J Wood Sci* 67(7), https://doi.org/10.1186/s10086-020-01938-4.

FittingKVdm example data file: "Janka hardness vs density of eucalypt wood.dta1".

4.1.16. Cooling down

Isaac Newton already discovered that warm objects cool exponentially to the ambient temperature. That makes sense: if they cooled linearly, they would keep getting colder and freeze to minus infinity, which is against the laws of thermodynamics. An exponential decrease means that the same percentage of the temperature surplus decreases per unit of time.

To check this, we can perform the following simple home experiment. We let some water come to the boiling point in an electric kettle with a built-in thermometer, as shown in Fig. 4.50. As soon as that happens, the power cuts off and the thing starts to cool down. We start the chronometer and regularly look at the thermometer. This is what I measured with the shown kettle:

t (minutes)	θ (°C)	$\theta-\theta_0$ (°C) (diff.with room temp.)
0	100	76
2	92	68
4	88	64
10	80	56
26	63	39
67	43	19
104	36	12
190	27	3
308	24	0
430	24	0

Fig. 4.50. Our home laboratory.

So, we expect: $\theta - \theta_0 = b \cdot a^t$, with θ_0 the ambient temperature, which was apparently 24 °C, because the temperature didn't drop any further. Obviously, this is valid for $t \geq 0$.

What do we expect for a and b?
• The parameter a must be positive and less than 1, otherwise there will be no cooling. The correct value will tell us how quickly the thing cools down, and that depends on which materials are used in the device, and how well they insulate, and also on the amount of water of course. For a thermos flask, we expect an a value close to 1; for a recipient with a thin wall, a smaller value. Zero is impossibly small: an amount of matter does not suddenly cool down immediately. Energy must be passed on to

the environment, and that always takes some time.
* The value of b should normally be around $100 - 24 = 76$ because that is what you get if you enter $t = 0$.

The "simple" way to find the a and b values is as follows: first subtract 24 °C from all temperatures to obtain the "temperature surplus" or the "excess temperature".
We could now estimate a by comparing different measurements. For example, $\theta - \theta_0$ went from 76 to 56 during the first 10 minutes. The "growth factor", as a is usually called, was therefore $56/76 \approx 0.7368421$. How much was that per minute? Don't make the mistake of dividing that number by 10 because that would mean the temperature would drop linearly. The correct way to find the per minute rate is by taking the 10th root of 0.7368421, namely, 0.9699234. If we compare minute 26 with minute 10, we get an estimate

$$a = \left(\frac{39}{56}\right)^{\frac{1}{16}} \approx 0.9776419$$

We could do this many times and average the obtained values, but fortunately we know there is such a tool as regression, which uses *all* the data at once!

So we dropped the measurements (and a few more) in the software to fit with the model "$y = ba^x + c$" where x is the time and y is the temperature. In FittingKVdm, there is no need to subtract the ambient temperature; the algorithm will find the best value for that too (c); in many simple software programs, you would need to do that, and you would have to omit the zero values because they make it crash! We can use MDLS here since (1) it makes sense to ask a question like "how much time does it take to cool down 60%?"; you could actually use this thermometer as a clock once you know the cooling rate; and (2) the measurements stayed nicely above the asymptotic value.

In Fig. 4.51, you see that the points are not very beautifully lying on the curve, but considering the primitiveness of the used equipment, it still looks reasonable. I entered the precision of the temperatures as ± 2 °C since it was really not more precise; the needle of the thermometer seemed to "hang" sometimes when the temperature was dropping too quickly. For the time imprecision I entered 0.05 minutes (3 s), because watching the chronometer and the thermometer more simultaneously was impossible.

The obtained parameters were $a = 0.9819 \pm 0.0053$, $b = 68.6 \pm 5$, $c = \theta_0$ $= 23.9 \pm 2.8$.

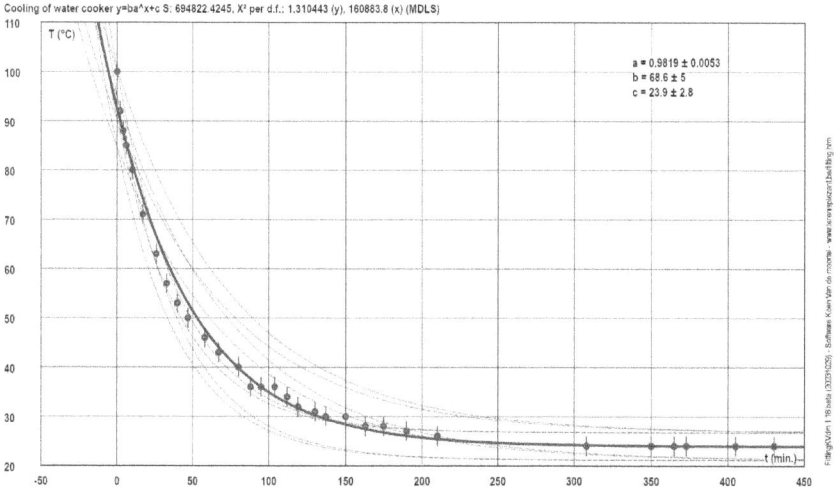

Fig. 4.51. The temperature measurements and the best fitting exponential curve

The error ranges, especially on c, are probably much less than the esti-mated ones. I guess the entered measurement error of $\pm 2\ °C$ must have been smaller toward the end because the temperature was stable. The problem with this thermometer seems to be that it is too slow, and that causes the biggest errors in the first minutes.

Remark: OLS produces $\theta_0 = 25.50\ °C$ which is quite impossible since the thermometer showed 24 °C for minutes without any change.

Now, of course I know there is a lot of room for improvement: using a fast thermocouple to measure (or even a kettle with a digital thermo-meter), and a data logger to record, would improve the precision, no doubt. But my point was: to show that, even with basic equipment, you can get usable results if you use regression!

FittingKVdm example data file: "Cooling_of_water_cooker.dta1".

4.1.17. A hot stone in water

What happens if you drop a hot stone in a glass of water? Of course, the temperature of the water will rise until all the heat is transferred from the stone and there is a new thermal equilibrium. A function describing this must look like a sigmoid, for example, the logistic, starting from room temperature and with the end temperature as the second horizontal asymptote. But, a glass is not a good heat insulator, so its contents will start to radiate heat away as soon as the temperature is above that of the surroundings. A candidate to describe the combination of this warming and cooling might be the one I called "double logistic"; see p. 36. Anyhow, it will be only an approximation, since heat flow is a complex phenomenon, a mixture of radiation, convection, conductance, dependant on the exact form of the recipient, the position of the thermometer probe, etc.

The proof of the pudding is in the eating... So I did the test for you. Here, x = time (t, in minutes, starting when the hot stone was dropped in a bowl of water at room temperature) and y = temperature of the water (T or Θ, in °C), measured with a thermocouple on a simple multimeter (Ohmeron 488B). No stirring was done. For the precision on x, ± 0.1 minutes was assumed, and ± 0.5 °C for y.

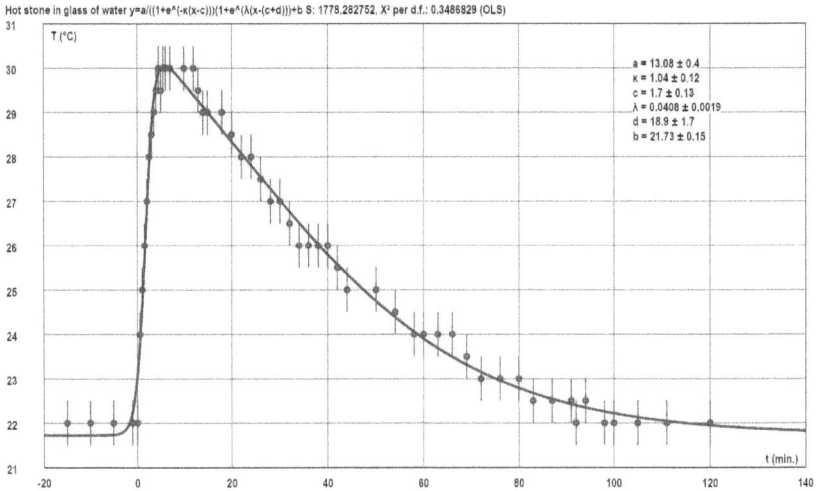

Fig. 4.52. Temperature of water, at t = 0 a hot stone was dropped in it. Best fitting "double logistic" curve.

Fitting with this function

$$f(x) = \frac{a}{\left(1 + e^{-\kappa(x-c)}\right)\left(1 + e^{\lambda(x-(c+d))}\right)} + b$$

yielded:
$\kappa = 1.04 \pm 0.12$ and $\lambda = 0.0408 \pm 0.0019$; $\kappa > \lambda$, so the ascending phase, reflecting the heat conductance in the water, was much faster than the descending phase, reflecting the heat radiation into the air.
Parameter $b = 21.73 \pm 0.15$ reflects the ambient air temperature, and $a = 13.08$ is a bit more than the peak height.

The pattern looks as expected, and $\chi_y^2 \approx 0.34 < 1$ means that the deviations are on average smaller than the measurement error, so that is very good. **Note:** χ^2 was only calculated in the vertical direction here because multidirectional regression was not possible here because there is a maximum in the pattern.

The "skewed peak 2" model (see p. 38) even fits a little bit better ($\chi_y^2 \approx 0.30$) but that also has one more parameter.

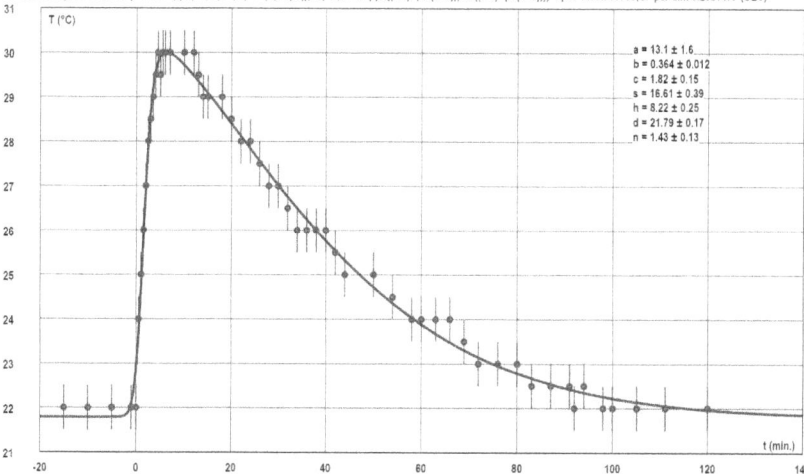

Hot stone in glass of water y=h/m·e^(-(ln(e^(-a(x-c)/s)+e^(b(x-c)/s)))^n)+d, m=e^(-(ln((a/b)^(-a/(a+b)))+ln((a/b)^(b/(a+b))))^n) S: 1475.738689, X² per d.f.: 0.2951477 (OLS)

a = 13.1 ± 1.6
b = 0.364 ± 0.012
c = 1.82 ± 0.15
s = 16.61 ± 0.39
h = 8.22 ± 0.25
d = 21.79 ± 0.17
n = 1.43 ± 0.13

Fig. 4.53. Same data, but with best fitting "skewed peak, type 2" model.

To improve the precision, of course you could use a data logger and a more expensive thermometer if you have that at your disposition.

FittingKVdm example data file: "Hot_stone_water.dta1".

4.1.18. Internal resistance of a battery

If you try to start a car with the lights on, you will notice that they become less bright. Why is that? You load the battery very heavily for a short time, i.e., with a small resistance R_L. The internal resistance of the battery (R_i) forms a voltage divider with R_L and as a result the voltage U that can be measured externally at the battery drops:

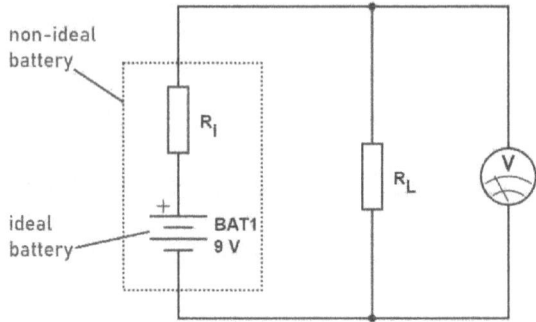

Fig. 4.54. Schematic of the experiment.

$$U = \frac{R_L}{R_L + R_i} U_B$$

where U_B is the voltage of the "ideal" battery (with $R_i = 0$) that is supposedly hidden in the real battery. In reality, the internal resistance is of course spread over the entire battery, but the system behaves more or less as if we can split the two.

Strictly speaking, R_L is here the replacement resistance of the parallel connection of the resistor we connected to the battery and the internal resistance of the voltmeter. The latter is usually so high (e.g., 10 MΩ) that it has no measurable influence. So, U_B can be measured directly to a very good approximation if we omit R_L and use a voltmeter with a very high internal resistance. After all, the fraction in the formula then becomes practically 1, except for roughly one millionth.

So, to determine R_i, we simply need to measure how much the voltage drops by loading the battery, and then transform the formula:

$$R_i = \frac{U_B - U}{U} R_L$$

I tried it with a regular 9 V battery (rechargeable Varta-NiMH, type 56722). When measured without load, this gave a voltage of 8.81 V.

To have a serious effect on the voltage, R_L must be of the same order of magnitude as R_i, or less. For example, if both are equal, the voltage is divided in half. But what if we have no idea how big R_i is? Of course, we can simply directly connect a very small resistor of about 1 Ω, and the voltage drop will certainly be noticeable. However... this gives us the problem that the battery will drain very quickly, and if U_B changes during the measurement, our formula no longer applies!

So I had to proceed a little more cautiously, first start with a high resistance and then lower it, and especially with smaller resistances, measure quickly! With a resistance of 100 kΩ (according to the manufacturer, but measured 98.7), nothing was noticeable. Only from less than 1 kΩ we started to see some change. With the last one, at 39.6 Ω, I saw the voltage drop quite quickly, so it was time to stop.

Caution, before we start calculating, we should note that there is a non-negligible error on those smallest resistance values, because the wire from the multimeter to the test resistor also has some resistance itself; in our case, 0.2 Ω. If you use alligator clips, this can quickly become 2–3 Ω. You measure this by simply shorting the wires.

The precision of the used Ohmeron 488B multimeter was 0.8% + 4 digits in most ranges and 0.8% + 5 digits in the lowest (0–400 Ω). For the voltage, it was 0.5% + 0.04 V.
The results are as follows:

R_L (Ω) raw	98.7k	10.81k	985	623	395	336	149	114.5	39.6
R_L (Ω)	98.7k	10.81k	984.8	622.8	394.8	335.8	148.8	114.3	39.4
σ_{RL} (Ω)	±1.2k	±0.13k	±12	±9.0	±3.7	±3.2	±1.7	±1.4	±0.82
U (V)	8.81	8.81	8.77	8.74	8.7	8.67	8.48	8.35	7.2
σ_U (V)	±0.08	±0.08	±0.084	±0.084	±0.084	±0.083	±0.082	±0.082	±0.076 ??

We can now calculate R_i, for example, from the first and the eighth measurement (the ninth might be less reliable):

$$R_i = \frac{8.81 - 8.35}{8.35} \cdot 114.3 \approx 6.296766467\Omega$$

How precise is this? We can do a lengthy calculation with the partial derivatives, but we can also make a quick estimation. Which part is the most error-prone? Probably the numerator of the fraction, since that is a difference of two numbers not so far apart.

Let's look at two "worst case" scenario's:

$$R_{i,1} = \frac{8.73 - 8.432}{8.432} \cdot 114.3 \approx 4.03954\,\Omega \quad \text{and}$$

$$R_{i,2} = \frac{8.89 - 8.268}{8.268} \cdot 114.3 \approx 8.59877\,\Omega$$

That's a big uncertainty!

The way to reduce that uncertainty is again regression. Put the data in a homographic model through the origin ($y = ax/(x + b)$, where $x = R_L$, $y = U$, then $a = U_B$ and $b = R_i$). The better estimate is now: $R_L = 7.4 \pm 1.4\,\Omega$.

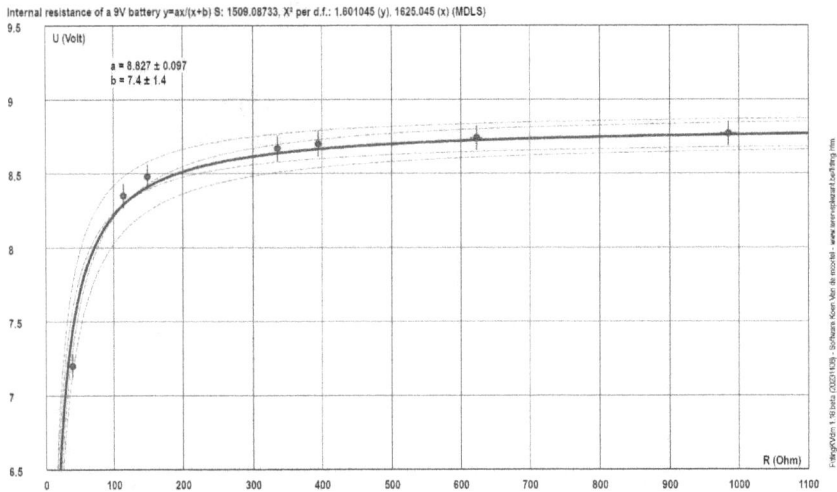

Fig. 4.55. Most of the measurements and the best fitting homographic function through the origin.

By the way, this is actually the experiment that the German physicist Georg Ohm (1789–1854) did to discover his famous law.

FittingKVdm example data file: "Internal_resistance_9V_battery.dta1".

4.1.19. Determining the capacitance of a capacitor

A capacitor, what kind of thing is that? You'll find many of them in every electronic device! Simply put, it is an electronic part that consists of two metal plates that are close to each other (hence the symbol ⊣ ⊢). In practice, those plates are usually rolled up with a thin layer of insulator in between, and then wrapped up in a package, with two wires coming out. If you connect them to a battery, that will pump electrons to one plate and suck them away from the other. So, one plate gets positively charged and the other negatively charged. Apparently, there is a current flowing through the capacitor, for some time. After a while, a lot of charge is piled up on each plate, and therefore the repulsion forces within each plate become bigger. This reduces the current and puts a limit on the charge you can store on the plates. The greater the voltage, the more "pressure" the electron pump provides, and therefore the more charge the capacitor can hold. The so-called "capacitance" C of the capacitor is therefore defined as the maximum charge Q (in coulombs) per unit of voltage U (in volts). A coulomb per volt is called a farad, after the British physicist Michael Faraday (1791–1867). Due to the attraction between the two plates, the charge remains even when the battery is disconnected. A capacitor can therefore serve as a storage vessel for electrons. One of the applications is, for example, to absorb short-term voltage drops. You see this with many devices when you turn them off: the control light remains on for a while until the capacitor in parallel with the power supply has discharged. A flash lamp also has one: when it is charged, it can release a lot of energy in a very short time, much more than a battery could. That is why you have to wait a while after each flash before you can take the next photo.

By applying an insulating substance between the plates whose molecules can be polarized (a so-called "dielectric"), the capacitance can be seriously increased.

How do you measure the capacitance of an unknown capacitor?
Well, you can charge it slowly by limiting the current with a resistor R in series, as shown in the schematic (Fig. 4.57).

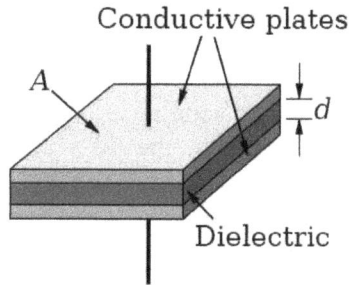

Fig. 4.56. A simple capacitor.
Image: Wikimedia Commons

The larger R, the smaller the maximum current, and the longer the charging takes, of course. As soon as you connect the battery by closing the switch, a current starts to flow and it drops exponentially to zero as

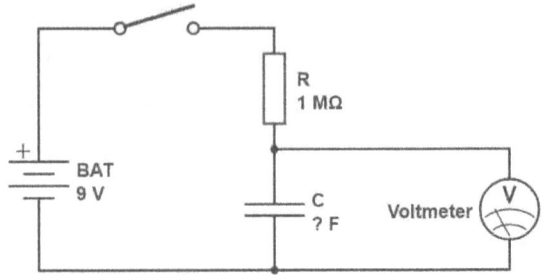

Fig. 4.57. Charging a capacitor over a resistor.

the capacitor fills. Consequently, the voltage across it (U) rises in the limit up to a maximal value U_{max}. Theoretically, that would be the battery voltage (U_B) if the internal resistance (usually called "input impedance") of the voltmeter were infinite. Realistically, a good voltmeter has an input impedance of about 10 MΩ, so the voltage over the capacitor can never exceed 10/11 of the battery voltage. Even the (non-ideal) capacitor may have a very small leaking current which decreases the maximum voltage a little bit more. This is what happens:

$$U = U_{max}\left(1 - e^{-\frac{t}{RC}}\right)$$

If you want to do this experiment at home or in a classroom, you should make sure the capacitor is big enough. The time span RC should have an "observable" length, since by then the capacitor is approximately 63% full; just replace t by RC to find this ($U/U_{max} = 1 - e^{-1} \approx 0.6321$). With a combination of $R = 1$ MΩ and $C = 1$ μF, $RC = 1$ s, which is too quick. A value of a few hundred μF is ideal for a demonstration. With $R = 10$ MΩ (the maximal value in the commercial range) you might lower C to the order of 10 μF. For higher capacitances, you can use a lower R, but for smaller capacitances you will need an electronic switching system to charge and discharge C regularly and an oscilloscope to visualize this.

I did the test with a so-called "electrolytic" capacitor on which was printed 220 μF. With a serial resistance of 1 MΩ (mentioned tolerance 1%, but since it was quite old, I measured it: 1.03 ± 0.062 MΩ), with $RC \approx 220$ s, I was in a comfortable measuring range to be able to use a normal chronometer on a watch or mobile phone.

Note that before starting such a measurement, you should of course discharge the capacitor first! With capacities and voltages that are not too large, this can be done by simply short-circuiting the two connecting wires. But I wouldn't risk that with one of 1 farad, because the wires could melt. If in doubt, just use

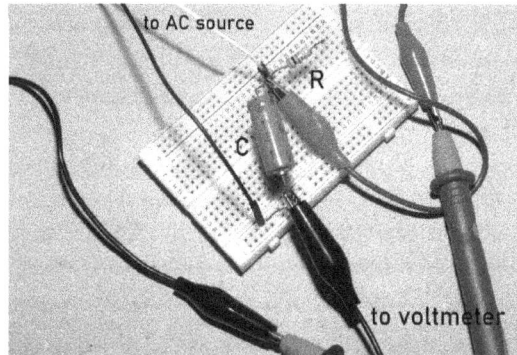

Fig. 4.58. The measurement set-up.

a small resistor of say 100 Ω, then the current cannot exceed 90 mA at, for example, 9 V.

It is also advisable to use a more stable voltage source than a battery, since it might discharge a bit during the experiment, even with such a low current. Especially rechargeable batteries discharge quickly a bit when they are just filled up; after a while they stay more stable for some time. I used a charger from a phone that produced a constant 8.54 V.

The measurements are as follows (*t* in s, *U* in V):

t	U_C	t	U_C	t	U_C	t	U_C	t	U_C	t	U_C	t	U_C	t	U_C
0	0.00	240	4.51	480	6.33	720	7.08	960	7.40	1200	7.54	1440	7.61	1680	7.65
30	0.81	270	4.83	510	6.45	750	7.13	990	7.42	1230	7.55	1470	7.62	1710	7.66
60	1.55	300	5.13	540	6.58	780	7.18	1020	7.45	1260	7.56	1500	7.62	1740	7.66
90	2.19	330	5.39	570	6.69	810	7.23	1050	7.47	1290	7.58	1530	7.63	1770	7.66
120	2.77	360	5.62	600	6.78	840	7.27	1080	7.49	1320	7.58	1560	7.64	1800	7.66
150	3.29	390	5.83	630	6.87	870	7.31	1110	7.50	1350	7.59	1590	7.64	1830	7.66
180	3.74	420	6.01	660	6.95	900	7.34	1140	7.52	1380	7.60	1620	7.65	1860	7.67
210	4.14	450	6.18	690	7.01	930	7.37	1170	7.53	1410	7.61	1650	7.65	1890	7.67

To determine C, I did a regression with an exponential model:

$$y = a \cdot \left(1 - e^{-\frac{x}{b}} \right)$$

with $x = t$, $y = U$, so $a = U_{max}$, $b = RC$.

Now, if you use the "real" uncertainties on the voltages as they are mentioned in the manual of the voltmeter, you get a weird result: the fitted values look okay, but the error on b (= RC) looks way to big: $b =$ 280 ± 46 s (Fig. 4.59). Why is that?

Well, in this case, according to the manufacturer of the instrument (Ohmeron MT 488B), the uncertainty on a measurement like 7.60 V, for example, was 0.078 V. That means the real voltage is "most likely" (99%? We don't know.) between 7.522 and 7.678 V. If that would be the real random fluctuation every measurement could have, we would see a lot of going up and down in the scatterplot. But that is not the case. This "± 0.078 V" should be interpreted as the maximal *absolute* error that can occur when the environment temperature, humidity, etc. change (within given limits). Now, during the short time span of an experiment, the conditions don't change much. The voltmeter may have a small systematic error because of a not-so-perfect calibration, but that won't change suddenly. In fact, to find the b parameter, it doesn't matter at all! Even if all the voltages were 50% wrong, the fitted b value should be just the same, as long as they are 50% wrong consequently in the same direction.

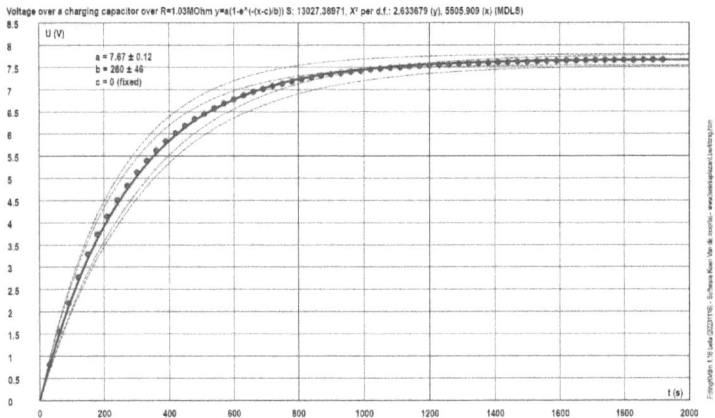

Fig. 4.59. The measurements with "real" absolute uncertainties on the voltages, causing a big uncertainty on the b parameter.

So, assuming a good stability of the instrument, we can use the *minimal* error of ± half of the least significant digit. At least, if the readings were stable… They did have some fluctuations though: the last digit went up and down sometimes, so I used a fixed uncertainty of ± 0.01 V for U. For t, I used ± 1 s. Of course with a data logger, this could be made smaller.

Voltage over a charging capacitor over R=1.03MOhm y=a(1-e^(-(x-c)/b)) S: 458577.4002, X² per d.f.: 26.99349 (y), 4033.128 (x) (MDLS)

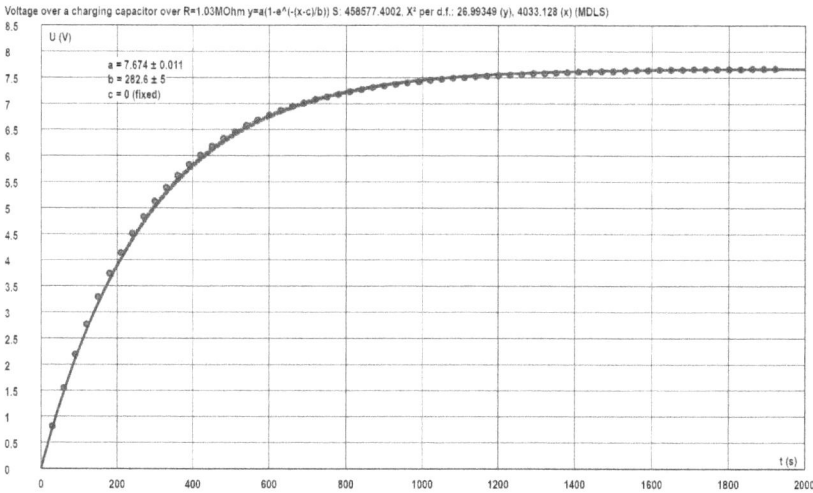

Fig. 4.60. The measurements with realistic uncertainties and the best fitting exponential curve.

This produced a more realistic $b = 282.6 \pm 5.0$ s and a perfect $a = 7.674 \pm 0.011$ V using MDLS. OLS produced $a = 7.6435 \pm 0.0019$ V which is impossible, since the last measured voltages were 7.67 V, so U_{max} cannot be lower. The produced value of $b = 271.4 \pm 0.41$ s is therefore also less reliable. MDLS is preferable here anyway since you can use a capacitor as a clock: you should be able to calculate the time from reading the voltage, using the same formula.

We can now calculate $C = b/R \approx 282.6$ s/1.03 M$\Omega \approx 274.36893$ µF, which is surprisingly more than the value printed on it. How precise is this?

$$\sigma_C = \sqrt{\left(\frac{\sigma_R}{R}\right)^2 + \left(\frac{\sigma_b}{b}\right)^2} \cdot C = \sqrt{\left(\frac{0.062}{1.03}\right)^2 + \left(\frac{5.0}{282.6}\right)^2} \cdot 274.36893 \approx 17.214 \mu F$$

So, we can conclude $C = 274 \pm 17$ µF. For more precise results, use a precision resistor! Some multimeters also measure capacitance directly, but then you miss the fun!

FittingKVdm example data file: "Charging_capacitor.dta1".

4.1.20. RLC filters

What happens if you put an alternating voltage (with frequency f) over a capacitor (with capacitance C), a coil (or "solenoid", with inductance L), and a resistor (with resistance R) in series, and you measure the voltage over the resistor, as in Fig. 4.61?

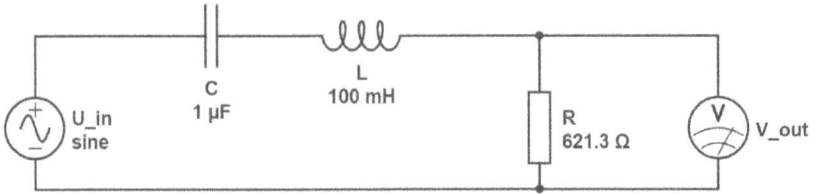

Fig. 4.61. An RLC bandpass filter.

The capacitor has an "apparent resistance" (impedance) that is inversely proportional to the frequency; higher frequencies will pass easier through it. The coil's impedance is proportional to the frequency; it gets "clogged" by high frequencies. The resistor's impedance is (ideally, and realistically in a very good approximation) independent of the frequency. For a "zero" frequency, the capacitor will be full quickly and it will be like it has an infinite resistance; while if the frequency goes to infinity, the coil will seem to have an infinite resistance, and again, the resistor will get nothing. So, you can understand there will be an "optimal" (or "resonance") frequency f_{res} for the signal transmission: not too low for the capacitor and not too high for the coil. Of course, there is a whole band of frequencies around f_{res} that gets through without too much reduction; that's why it's called a **"bandpass" filter**.

The output voltage can be calculated exactly from the theory about these components. Suppose ω is the pulsation of the incoming signal (= $2\pi f$). The part of the incoming voltage (U_{in}) that can be measured over the resistor (U_{out}) is equal to the impedance of the resistor (R) divided by the total impedance (Ohm's law). Here, the impedances of the coil ($L\omega$) and of the capacitor ($1/(C\omega)$) have to be treated as imaginary numbers, and R as a real number, so the total has to be calculated as the modulus of the complex sum. Also, the internal resistance of the coil (R_L) cannot be neglected.

Hence,

$$\frac{U_{out}}{U_{in}} = \frac{R}{\sqrt{(R+R_L)^2 + \left(L\omega - \frac{1}{C\omega}\right)^2}} = \frac{1}{\sqrt{(1+r)^2 + \left(af - \frac{b}{f}\right)^2}}$$

with

$$r = \frac{R_L}{R}, \quad a = \frac{2\pi L}{R}, \quad b = \frac{1}{2\pi RC}$$

The peak height (reached when $af = b/f$) can only be 1 if the coil is "ideal" (zero internal resistance, so $r = 0$).
This is the RLC transmission function mentioned on p. 40, with $x = $ frequency.

If you want to check how this works in the real world, you need a signal generator, but if you don't have laboratory equipment, the headphone exit of a laptop pc will do just fine; that's how I did it. The signal was generated with the Windows beep function (e.g., in Delphi Pascal, the command "beep(500, 10000);" generates a signal with a frequency of 500 Hz during 10000 ms = 10 s). The errors on the frequencies can be assumed to be small (say 0.1 Hz) since they are derived from the clock of the computer. The voltages were measured with a simple multimeter (OWON OW18E). The errors according to the manufacturer were 0.05% + 10 digits (see p. 64).
Then, for each measurement, a "y" value and its error had to be calculated:

$$y := \frac{U_{out}}{U_{in}}$$

$$\sigma_y = \sqrt{\left(\frac{\sigma_{U_{out}}}{U_{out}}\right)^2 + \left(\frac{\sigma_{U_{in}}}{U_{in}}\right)^2} \cdot y$$

This is easy to do with a spreadsheet program.

The values of L (100 mH ± 10%?) and C (1.0 µF ± 10%?) shown on the schematic, were given by the manufacturer; R was measured as 621.3 ± 1.9 Ω (accuracy ± (0.3% + 3 digits) according to Owon).

If they were correct, we could expect a peak at:

$$f_{res} = \sqrt{\frac{b}{a}} = \frac{1}{2\pi\sqrt{LC}} = \frac{1}{2\pi\sqrt{100 \cdot 10^{-3} \cdot 1 \cdot 10^{-6}}} \approx 503.29 \text{ Hz}$$

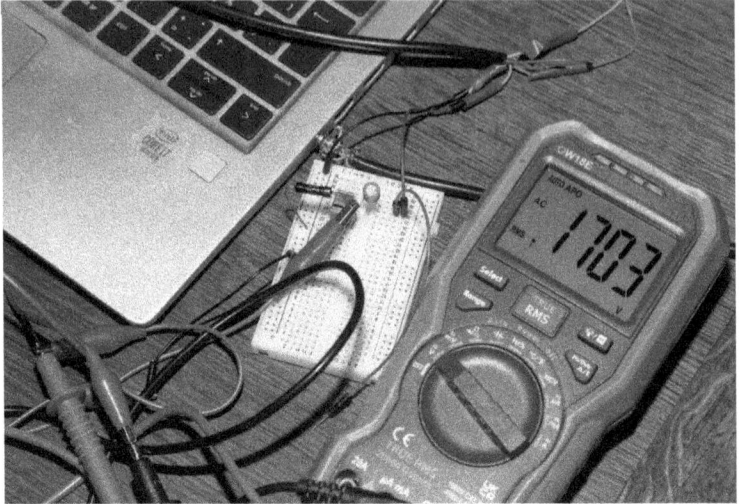

Fig. 4.62. The set-up for the measurements.

The results from 17 measurements with $f = 10\text{–}4000$ are shown in Fig. 4.63.

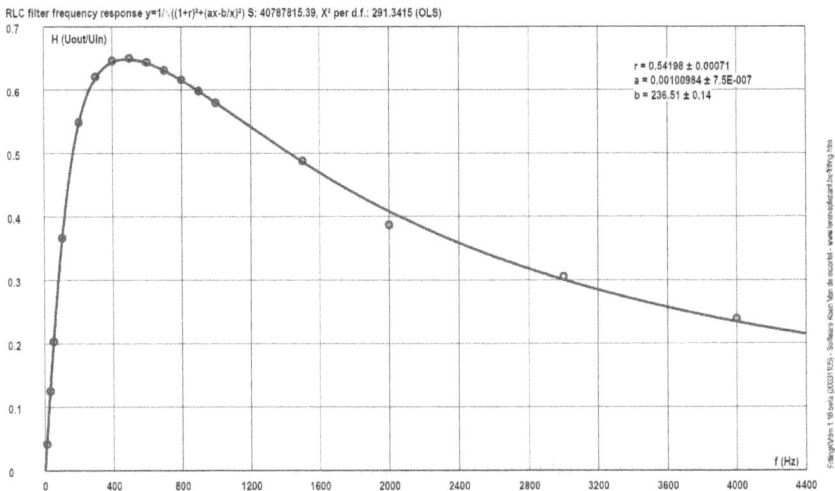

Fig. 4.63. Measured responses through the filter, and the best fitting theoretical curve.

From the fitting, we get
$r = 0.54198 \pm 0.0071$
$a = 0.00100984 \pm 0.00000075$
$b = 236.51 \pm 0.14$
$f_{res} = \sqrt{(b/a)} = 483.9 \pm 1.2$ Hz

My multimeter didn't have an option to measure L. But now we can calculate it!

$$L = \frac{aR}{2\pi} \approx 99.855974 \text{ mH}$$

$$\sigma_L = \sqrt{\left(\frac{\sigma_a}{a}\right)^2 + \left(\frac{\sigma_R}{R}\right)^2} \cdot L \approx 0.8020 \text{ mH}$$

So, we should say $L = 99.86 \pm 0.80$ mH, which is pretty much confirming the manufacturer's specification. The value for C was a bit further off; check this as an exercise.

This shows that you can use this setup to determine L (and C) if you don't have the equipment to measure it directly.

In fact, it's not necessary to put a coil in the circuit if you want to find out C, or a capacitor if you want to find L. In the first case, set $a = r = 0$, in the second, $b = 0$.

Test results with the same components are as follows:

(1) With only R and C, we have a so-called **"high-pass" filter**, since the capacitor allows the higher frequencies to pass easier.

Fourteen measurements were done from $f = 10$ to 3000 Hz. The fitted $b = 236.9 \pm 1.2$. Hence,

$$C = \frac{1}{2\pi R b} \approx 1.0813187 \,\mu\text{F}$$

$$\sigma_C = \sqrt{\left(\frac{\sigma_R}{R}\right)^2 + \left(\frac{\sigma_b}{b}\right)^2} \cdot C \approx 0.006398 \,\mu\text{F}$$

So, $C = 1.0813 \pm 0.0064$ μF.
With $\chi_x^2 \approx 0.11$, the fitting is quite precise.

Directly measured with the multimeter, the reading was: $C = 1.0908$

± 0.034 μF (accuracy: ± (3.0% + 10 digits) according to Owon), so that's not a bad result.

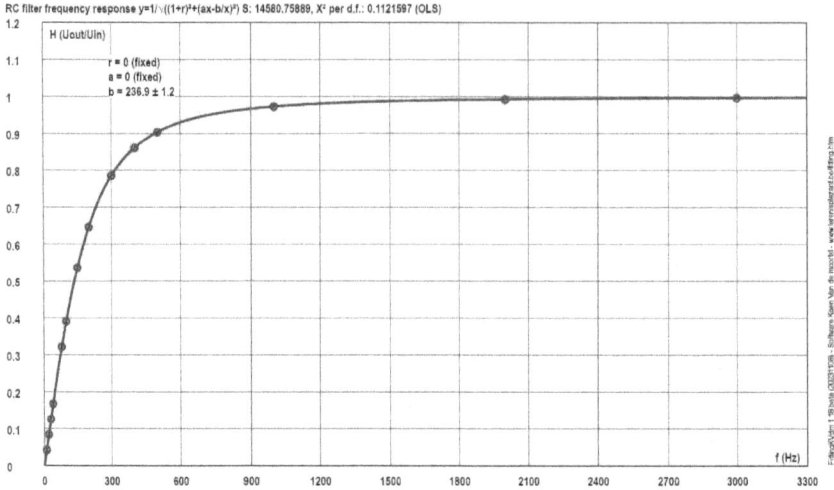

RC filter frequency response y=1/√((1+r)²+(ax-b/x)²) S: 14580.75889, X² per d.f.: 0.1121597 (OLS)

Fig. 4.64. Transmission through an RC filter.

(2) With only R and L, we have a **"low-pass" filter**, since low frequencies are less hindered by the coil. Seventeen measurements were done from f = 10 to 5000 Hz.

RL filter frequency response y=1/√((1+r)²+(ax-b/x)²) S: 44320211.76, X² per d.f.: 295.4681 (OLS)

Fig. 4.65. Transmission through an LR filter.

The fitted parameters are:
$r = 0.53679 \pm 0.00064$
$a = 0.00102609 \pm 0.00000065$
So, $L = aR/(2\pi) \approx 0.101462822$ H

$$\sigma_L = \sqrt{\left(\frac{\sigma_a}{a}\right)^2 + \left(\frac{\sigma_R}{R}\right)^2} \cdot L \approx 0.00031687$$

Or $L = 101.46 \pm 0.32$ mH.

$\chi_x^2 \approx 295$ means that the measurement errors might be bigger than promised by Owon, and therefore the error on L is probably also a bit bigger.

If you look at the scatterplot, you see that there might have been a small mistake in the measurement with $f = 2000$ Hz. It's not sure, but it looks suspicious. If we refit without this point, we get: $a \approx 0.0010135486$ which would cause L to go down a little bit too, to coincide better with the first measurement.

Remarks:
(1) If you want to do this experiment, make sure R is such that a big enough difference in transmission can be measured in the frequency range that can be produced by your generator (PC: only about 10–10000 Hz). And R should also stay much lower than the input impedance (internal resistance) of your voltmeter (or you should use the total resistance of R and the meter in parallel), but also not too low to burn the output of the generator; to stay safe I would use at least 100 Ω.

(2) In fact, you could derive the value of C from just one measurement, yes, but the error margin would be much bigger.

FittingKVdm example data files: "Filter 1 RLC.dta1", "Filter 2 RL.dta1", "Filter 3 RC.dta1"

4.1.21. Diode characteristic

A diode is a semiconductor device with a non-ohmic resistance: the current (I) through it is not directly proportional to voltage (U) over it. The ideal I vs. U curve is not a straight line but an exponential function shifted through the origin, if the temperature stays constant, according to the theory of the American physicist William Schockley (1910–1989):

$$I = a\left(e^{bU} - 1\right)$$

The parameter a in the formula is minus the maximal current if the diode is reversed (before it burns to ashes). Parameter $b = q/(nkT)$, with $q =$ the elementary charge, $k =$ Boltzmann constant, $T =$ absolute temperature, and $n =$ ideality factor between 1 and 2.
See https://en.wikipedia.org/wiki/Shockley_diode_equation.

This theory can easily be tested at home or in a classroom with basic equipment; see the schematic. (It's wise to add an additional resistor in series, to keep the current below the maximum allowed value.)

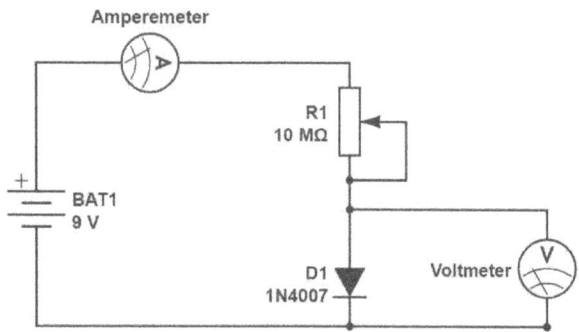

Fig. 4.66. Measurement of the diode characteristic.

I did the measurements with a classic silicon diode of type 1N4007.
The voltage was measured with an Ohmeron MT488B multimeter in the range 0–4.000 V. The accuracy according to the manual should be 0.5% + 4 counts = 0.5% + 0.004 V.
The current was measured with an Owon OW18E multimeter. Its accuracy in the 200.00 µA range should be 0.5% + 10 counts = 0.5% + 0.1 µA. In the 2.0000 mA range, 0.5% + 1 µA; 20.000 mA range, 0.5% + 10 µA; 200.00 mA range, 0.5% + 100 µA. The "10 counts" are certainly not exaggerated since the displayed values went up and down quite a lot.

As you can see in Fig. 4.67, the model fits quite well.

It predicts a maximal reverse current of 0.0146 µA (fitted with MDLS), which is better than the datasheet gives as "worst case" at 25 °C: 5 µA when 1000 V is applied. Direct measurement with a reverse voltage of 12 V (without voltmeter) gave a value of less than 0.01 µA (unmeasurable). See, www.alldatasheet.com/datasheet-pdf/pdf/58830/DIODES/1N4007.html.

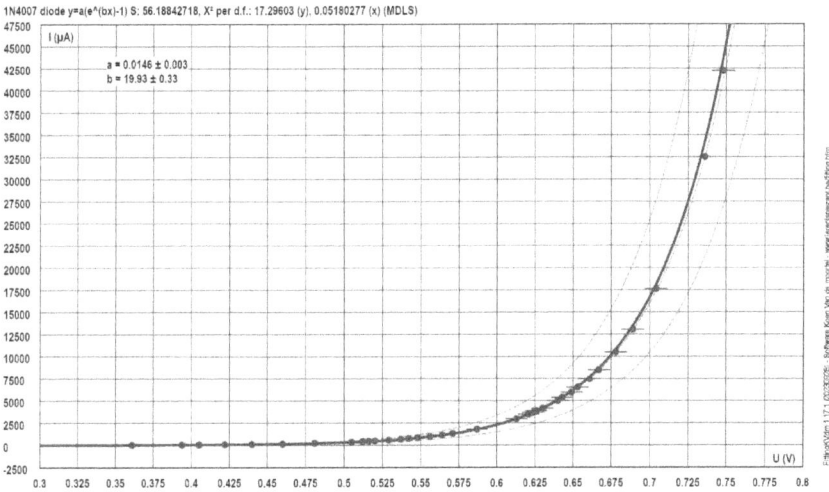

Fig. 4.67. 1N4007 diode measurements and the fitted "ideal" curve.

By the way, fitting with OLS gives $a = 0.0160$ µA. Parameter b is practically the same: 19.78.

Remark: If you do the measurements with reversed voltage and the shown schematic, you will see some current in the order of 1 µA, but that is due to the internal resistance of the multimeter (usually about 10 MΩ), so don't do that.

The same measurements were done with a germanium diode of type OA72 (Fig. 4.68) and a white LED, but they don't fit well in the "ideal diode" model. If the fitting is done without the last three points, the curve follows the points to about 0.8 V, and the next points fall way below the model predictions.

A possible explanation might be that the internal temperature at the p-n junction raises significantly, since Ge diodes and LEDs have a higher conductivity, so the parameter b becomes smaller when U rises, so the current is lower than we expect. This temperature is difficult to measure with simple equipment. Keeping the diode temperature as constant as possible (using cooling equipment) might possibly help obtain a more "ideal" curve.

But, a more important problem is that a significant internal resistance can exist in the diode, which causes a voltage drop, and makes it impossible to calculate U explicitly from I, only numerically by iteration. More details and a calculator can be found at this site from the Technical University of Graz , Austria:

https://lampz.tugraz.at/~hadley/psd/L6/pnIV.php

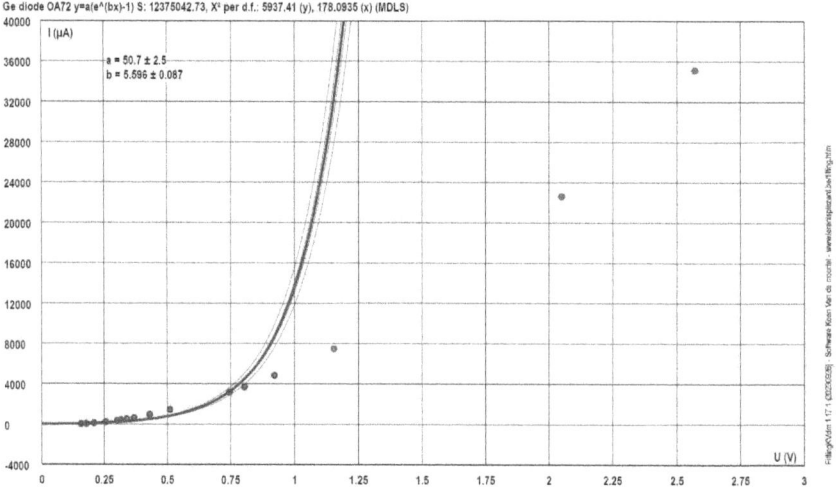

Fig. 4.68. Measurements from a Ge diode and fitted curve without the last three points.

FittingKVdm example data files:
"Diode Si 1N4007.dta1", "Diode Ge OA72.dta1", "Diode White LED.dta1"

4.1.22. Calibrating a salinity probe

How much salt is there in a river, lake, sea, etc.? This question is often asked in environmental research.

The most direct way to find it out is by taking a sample; let the water evaporate and study the residuals. But that's quite time consuming, so an indirect approximation is often preferred: measure the conductivity and deduce the salt concentration from that. The more substances are dissolved in water, the better it conducts electricity (well... to a certain extent), and most of the ions in natural water come from salt.

Measuring conductivity is fast and relatively easy but not trivially easy... You might think that just putting the probes of a simple multimeter in the water and switching it to the "ohms" position will do it, but no, it's really a lot more complicated.

First of all, an ohmmeter puts a small *direct* voltage on its probes and then measures the current, so it can calculate the resistance, which is simply the inverse of conductance ($G = I/U$). The problem with this direct current is that it causes electrolysis, and the current will keep changing because of the reactions happening in the water. To avoid that, a small *alternating* current has to be used.

Second, the current is not only dependent on the voltage, but also on the shape and dimensions of the probe. Therefore, standardized probes are used, but you can calibrate any shape of probe, as we shall see.

You can use many things as a probe, but, just keep in mind that
(1) The shape is fixed, not bendable, and the metal parts in the water should always have the same surface and distance. Just hanging two metal plates in the water is not correct.
(2) The metal parts should not be too sensitive for corrosion; a plain copper wire will be covered with an oxidation layer in no time.

Fig. 4.69. A suitable cheap probe.

Professional (expensive) probes are covered with platinum, but for occasional use, a normal power plug with a rubber cover will do fine (Fig. 4.69), or a smaller one from an old cell phone charger or whatever.

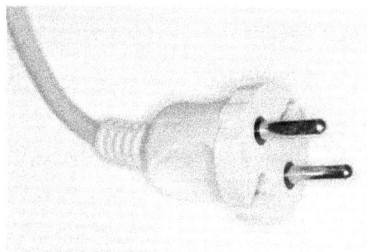

To obtain a small alternating current, you might use a transformer connected to the mains voltage, but that's not really the safest, and it produces a fixed frequency of 50 or 60 Hz. Any cheap low-frequency oscillator (sine or square wave generator) will do, and it will allow you to adjust the voltage and the frequency. If you are handy, you can make one yourself, see for example, www.nutsvolts.com/magazine/article/seven-common-ways-to-generate-a-sine-wave. Probably the simplest and cheapest source available for everyone, is the speaker exit of your computer. Use any sound program to generate a simple sine wave. You'll just need a plug and some alligator clips to make the connections with the volt and ampere meters and the probe. To generate some data for you, I used the windows beep function (available in any programming language).

If the output amplitude of the generator is very stable, you only need to measure the voltage once, and you can use the same instrument to measure the current. But most probably the voltage will drop when the water becomes more salty, and it's more practical to use separate instruments to measure U and I.

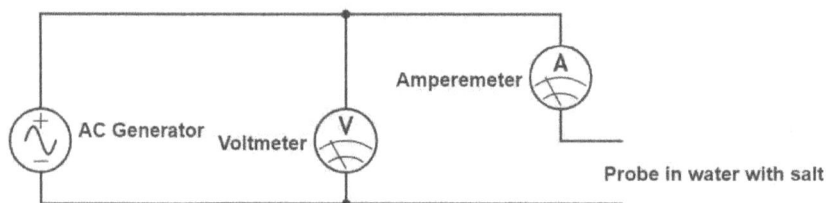

Fig. 4.70. Schematic of the measurement set-up.

Now, the *frequency* of the applied voltage,... does it matter? That is quite simple to find out: just measure it, with a fixed concentration. A PC can generate frequencies in the audio range easily from 50 to 10000 Hz. Using my probe, in water with 0.52 g/l salt (this is slightly more than in our local tap water), I got the strange result you see in Fig. 4.71: the conductivity ascended in the beginning, stayed more or less constant until about 3 kHz, and then apparently some kind of resonance phenomenon was going on, and after 8 kHz, the measurements started to become unstable and fluctuating. Between 1 and 8 kHz the voltage dropped from about 0.4 to 0.003 V and hence the relative measurement errors increased. The higher the frequency, the more important the role of the probe as a capacitor, and that's not what we want to measure, so it seemed plausible to me to do the rest of the measurements with a frequency in the beginning of the stable area, say 800 Hz.

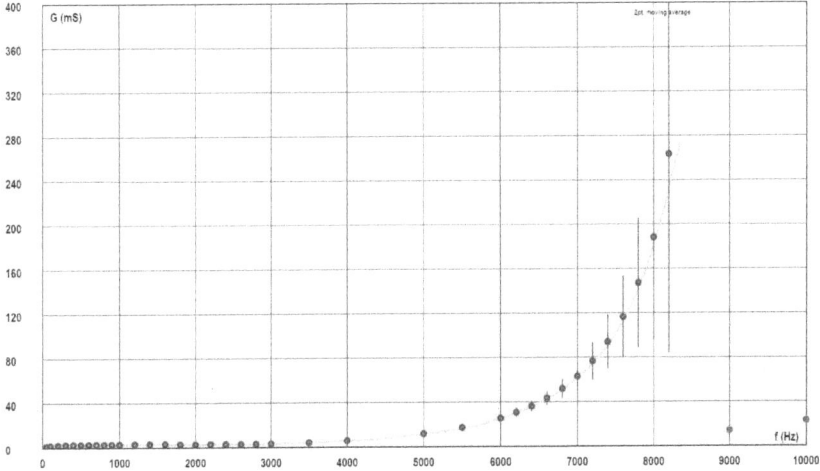

Fig. 4.71. Conductivity vs. frequency in a 0.52 g/l salt solution.

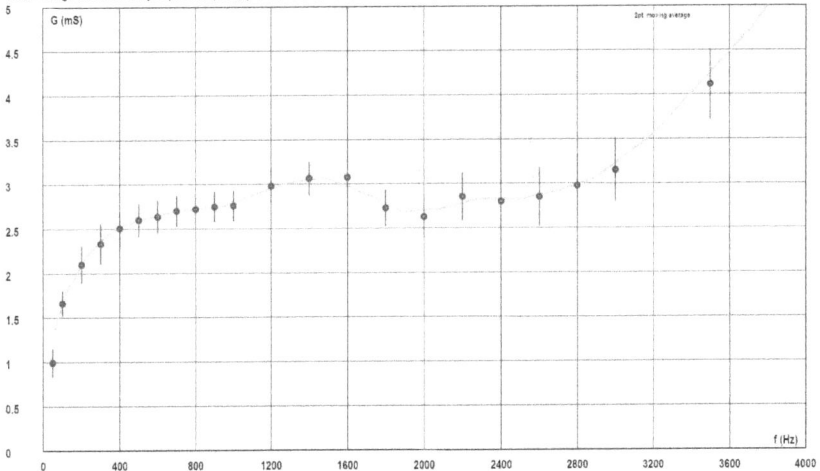

Fig. 4.72. The left part of the previous graph magnified.

What else is going on that explains this graph? I guess lots of doctoral theses have been written about all the possible physical and chemical processes happening in ionic solutions with an electrical current flowing through! You can find some interesting thoughts in this paper from the French instrument manufacturer "Radiometer Analytical SAS": www.tau.ac.il/~chemlaba/Files/Theoryconductivity.pdf.

Anyway, we don't need to know all the complex details here; we just want to find a mathematical function that describes the relationship between salt concentration and conductivity precisely enough for practical use.

Which range of concentrations should we measure? Probably everything between tap water and sea water (average is 35 g/l) is okay to start? According to the same source, weird things start happening with higher concentrations anyway, and a one-to-one relationship between C and G is no longer possible!

Now, if you want to repeat this calibration at home or in your classroom or laboratory, you need to make a series of water samples with different known amounts of salt in it. For example, you could start from scratch every time with a liter of demineralized (!) water and weigh, for example, 0.1 g of salt for the first, then 0.2, 0.5, 1, 5, 10, etc. for the next samples. You need smaller steps in the beginning, since the difference between no salt and a little bit of salt is more noticeable than "much" and "very much" salt. This is fine but it has disadvantages: you need a lot of distilled water, and you need a precise, sensitive, and stable scale. You can also add bits of salt a number (n) of times, but don't forget that your measurement error on the total weight will be the imprecision on the scale multiplied with the root of n.

It might be more practical to make one concentrated reference solution and dilute that. Use a syringe to take precise samples from it to mix with pure water, stir well, and wait a minute or so until the water is calm and the current is stabilized.

You also need to keep an eye on the temperature and keep it very stable since the conductivity can go up a few percent with every degree!

It's advisable to put the measurements in a spreadsheet program together with the measurement errors on U and I according to the instrument manufacturer, so you can easily calculate the errors on $G = I/U$:

$$\sigma_G = \sqrt{\left(\frac{\sigma_I}{I}\right)^2 + \left(\frac{\sigma_U}{U}\right)^2} \cdot G$$

The calculation of the errors on the concentration is similar.

Fig. 4.73 shows my results with French sea salt (with 2% NaI added) I found in our kitchen, at a room temperature of 21.1 °C.

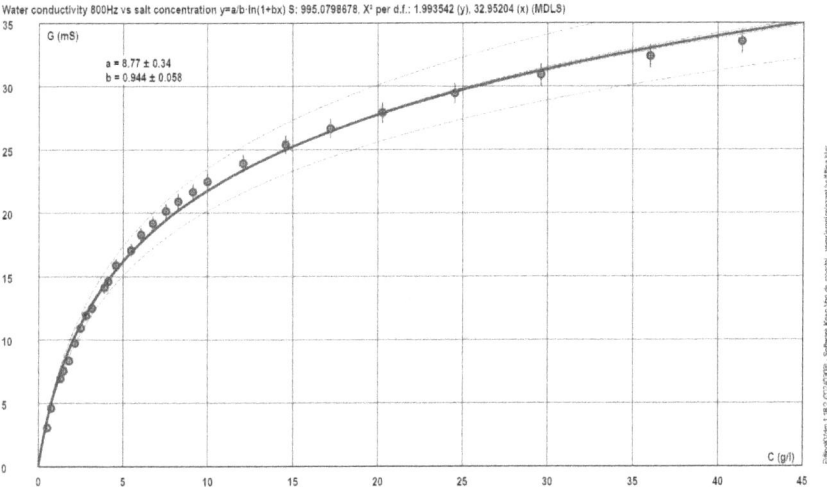

Water conductivity 800Hz vs salt concentration y=a/b·ln(1+bx) S: 995.0798678, X² per d.f.: 1.993542 (y), 32.95204 (x) (MDLS)

a = 8.77 ± 0.34
b = 0.944 ± 0.058

Fig. 4.73. The 800 Hz conductivity measurements for different concentrations, and the best fitting logarithmic function (shifted through the origin).

There is a clear pattern visible that looks like a square root function at first sight, but a power function doesn't really fit well. A logarithmic function shifted through the origin describes the data surprisingly well. A Weibull growth function, which has one additional parameter, fits even better, see Fig. 4.75. Since the whole point of a calibration is to use the inverted formula, to estimate x from y, MDLS is definitely recommended above OLS!

Theoretically, the conductivity will never be absolutely zero with pure water, since there are always some ions in water: positive hydrogen ions (protons) and negative oxygen ions, but as you can check yourself, it will be hard to measure.

The pattern looks a lot like the left part of a graph, shown in the paper I mentioned earlier indeed (Fig. 4.74). A calibration for the whole range of salinity (OE) is not possible since the function that covers this is not invertible.

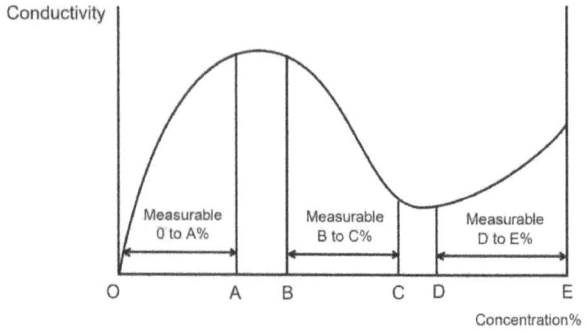

Fig. 4.74. G vs C according to "Radiometer Analytical SAS" company.

Can it be extrapolated to predict the conductivity of the Dead Sea (340 g/l)? I wouldn't do it... see above! Can it be used to calculate the salinity of an unknown solution? The curve runs nicely through the error flags, so yes, as long as you can be sure that the concentration is less than about 45 g/l, and the water temperature is the same as your calibration temperature! You might want to do different calibrations at a range of temperatures if you want to do field work.

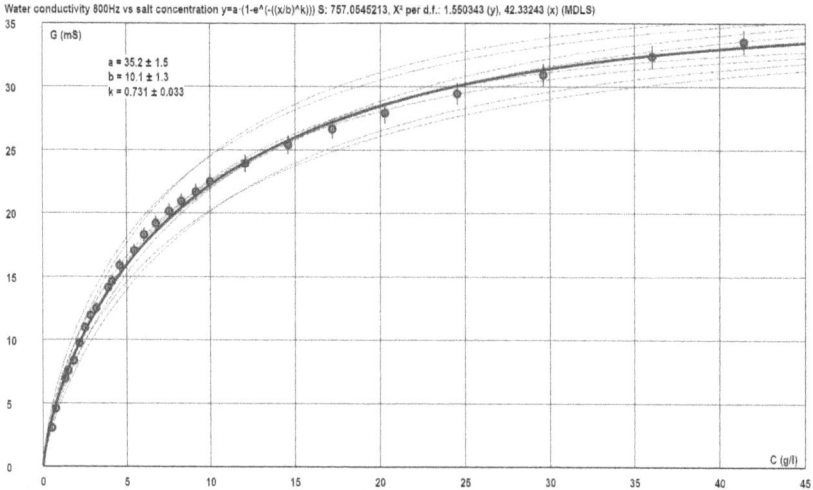

Fig. 4.75. The same measurements and the best fitting Weibull growth function.

FittingKVdm example data files:
"Salt_water_conductivity_800Hz.dta1", "Salt_water_G_vs_f.dta1".

4.1.23. Measuring concentrations with... your smartphone

Can a mobile phone be used as a laboratory instrument? Sure it can be! Maybe not top quality, but good enough for a fun home or classroom experiment!

If a liquid changes color by dissolving a substance in it, we can estimate the concentration visually: using our naked eye, we can clearly see the difference between weak and strong coffee, for example. With the help of a bit of modern toys, we can even do that a lot more accurately. We just need a setup to measure the light absorbed by the solution. As a light source, we can simply use diffuse stable ambient light (daylight that is reflected on a light, preferably white, wall, for example) that falls through a slit in a box. A serious camera can, of course, serve as a detector, but a simple mobile phone will more than suffice. A rectangular glass container would be ideal as a container, but an ordinary kitchen glass will work too. Place it against a hole in the box as close as possible to the lens of the camera of your "smart phone", in order to minimize reflections on the outside of the glass. During the measurement, it is best to close the lid of the box in order not to be bothered by all kinds of reflections in the glass, see Fig. 4.76.

Now, how do we *calibrate* our concentration meter? Suppose, for example, that we want to measure the strength of our coffee. The simplest way is to take a few pictures of a test liquid with a few known concentrations, e.g., first a glass of pure water, then water with a coffee spoon of very strong instant coffee, then with two, three, four, etc. spoons (and stir every time of course). It will undoubtedly also work with ink drops or red cabbage juice.

Here we have to be careful! After all, if you take a photo in automatic mode, the device will adjust the sensitivity (ISO) and the exposure time so that the photo always looks light. And that is just *not* what we want, of course. So you have to choose the "pro mode" in the camera options and then set the ISO to a fixed value (e.g., 100) instead of "automatic". If the aperture can be controlled, it should also be fixed, preferably at the largest opening (f/smallest number) to have a background that is as blurry as possible. In daylight, a shutter speed of 1/200 s will probably be fine, but you may need to make some adjustments. The lightest photo (pure water) should be light, but not overexposed white. So first try to set the shutter speed so that this works. Note that also the white balance is to be set to "manual", otherwise the device will adjust the colors because it thinks that the light source is changing! It doesn't matter if the white balance is set to cloudy, sunlight, tungsten, or whatever, as long as it doesn't change during the measurements. The flash must of course also be turned off because we only want to measure the transmitted light.

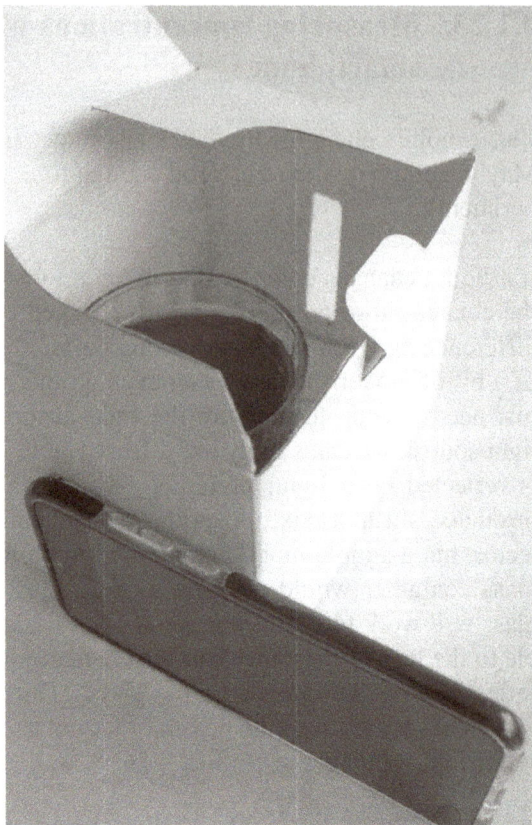

Fig. 4.76. The experimental set-up. A small hole has been made in the box for the lens of the mobile phone. (The box must be closed for the measurements, of course.)

In my test, I started with pure water and then added one to four spoonfuls of strong coffee. Fig. 4.77 shows you the result.

In order to find a relationship between the concentration and the colors in the photos, we first need to calculate the concentrations. This doesn't need to be done in mol/l or g/kg or anything like that; for the sake of simplicity, it will suffice here to use "spoons per glass" as a unit. I determined experimentally that "1 glass" (almost full, the volume we started

Fig. 4.77. Middle parts of the photos of pure water (left), and water with 1, 2, 3 resp. 4 spoons of strong coffee.

with) was 101 spoons. So if we add 1 spoonful of coffee to the glass, the concentration is actually 1 spoonful per $(1 + 1/101)$ glass, or 0.99 spoonfuls/glass. With 2 spoons that becomes $2/(1 + 2/101) = 1.96$, with 3, 2.91 and with 4, 3.85. Considering the way I measured the volumes, the precision will certainly not be better than 1%, except for the first value ("pure" water); theoretically, that should have 0% error (unless a few coffee molecules got in the water somehow), but that would give the point an infinite weight in the model fitting, so the x error could be set to a very small value here.

Now we still need to be able to paste numerical values on the photographed color tones. This is very easy with any photo editing program, e.g., "Digital Photo Professional" from Canon or "ViewNX" from Nikon (both free): just open the photo and move the mouse over the evenly colored area and you will see the color values. A common camera only registers three colors, red, green, and blue, and in a JPG file, these each get values between 0 (dark) and 255 (light). Since not every pixel in the center of the photo has the same values, we have to assume a measurement error of at least ± 1.

Remark: There are also smartphone apps, like "Physics Toolbox", that measure the RGB values directly without having to take a photo. You have to convert the measurements from the hexadecimal format though. See, www.vieyrasoftware.net.

Following are the obtained values from the photos above:

Conc. (spoons/glass)	R	G	B
0 ± 0.0001	249 ± 1	248 ± 1	232 ± 1
0.99 ± 0.01	216	153	40
1.96 ± 0.02	163	56	5
2.91 ± 0.03	96	16	1
3.85 ± 0.04	50	5	0

The numbers reflect what we see: the more coffee, the more light is absorbed, but the blue is absorbed the most, hence we see the remnants of the white light that comes through as brownish/reddish.

What pattern are we expecting? It would be nice to obtain a formula that would allow us to determine the concentration in an unknown sample. According to the *law of Lambert–Beer*, the transmission (1 – absorption) should decrease exponentially with the concentration.

And yet, the graph for the red values (Fig. 4.78) looks more sigmoid-like, and it fits quite nicely with a logistic function (baseline zero).

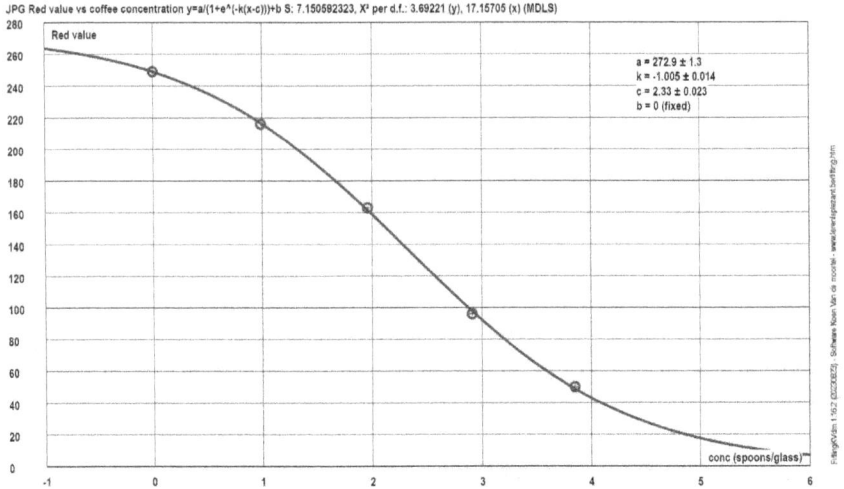

JPG Red value vs coffee concentration y=a/(1+e^{-k(x-c)})+b S: 7.150592323, X² per d.f.: 3.69221 (y), 17.15705 (x) (MDLS)

$$a = 272.9 \pm 1.3$$
$$k = -1.005 \pm 0.014$$
$$c = 2.33 \pm 0.023$$
$$b = 0 \text{ (fixed)}$$

Fig. 4.78. The logistic function that fits the "JPG red vs concentration" measurements.

The green and the blue channels look similar but steeper. That means the red values are the best suited to estimate a concentration, at least in this range. The blue values could be used to detect smaller concentrations since they react quicker.

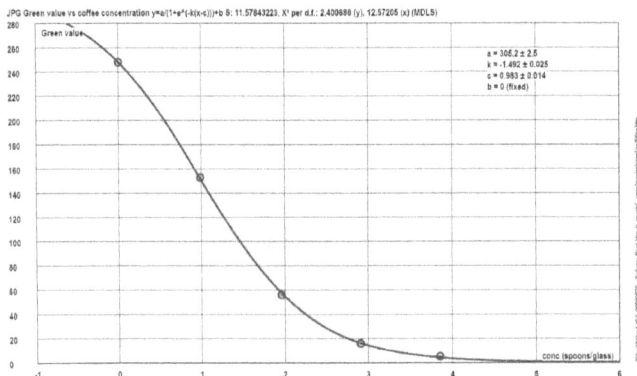

JPG Green value vs coffee concentration y=a/(1+e^{-k(x-c)})+b S: 11.57843223, X² per d.f.: 2.400688 (y), 12.57205 (x) (MDLS)

$$a = 305.2 \pm 2.5$$
$$k = -1.492 \pm 0.025$$
$$c = 0.983 \pm 0.014$$
$$b = 0 \text{ (fixed)}$$

Fig. 4.79. JPG Green vs concentration.

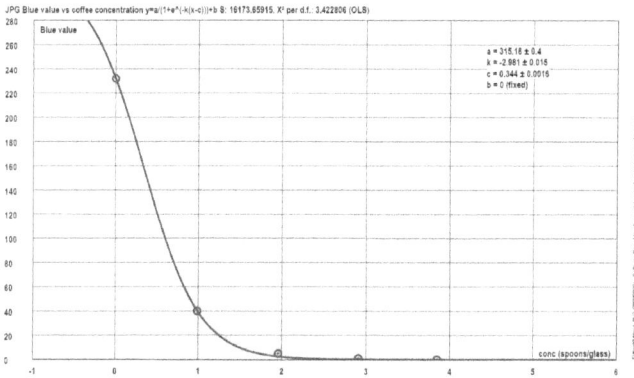

Fig. 4.80. JPG Blue vs concentration.

These remarkable results are too good to be coincidence. What could be behind this sigmoid pattern while we expected to see an exponential curve? Is there any reason to doubt the Lambert–Beer law? I guess not!

I have to admit, it also puzzled me for a few hours, but then my knowledge of photography came in handy. What does a camera do while creating a JPG file? You might expect that all the measured light values between a certain dark and a light threshold would be linearly converted to numbers between 0 and 255, i.e., all colors with less than a certain minimum value of red would have a red value of 0, and all those that are too bright a red value of 255. Well, it actually doesn't quite happen that way. The conversion of light intensity to number (the so-called *"tone mapping"*) is done approximately according to a logistic function: this way the dark colors still get a number that does not quickly become 0, and the lighter colors do not immediately become 255. The advantage is that the "dynamic range" of the photo (the difference between the lightest and the darkest that can be displayed) becomes a bit larger. You will be able to see some more details in the shadows and also in the lightest parts of the photo (clouds, for example). Fig. 4.81 illustrates this.

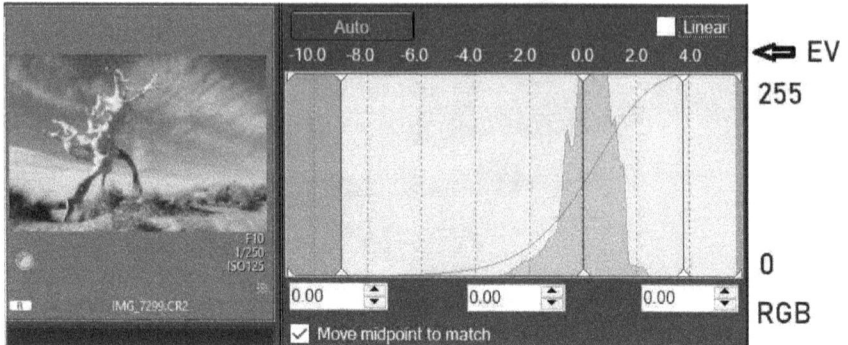

Fig. 4.81. A raw image, its histogram, and the tone mapping curve, as it can be adjusted in Canon's "Digital Photo Professional" software. The horizonal scale shows the exposure value (EV) and the vertical one the RGB values in the final JPG image. The conversion is usually not linear.

This must undoubtedly explain our observations. We know now that we need a logistic function to calibrate this kind of set-up, for optimal accuracy! And since we want to use the inverse function to calculate a concentration from a photograph, MDLS will certainly give the most accurate results, in this case hardly noticeable though, because the points fit so well on the curve (in the limit, if there are parameters that make the curve fit exactly, OLS and MDLS give the same results of course).

Example: If an unknown concentration gives a red value of 200 (always an integer value, which means it's rounded from a number between 199.5 and 200.5), we can conclude that the concentration must be $f^{-1}(200) \approx 1.3266$.
How precise would this be? $f^{-1}(200.5) \approx 1.3173$, and $f^{-1}(199.5) \approx 1.3359$. $(1.3359 - 1.3173)/2 \approx 0.0093$, so the concentration would be about 1.327 ± 0.009 spoons/glass. Not bad for such a primitive setup, isn't it?

Another question we can ask ourselves is as follows: "How can we improve the *measuring range* of our system?" In other words, what should we do to measure higher concentrations? Right now, they would give RGB values too close to zero. My first idea was to just double the shutter speed of the shot, for example, and then halve the RGB values, but given the non-linearity of the tone mapping, that's not going to be reliable.
We could, however, do a re-calibration with, for example, a 10 times slower shutter speed, which will then be worse for small concentrations

(overexposed photos) but better for bigger concentrations. To determine an unknown concentration, we could take two photos with the different shutter speeds, and use the photo with color values closest to 127 (half-way between 0 and 255). As a result, our measuring range would already expand somewhat.

Or... instead of using a mobile phone, we could also stick different color filters and a simple photocell (LDR: Light Dependent Resistor, cost: ca. 0.5 €) in front of the hole in the box, measure its electrical resistance with a simple multimeter, and recalibrate. The calibration curve should look more like an exponential function then because there is no "tone mapping" happening.

Remarks:

- To avoid the annoying "zero concentration" measurement, it might be more convenient to start with a "pure" solution (very strong coffee, juice, milk, ink, etc.) and then add small amounts of water for each measurement.
- This set-up will only be good to measure concentrations of this type of coffee. If you want to detect two different substances (mixed but not interacting), you will have to use at least two different color measurements.
- We have in fact made a primitive *spectrometer* with this set-up. In the industry and serious laboratories, spectrometers that can distinguish many more than three colors are used. Of course, they work with more stable calibrated light sources and detectors, etc. In the production of petrol, for example, these color spectra are continuously monitored to make sure that the composition doesn't deviate too much from the desired one. Each component of the mixture absorbs certain specific colors, and if you know them, you can accurately estimate each percentage, even with a multi-linear approximation if the concentrations don't vary too much.

For more information, google "absorption", "Lambert–Beer", "tone mapping", "spectrophotometry", etc.

FittingKVdm example data files: "JPG-Red_vs_coffee_concentration.dta1", "JPG-Green_vs_coffee_concentration.dta1", "JPG-Blue_vs_coffee_concentration.dta1"

4.1.24. The refractive index of a CD box

Light does not follow the *shortest* but the *fastest* path from one point to another. If a light ray enters a material in which it can move less quickly, it will therefore refract toward the normal to the surface. You can compare this with the path you will choose if you have to walk a distance from one point to another and then swim a distance. Swimming is slower, so you will prefer to shorten the stretch in the water. There is an "optimal" path, which is more broken the greater the speed difference. Determining this path is a so-called extremum problem, which was solved by the Dutch scientist Willebrord Snel van Royen (1580–1626), better known by his Latin name Snellius (in English mutilated to "Snell"). If light travels n times slower in a given medium than in a vacuum, the light will refract in this way when going from medium 1 to medium 2:

$$n_1 \sin \alpha_1 = n_2 \sin \alpha_2$$

where, α_1 and α_2 are the angles between the "normal" line (perpendicular) to the parting surface, on the incoming side (substance, or "medium" 1) and the outgoing side (medium 2). The material constant n is called the "refractive index". It is exactly 1 for vacuum (light has its maximal speed if it can move freely), but in air, it's not much more, 1.000293. So if we follow a ray of light going from air to a much denser material, n_1 can usually be neglected.

How can we measure the refractive index of a certain substance?
Well, all we need is a thin beam of light that we can easily trace, a piece of the material to be examined with a straight edge, a sheet of paper, a pencil, and a protractor.

I tried it with something that everyone has at home (well, for a few more years probably): the lid of a CD case, which according to Dr. Google is supposed to consist of polystyrene. You could tinker with a flashlight and two slits to construct a fine light source, but a laser pointer is easily available these days.
Aim the laser beam at the box, the bottom of which rests neatly against the edge of the sheet, and then mark points A, B, and C with a pencil as shown in Fig. 4.82.

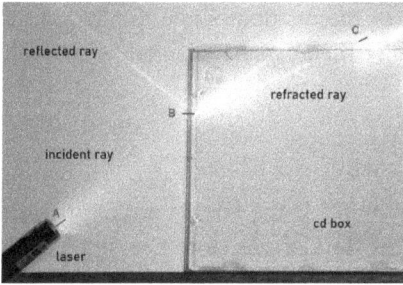

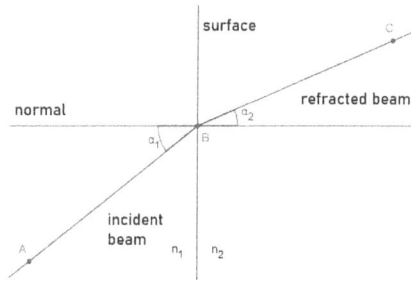

Fig. 4.82. Measurement of the refractive index of polystyrene. Fig. 4.83. The optical path, schematically.

Then draw the line segments AB and BC and the normal, and measure both angles.

I thus measured $\alpha_1 = 30°$ and $\alpha_2 = 18°$, from which we can calculate

$$n_2 = \frac{\sin(30°)}{\sin(18°)} \cdot 1.000293 \approx 1.618508$$

Now, that number of digits after the decimal point is probably a bit exaggerated. How accurate could the result be?

The simplistic school rules for significant figures are of no use to us now. With a little common sense (in Flemish we call this "farmer's logic"), we can figure something out.

What's the worst that could go wrong? The numerator of the fraction being less and the denominator being more, or vice versa. By marking the indications and the lines a little bit wrong, or by placing the protractor a little bit wrong, a mistake of one degree is easily made. The sine function is increasing for angles between 0 and 90°, so the worst cases are (since the error on $n_1 = n_{air}$ is negligible)

$$n_{2,max} = \frac{\sin(31°)}{\sin(17°)} \cdot 1.000293 \approx 1.76210$$

and

$$n_{2,min} = \frac{\sin(29°)}{\sin(19°)} \cdot 1.000293 \approx 1.48955$$

So we would have a maximum deviation of about 0.14 upward or downward, which is quite a lot for a refractive index.

But we can calculate the error propagation a bit neater, using derivatives:

A small deviation in the measurement of the incident angle $\Delta\alpha_1$ will give a deviation in the refractive index Δn_2 that can be estimated as (see p. 72):

$$\Delta n_2 \approx \frac{\partial n_2}{\partial \alpha_1} \cdot \Delta\alpha_1 = \frac{n_1}{\sin(\alpha_2)} \cdot \cos(\alpha_1) \cdot \frac{\pi}{180} \Delta\alpha_1$$

Note the $\pi/180$ that you must multiply with, if the angles are expressed in degrees! (After all, the cosine is just the derivative of the sine if the angles are in radians, so you have to use the chain rule!)
If $\Delta\alpha_1$ is the measurement error on α_1 ($\sigma_{\alpha 1}$), i.e., of the order of one degree, then Δn_2 becomes approximately 0.048927.

Analogously, a variation $\Delta\alpha_2$ gives the effect:

$$\Delta n_2 \approx \frac{\partial n_2}{\partial \alpha_2} \cdot \Delta\alpha_2 = -\frac{n_1 \cdot \sin(\alpha_1)\cos(\alpha_2)}{\sin^2(\alpha_2)} \cdot \frac{\pi}{180} \Delta\alpha_2$$

And that becomes about -0.033208.

The precision of n_1 is a lot better, hence the influence on n_2 is negligible. If we may assume that the errors at both angles are independent of each other, and that the probability that they both act equally strongly in the same sense is therefore not very great, we may add both effects as if they were perpendicular to each other, and we get as "statistically expected precision" of n_2:

$$\sigma_{n_2} \approx \sqrt{0.048927^2 + 0.033208^2} \approx 0.059132$$

That is more reassuring than the earlier rough worst estimate of ± 0.14. But still, we can do better. We can repeat the measurement with other angles. So that's what I did:

a_1 (°)	a_2 (°)	n_2
19.5	11.5	1.67483
30	18	1.61851
35.5	22	1.55062
48.5	27.5	1.55062
54.5	29.5	1.65377

You see, the results are all well within the limits of "worst case scenarios"! But which of the five measurements could enjoy the most confidence? Can our common sense also say something about that? Well yes. The larger the angle, the less the sine rises, so the smaller the deviations in numerator and denominator become, so the measurements with larger angles should in principle be slightly more reliable.

Of course,. we can also just calculate the average value: $n_2 = 1.62358$. The standard deviation of these five outcomes is 0.0471, so the error on the mean is $0.0471/\sqrt{5} = 0.0211$, which means that $n_2 = 1.624 \pm 0.021$ with 68% confidence, or $n_2 = 1.624 \pm 0.042$ with 95% confidence. Strictly speaking, we should take the *weighted* average, because the relative precision is better with the bigger angles! The n_2 values with bigger angles should weigh more.

But, we can also get a more accurate value by using regression, which uses all the collected information and the right weights. In order to use that, we have to express the angle of the broken beam as a function of the angle of the incident beam:

$$\alpha_2 = f(\alpha_1) = \text{Arc}\sin\left(\frac{\sin\alpha_1}{n}\right)$$

with $n = n_2/n_1$ ($\approx n_2$) the *relative* refractive index of polystyrene in relation to air. Now we can determine n by putting the measurements in a regression program that can handle this kind of function, like FittingKVdm. Since the variables α_1 and α_2 have a very symmetrical role, it shouldn't matter which one we use as x or y, so we certainly should use MDLS.

The result is $n = 1.617528 \pm 0.035$, so $n_2 = 1.618 \pm 0.035$. OLS would give 1.623147 ± 0.035, so $n_2 = 1.624 \pm 0.035$.

In any case, regression will give more leverage to the measurements with bigger angles, which is in correspondence with their better relative precision.

Another major advantage of such a regression analysis is clear on the graph: measurements that deviate somewhat from the general pattern (the model) stand out more, as in this case the middle one. All measurements are close enough to the curve if you look at the error flags. This is reflected in the χ^2 values being in the order of magnitude of 1.

Refraction air-->polystyrene y=Arcsin(sin(x)/n) S: 4.556074628, X² per d.f.: 0.403731 (y), 1.656738 (x) (MDLS)

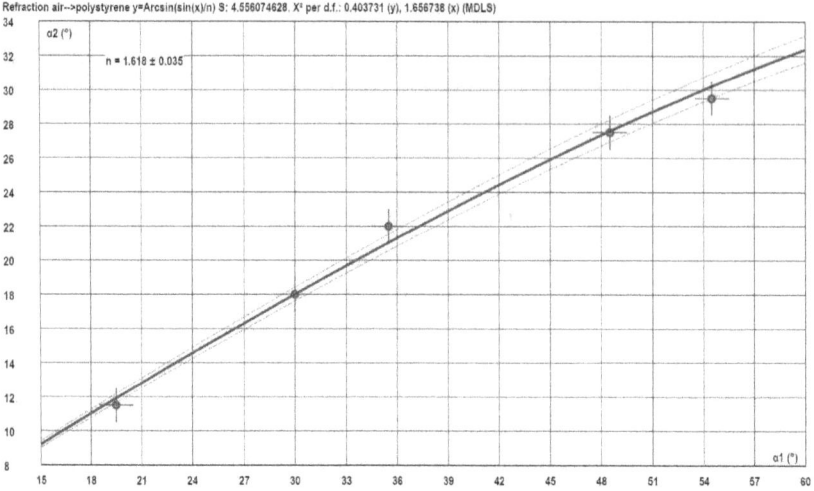

Fig. 4.84. Best fitting theoretical curve through the data.

Are we happy with this result?
Our laser, according to its label, had a wavelength of 532 ± 10 nm. According to https://refractiveindex.info, a website where you can find the optical properties of a lot of substances at different wavelengths, the refractive index at 532 nm should be 1.5983 (and 1.5997 at 522 nm, and 1.5969 at 542 nm). So, our estimate, using simple means, was not so bad! And MDLS gave a better result than OLS.

Using a solid block of test material, and more precise angle measurements, it's possible to improve the accuracy, of course. But that would require expensive optical equipment. The goal of the story here was to show what one can do with simple home material, and how the best results can be extracted using regression.

FittingKVdm example data file: "Refraction_polystyrene.dta1".

4.1.25. Radioactive decay

The changing rate of radioactivity of an unstable isotope is a classic text-book example to illustrate exponential decay.

Prof. Frank Fokkema, from the University of Amsterdam, NL, was so kind to provide me with some measurements done in his laboratory on a short living radioactive source (^{220}Radon). The variables here are as follows: $x = t$ = time in seconds; $y = N$ = counts from a Geiger counter near the sample. $\sigma_x = 2$ because the count intervals were 4 s. $\sigma_y = \sqrt{y}$ since it is a count.

As expected, the exponential model (with baseline!) fits well. Parameter c represents the background radiation, the horizontal asymptote ($y = c$) of the curve. Radioactive decay of an atom occurs with a statistical probability, not with a precise time schedule; that's why there is a lot of "noise" in the measurements.

The derived parameter "half-life" is 52.2 ± 1.2 s. According to Wikipedia (Dutch version, consulted 18 Jan. 2024), this should be 55.6 s, but Britannica says 51.3, see,
www.britannica.com/science/radon#ref237175.
Since the data contain many points above and below the asymptote, OLS had to be used.

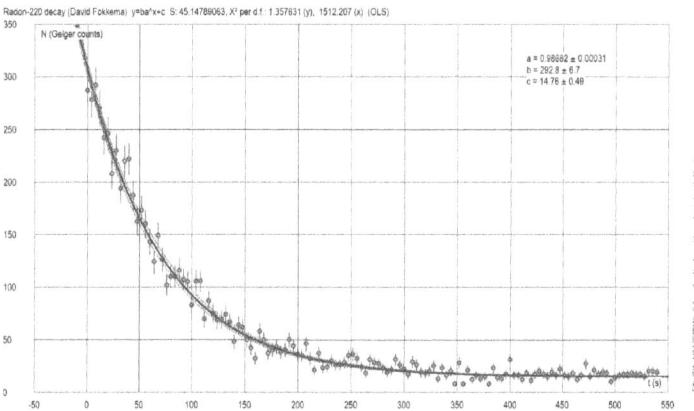

Fig. 4.85. Radioactivity counts from a Radon-220 sample and the best fitting exponential function.

FittingKVdm example data file: "Radon220 decay.dta1".

4.1.26. The lifetimes of pressure vessels

How long does a device, a tool, a car, etc. last? That depends a lot on
chance, like radioactive decay, but the "chance" is moderated by many
factors here: how well it was maintained and not abused in rough circumstances, how often it was used, etc. Maybe there was even a factory-
built-in timing to sabotage the device, as it seems to be done with some
printers, cell phones, washing machines, etc.? I know for sure it is done
by at least one company that produces insulating windows because I was
once working for them, and they fired me for not agreeing with this
policy. So, the function to describe what happens with a certain type of
products after usage is to be determined empirically.

I found an interesting example in a book by Annette J. Dobson: *An introduction to generalized linear models*, 2nd ed. (Chapman & Hall/CRC,
2001), chapter 4.2: *Example: Failure times for pressure vessels*. This
can be found at
www.academia.edu/18454337/AN_INTRODUCTION_TO_GENERALI
ZED_LINEAR_MODELS.

The original data were 49 ages of Kevlar epoxy strand pressure vessels
when they broke down (in hours): 1051, 1337, 1389, 1921, 1942, 2322,
3629, 4006, 4012, 4063, 4921, 5445, 5620, 5817, 5905, 5956, 6068,
6121, 6473, 7501, 7886, 8108, 8546, 8666, 8831, 9106, 9711, 9806,
10205, 10396, 10861, 11026, 11214, 11362, 11604, 11608, 11745,
11762, 11895, 12044, 13520, 13670, 14110, 14496, 15395, 16179,
17092, 17568, and 17568.

In the source text, a distribution was calculated by grouping them into
seven classes. Then a Weibull distribution was fitted through these
seven numbers. However, doing this, a lot of information is lost. What I
did, conserves all the information: x_1 = the time when the first vessel
broke, y_1 = 1; at time x_2, the second vessel broke, y_2 = 2, and so on: y_i =
the total number of broken vessels at time x_i.
I assumed the same error (1 hour) for each time value. For the y values, I
used the usual square root. Now we have to fit with the *cumulative*
distribution of course. Anyway, it should be a sigmoid-like function.

The text above recommends to use the Weibull distribution with k = 2.
Why 2? Is there a theoretical preference for that value? Apparently not;
I quote: "Although there are discrepancies between the distribution and

the data for some of the shorter times, for most of the observations the distribution *appears* to provide a good model for the data.".

Since the last data points are probably in the horizontal asymptote area, I used OLS here. MDLS produces a slightly higher value for the limit value (parameter *a*). The fitted curve seems reasonable: it goes through most of the error flags.

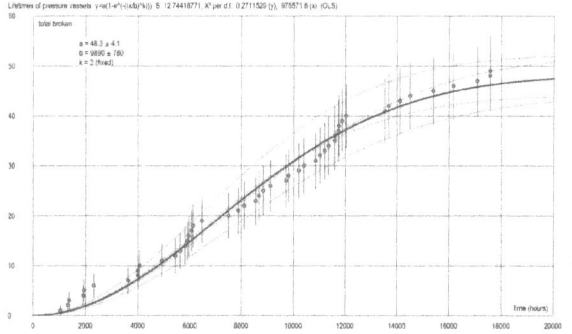

Fig. 4.86. Times before a number of vessels break down, and the best fitting Weibull growth model, with the "recommended" *k* = 2.

But, in my opinion, there is more need to fix the parameter *a* to 49, since that is the *real* limit of possible breakdowns, and let the value of *k* open to be adjusted to the empirical reality! Doing that we get *k* = 1.73 ± 0.13.

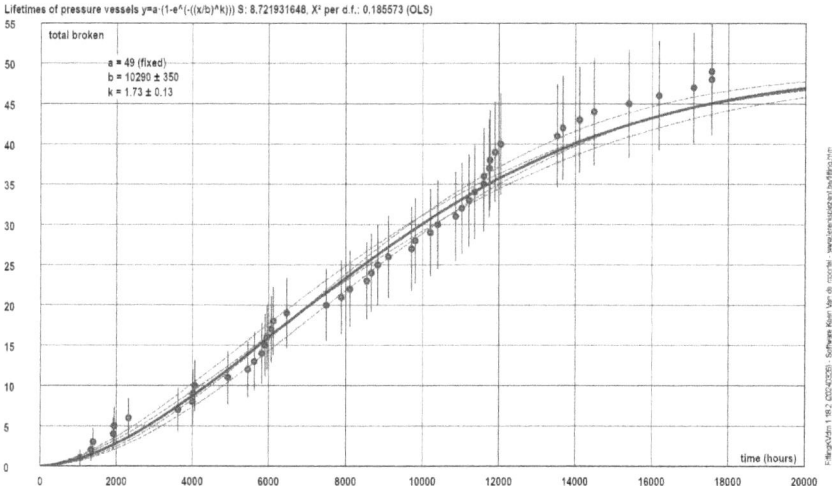

Fig. 4.87. Times before a number of vessels break down, and the best fitting Weibull growth model with the theoretical *a* = 49 and the empirical *k* value.

FittingKVdm example data file: "Lifetimes_pressure_vessels.dta1".

4.1.27. The flow rate of a powder

Fine powders and granular materials are used in numerous industrial applications. For optimal processing, their physical and chemical properties need to be determined.   One of these is how easy a powder "flows" when it has to be poured in, e.g., a capsule or any other container. At first sight, you might think it's enough to measure how much powder flows through a certain area per time unit, and you can recalculate it for any other area. The reality is a little bit more complicated: if a hole is too narrow, no powder will fall through it at all and according to the so-called Beverloo law [e.g.: Jianhua 2022], once the hole is wider than a minimal size d_0 (a bit more than the size of the grains), the flow rate Q increases not with the square of the diameter d (~surface) but with the power 2.5:

$$Q = a \cdot (d - d_0)^{2.5} \qquad \text{if } d > d_0, \text{ else } Q = 0$$

The proportionality parameter a is dependent on the density, gravity, and other factors.

Let's check if that is true!
Some data that are openly available, are from a kind of lactose powder, called "Flowlac90" from Meggle Pharma. The measurements were done with a device from the Belgian company Granutools, called "Granuflow". This instrument has seven holes with very precisely known diameters, through which the flow can be observed.
These are their measurements; see,
www.news-medical.net/whitepaper/20190910/A-Guide-to-the-Flowabilit
y-Classification-of-Lactose-Powders.aspx.

d (mm)	4	6	8	10	12	14	16
Q (g/s)	0.78	2.34	4.92	9.21	14.34	20.52	27.07

These data were put in FittingKVdm ($x = d$, $y = Q$) and the best fitting "Power with horizontal shift" model was asked. Estimated precisions are $\sigma_x \approx 0.01$ mm and $\sigma_y \approx 0.01$ g/s.

In a reasonable approximation, the data seem to fit, but... the shift parameter d_0 (= c in the software) is negative (-0.318 mm with MDLS, and even more negative with OLS), which is against the Beverloo theory. In the article, they say: 0, which is also impossible.

The strange result might be explained by the simplistic model assumptions that the particles are round and without mutual attraction (cohesion). Improved models have been suggested [e.g., Garcimartín].

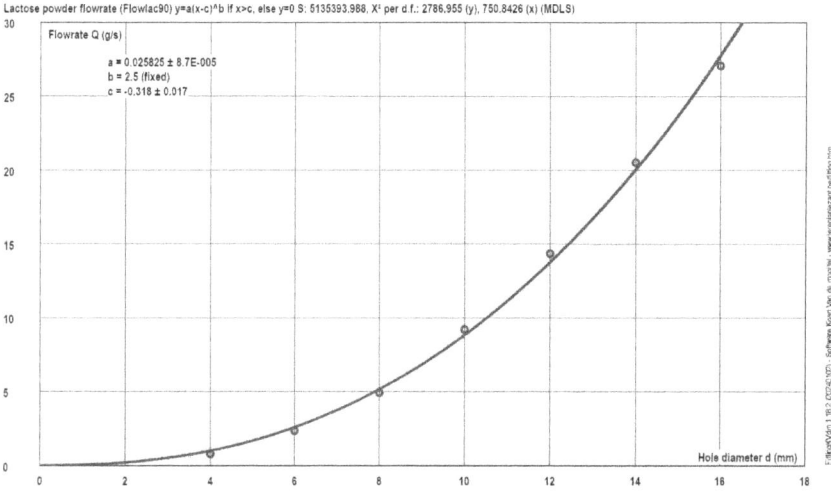

Fig. 4.88. Lactose flowrate measurements and best fitting shifted power function (exponent 2.5).

Or... and this makes also sense to me: the flow rate could just be proportional to the square of the diameter. It fits 10 times better ($S \approx 4.9 \cdot 10^5$ instead of $5.1 \cdot 10^6$), and d_0 ($=c$) becomes positive.

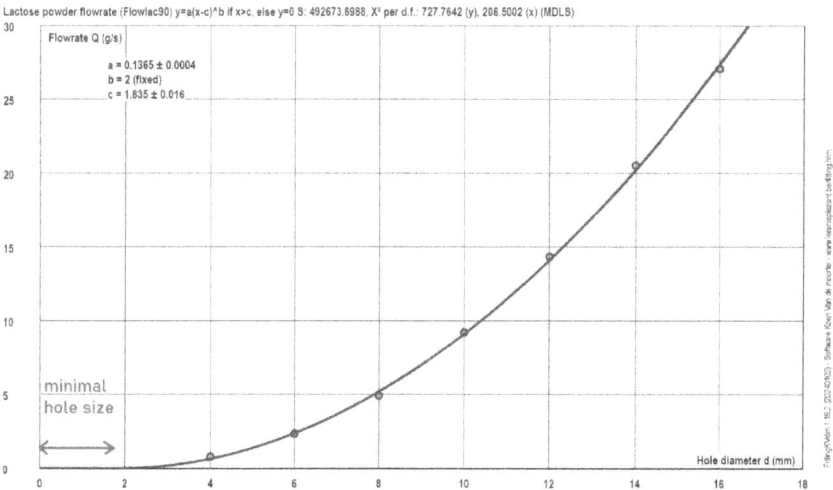

Fig. 4.89. Same data, but with a fitted shifted quadratic function.

References:

• Garcimartín, Angel *et al.*, *Flow and Jamming of Granular Matter through an Orifice*, Universidad de Navarra (Pamplona, Spain), see, www.unav.edu/documents/15083165/15313444/agm1_TGF07.pdf.

• Jianhua Fan, Li-Hua Luu, Pierre Philippe, Gildas Noury (2022), "Discharge rate characterization for submerged grains flowing through a hopper using DEM-LBM simulations", *Powder Technology*, Vol. 404. See, https://doi.org/10.1016/j.powtec.2022.117421.

FittingKVdm example data file: "Powder flowrate (Flowlac90).dta1".

Fig. 4.90. The measurement device.
Source: https://www.granutools.com/en/.

4.1.28. Sunspots

Dark spots on the Sun have been observed for at least two millennia. We know now that they are relatively "cool" places due to magnetic fields hindering the movement of plasm, and that their number varies periodically with maxima when the magnetic poles switch. This rhythm is also correlated with small climate variations on Earth.

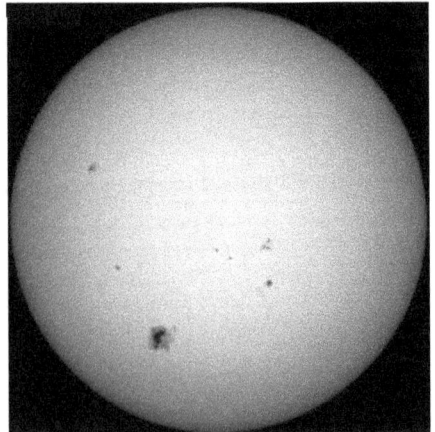

Fig. 4.91. Sunspots. *Source:* NASA 7 June 1992.

To illustrate a periodic model, I used some data from WDC-SILSO, the Royal Observatory of Belgium, in Brussels.
See, www.sidc.be/silso/infosnytot.

The variables here are as follows, x = time in years (1818–2021) and y = monthly average of the daily counted sunspot number (and standard deviation of that average).

This is a tricky one! You clearly see a "periodic" pattern, but it is not very constant: the period and the amplitude are slightly varying.

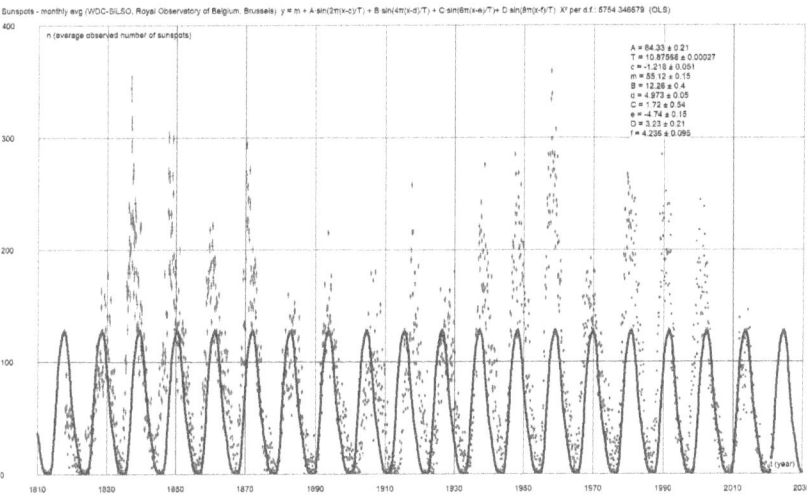

Fig. 4.92. Sunspot counts (monthly avg.) and best fitting sine wave with three additional harmonics.

A **sine wave model with a few harmonics** (possible with FittingKVdm) will fit reasonably, at least the bottom part of the dots, but only if you start with a good estimation of the period $T \approx 10.9$ years and set the limits to minimum 10.7 and maximum 11.1.

For the amplitudes of the main wave and the first harmonics A, B, C and D, you can enter start values 1 and limits 0–100 for example; for the phase shifts c, d, e, and f you can start with 0 and limits −6 and 6 since adding an integer multiple of periods makes no difference. It takes a few thousand iterations to get a stable fit, and a long calculation time for the precision estimation of the parameters.

There is a certain asymmetry in the fitted curve, but that might be an artifact since this periodic function cannot adapt its parameters to fit with a varying period or amplitude.

The period converges to 10.87568 years, but the sum of squares S has many other minima near that are almost as deep as this one. So the slightest change in initial value can cause T to converge to one of the other minima. This can be made visible in FittingKVdm, see Fig. 4.93.

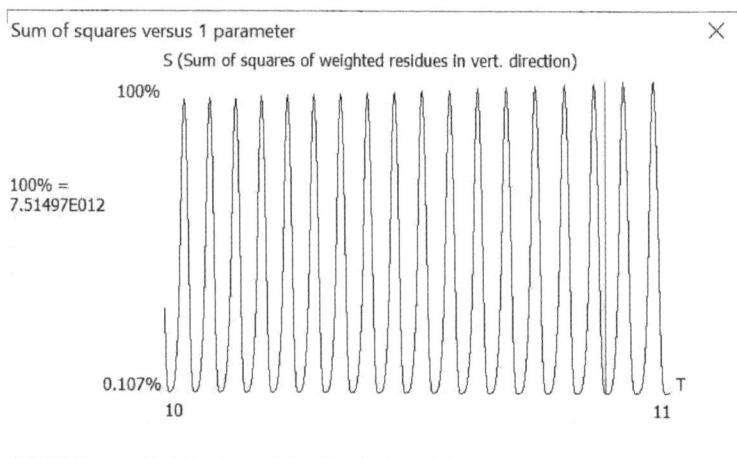

Fig. 4.93. S vs. T for the given dataset. The absolute minimum is indicated by the red line.

This is a good example to show the advantage of subtracting the biggest part of the time. In the original file, the time was given in years AD. In that case, a very small deviation will give an enormous difference after 1900 years, since the influence of a small error ε will be blown up. That risk is much less if you subtract for example 1800 from each year:

$$\left|\sin\left(\frac{2\pi\cdot1900}{T+\varepsilon}\right)-\sin\left(\frac{2\pi\cdot1900}{T}\right)\right|\gg\left|\sin\left(\frac{2\pi\cdot100}{T+\varepsilon}\right)-\sin\left(\frac{2\pi\cdot100}{T}\right)\right|$$

Another candidate for a model is the **"periodic peak function"** (see p. 49), simpler with fewer parameters, no asymmetry allowed.

This time, the time values were reduced to "number of years starting from 1800", in order to reduce the number of minima in the S vs. T curve.

Sunspots - monthly avg (WDC-SILSO, Royal Observatory of Belgium, Brussels) y = A·(k(k+1)/(sin²(π(x-c)/T)+k)-k)+m X² per d.f.: 7001.067898 (OLS)

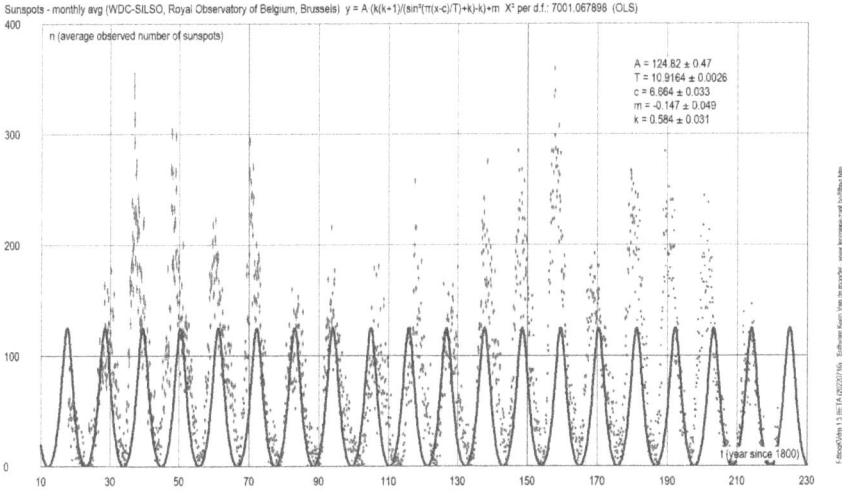

Fig. 4.94. Same data, but now with a simpler model with symmetric peaks and years since 1800.

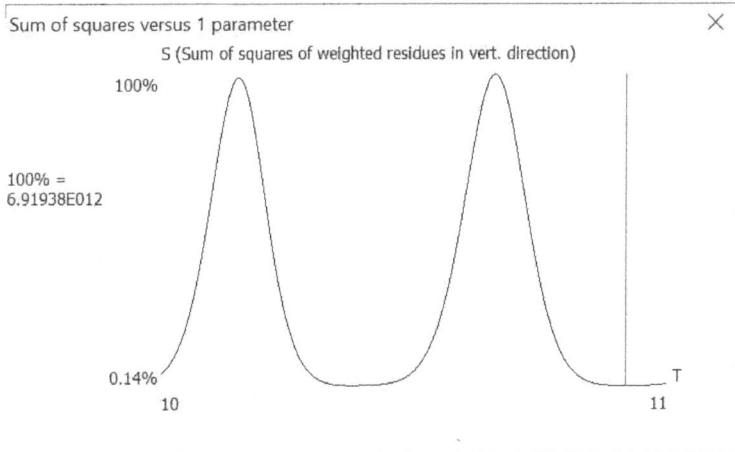

Fig. 4.95. Again S vs T, but now 1800 is subtracted from the years.

Now, there are only two minima in the S vs. T curve instead of 19, which makes the choice of the initial value for T less critical. Note that it is unavoidable to have multiple minima in S vs. T if the data are more or less periodical, but the deepest minimum should be the "real" period.

The first and second models have slightly different periods, which is not a good sign. The reason is that the periodicity is not strict. There might also be two cycles mixed up (explaining the different peak heights), or there is just some chaotic instability in this phenomenon.

Astronomers found ways to predict the height of the next peak once a cycle started, but predicting the exact moment of the 100th peak from now is probably quite unrealistic. See, for example,
Podladchikova, Tatiana *et al.* (2022), "Maximal growth rate of the ascending phase of a sunspot cycle for predicting its amplitude", *Astronomy & Astrophysics*, 663. Online
www.researchgate.net/publication/361568541_Maximal_growth_rate_of _the_ascending_phase_of_a_sunspot_cycle_for_predicting_its_amplitud e.

If you like to follow current predictions of astronomers, have a look at www.swpc.noaa.gov/products/solar-cycle-progression.

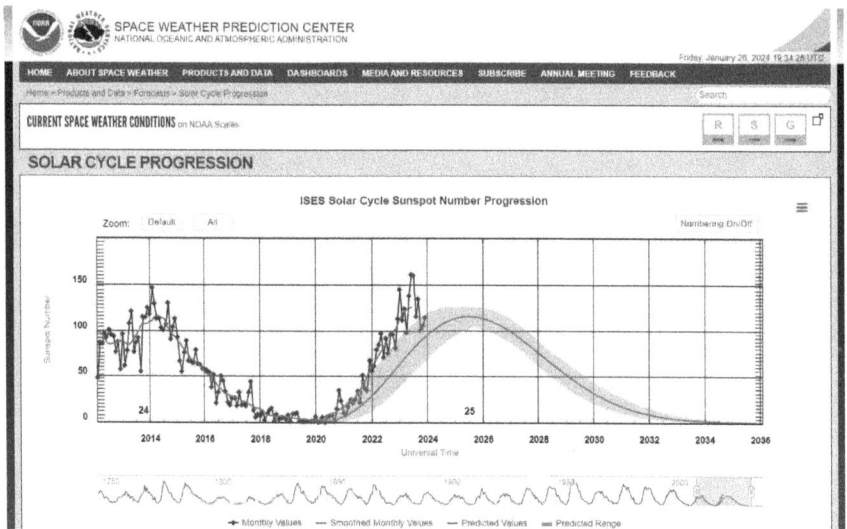

Fig. 4.96. Measurements and predictions of sunspots by NOAA.

FittingKVdm example data file: "Sunspots-monthly avg.dta1".

4.2. Geography, Climate, etc.

4.2.1. The simplest climate model

The first thing you want to know when you travel to a place you never went to is what kind of clothes should you pack in this time of the year. Of course, no year is the same, but with a periodic function you can make a very simple model to estimate the temperatures when you arrive. To provide you with an example, I took a random place on the globe: Chatanga (sometimes also transcribed from the Cyrillic alphabet as "Khatanga") in the arctic area of Russia (71.79°N, 102.50°E). Quite chilly winters, reasonable summers.
I found average monthly temperatures from January 2009 to May 2020 at
www.worldweatheronline.com/chatanga-weather-history/taymyr/ru.aspx.
For the uncertainty on the time, I took ± 0.5 month. For the "errors" on the temperature values (given in °C rounded to integer numbers), I assumed ± 1°C since no standard deviations were given.

The simplest model imaginable to connect the dots more or less is a sine wave (p. 45).

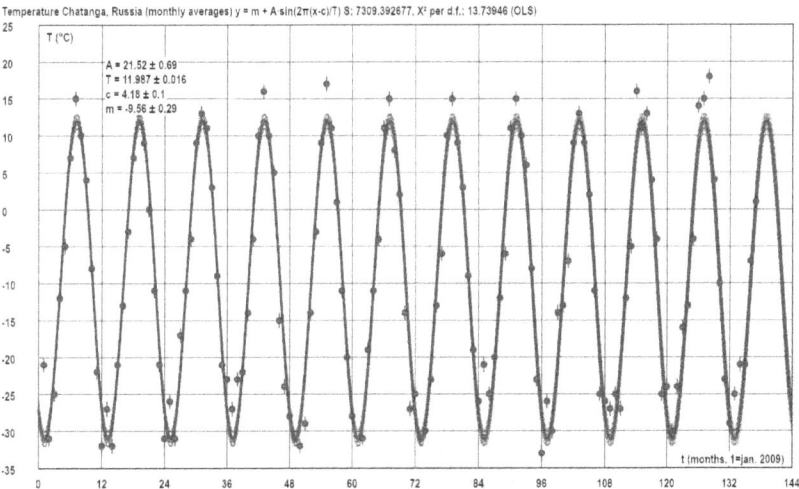

Fig. 4.97. Average monthly temperatures in Chatanga, Russia, for almost 12 years, and the best fitting sine function.

The expected period is of course 12 months, and the algorithm produces 11.987, which is not bad. For more precision, we should use the exact number of days since a certain start date since months don't have equal lengths and some years have more days than others! And also more precise average temperatures and their standard deviations would help! The average yearly temperature turns out to be -9.56 °C.

If you want to refine the model a bit, you can use the *skewed wave* function (see p. 52). That reveals a systematic asymmetry: a slightly slower "heating up" phase and a faster "cooling down": parameter $k = -0.105$ ± 0.048 ($k = 0$ would mean symmetry).

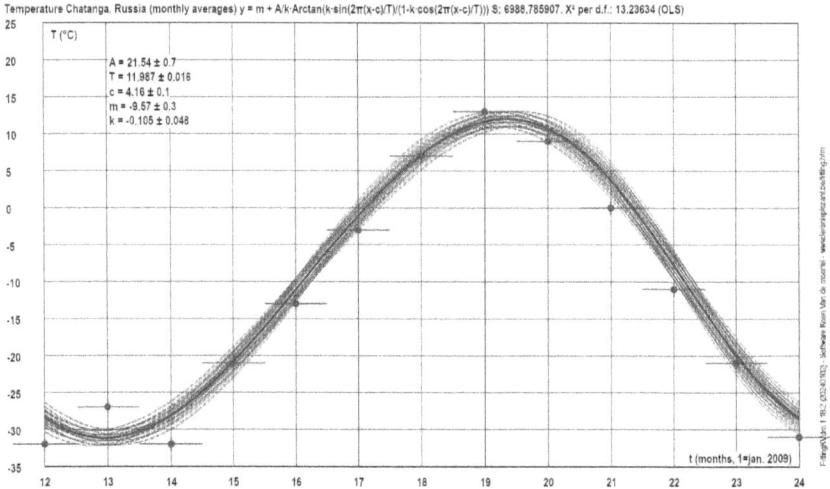

Fig. 4.98. The best fitting "skewed wave" function through the same data, shown for one year.

Another test you can do with the same data is to calculate the best fitting *linear model* here to find out if there is a trend toward warming or cooling.

But, be careful to avoid a bias here!

- Make sure the data cover an integer number of years (a multiple of 12 months) then by setting the first or the last data points to "inactive"!
- Also, this time, use OLS since there is no direct causal relationship between the time and the temperature; it doesn't make sense to switch the independent and the dependent variable here! In the given time span, a warming trend of about 0.02 °C per year can be seen, probably highly appreciated by the locals. Beware, extrapolation of this trend is very speculative!

Temperature Chatanga, Russia (monthly averages) y=ax+b S: 131076.4727, X² per d.f.: 252.0701 (OLS)

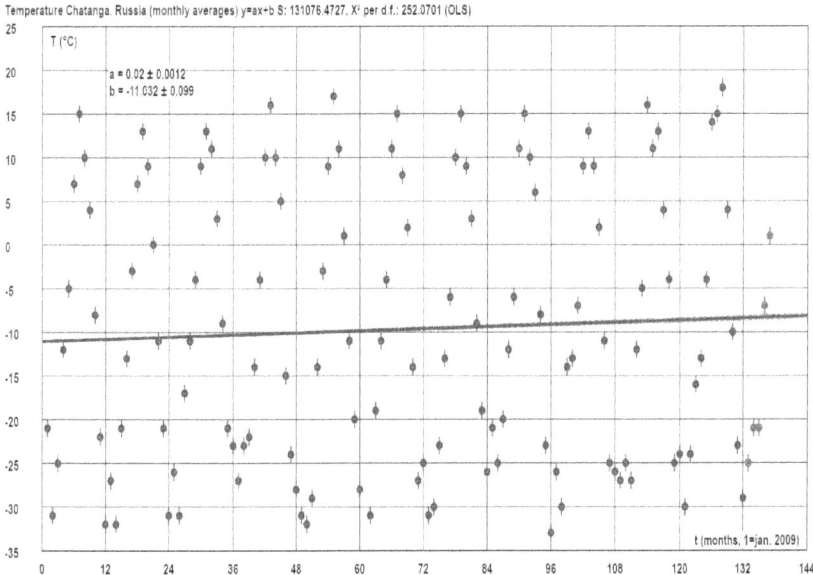

Fig. 4.99. Linear trend line through the same data (multiple of 12 months only!).

FittingKVdm example data file: "Temperature Chatanga.dta1".

4.2.2. How does the temperature vary in one day?

This is a simple home experiment: put a thermometer in shady place and register the temperature throughout the day. The idea was to watch how the cycle of warming and cooling went, so I chose a stable sunny day with no significant disturbances from wind or variable clouds or anything. The measurements were done approx. 1 m above the ground, in the hamlet of Popovica, in the hills a few km south of Novi Sad, Serbia, ca. 110 m above the Danube river, on a nice summer day (4 August 2023).

So, our variables here are as follows: x = clock time (decimal hours, precision ca. 1 minute); y = temperature (°C). The thermometer had a resolution of 0.1 °C. The absolute precision was not known, but by comparing with other thermometers, we could assume a precision of 0.1 °C.

Which function can describe the observations? Since we expect a more or less cyclic phenomenon, at least for a few days while the weather stays the same, a plain sine wave could be a good approximation. For the fitting process, T should be fixed to 24 since not enough data points were collected to auto-detect the period precisely.

Apparently, there were some fluctuations, sometimes a very light wind, but the sine wave fits more or less, and it "predicts" a maximum of 32.5 °C around 16 h while it actually was 33. The predicted minimum of 23.9 °C might have been a little bit high but not more than a degree off, I estimate.

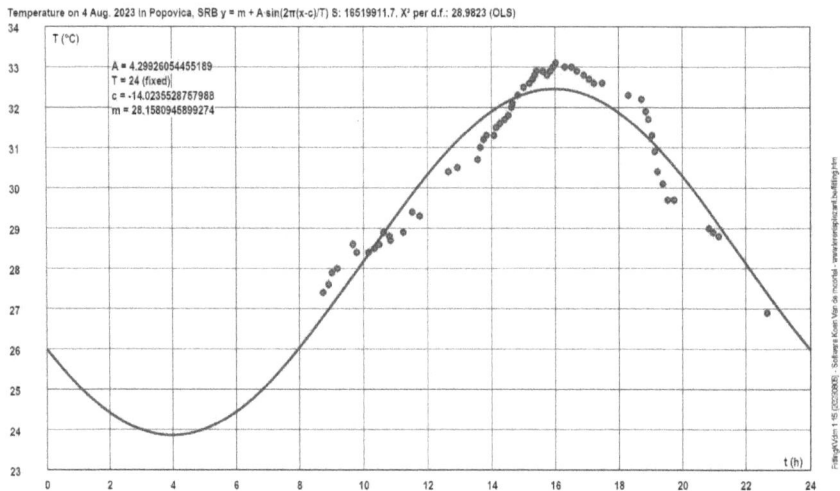

Fig. 4.100. A simple sine wave fitted through the measurements.

There seems to be an asymmetry though: the warming up goes slower than the cooling down, so maybe a sine wave with a few harmonics ("Sine wave with harmonics" model in FittingKVdm) might fit better with the data?

Well, yes it does, as you can see in Fig. 4.101 … (I had to adjust the vertical scale.)

The iteration went very slow toward convergence, which is a bad sign; it means the pattern isn't very clear. Eventually, this model will always fit the data better, since it has six more parameters. But … it predicts a minimal night temperature of about 1°C! That's totally absurd.

Temperature on 4 Aug. 2023 in Popovica, SRB y = m + A·sin(2π(x-c)/T) + B·sin(4π(x-d)/T) + C·sin(6π(x-e)/T)+ D·sin(8π(x-f)/T) S: 3351272.629. X² per d.f.: 6.571123 (OLS)

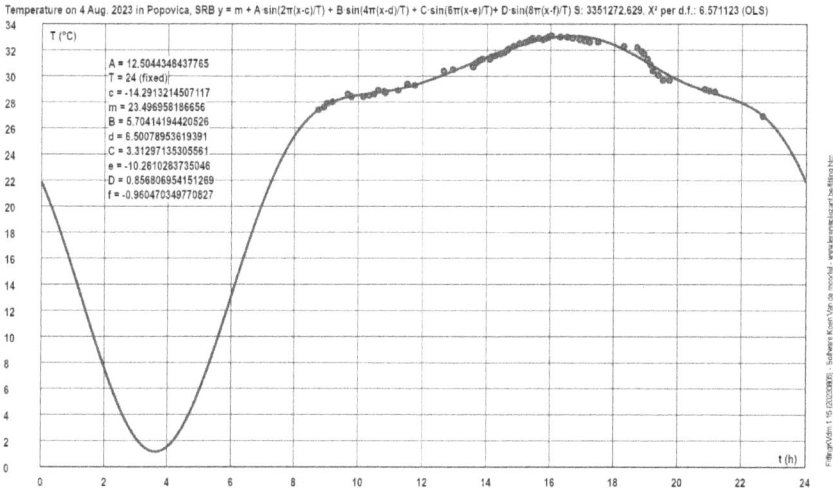

Fig. 4.101. A sine wave with its first three harmonics fitted through the measurements.

This is a typical example of "**overfitting**": letting the model adapt to the small fluctuations, by fitting more parameters, is the same as believing that these random ups and downs are part of "the general pattern", which is absurd. The consequence is that extrapolations become much less reliable! The "best fitting" model is not necessarily the best!
You might understand now why one has to be extremely careful when interpreting predictions of climate models with thousands of parameters! This is only the beginning of an embryo of a climate model!

If we want to detect the *asymmetry* in the pattern, we better use a simpler model, like the "Skewed wave". That has only one additional parameter, so the risk of overfitting is smaller.

The high and the low in the curve are still very realistic, and indeed some asymmetry shows up: $k = -0.249$, see Fig. 4.102.

Some afterthoughts …
Why is the maximum that late? Okay, the soil needs some time to warm up, so the maximum is always reached a few hours after noon, but the Sun culminated that day on this location at 12h47. The maximal temperature was observed at 16h01, that's 3 hours and 14 minutes later! Is this purely by chance or is it normal here?
And, is this really a typical pattern, the asymmetrical wave?

To answer both questions, we would need to do repeated measurements on similar days (stable and sunny). A hypothesis that might explain the pattern: maybe the temperature would normally start dropping at a more usual time, say around 15h, but the hot air from the asphalt and the concrete in the city keeps ascending uphill for sometime? This could be checked by comparing with measurements from the other side of the city, in the plains.

Of course, if you have an electronic thermometer and a data logger, please use it to make these kinds of measurements easier...

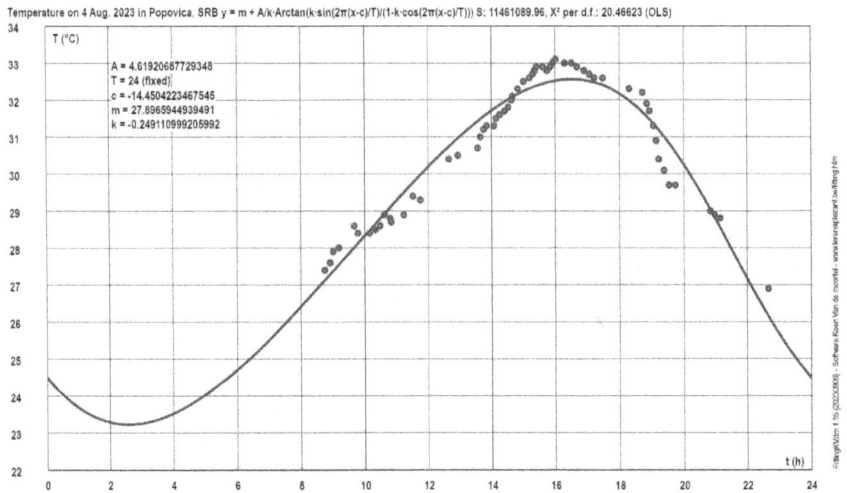

Temperature on 4 Aug. 2023 in Popovica. SRB y = m + A/k·Arctan(k·sin(2π(x-c)/T)/(1-k·cos(2π(x-c)/T))) S: 11461089.96, X² per d.f.: 20.46623 (OLS)

A = 4.61920687729348
T = 24 (fixed)
c = -14.4504223467545
m = 27.8965944939491
k = -0.249110999205992

Fig. 4.102. A "skewed wave" function fitted through the measurements.

FittingKVdm example data file: "Temp Popovica 20230804.dta1".

4.2.3. Temperature versus geographical latitude

The average air temperature near the equator is much higher than near the poles, obviously. But how does it vary? Can we "predict" the "normal" temperature for latitude x reliably? Of course, that will depend on many other factors too, like prevailing wind directions, cloudiness, proximity of the ocean, altitude, etc.

Places far inland usually have a more severe climate than coastal places, so let's limit ourselves to the last ones, at sea level, just to complicate things as little as possible.

Of course, we could study satellite data to get the most complete information, but it's a fun classroom task to look up a number of places in the atlas (or Google Maps or whatever), their geographical latitudes (decimal!), and their average temperatures.
The climate data can often be found on www.worldclimate.com and https://en.wikipedia.org/.
For my example dataset, I collected information from 73 places all over the world. Most of them were from airports near those places. Avoid data from city centers because they are always warmer than their surroundings. The measurement imprecisions were assumed to be half of the least significant digit since they were not given. The time span for averaging the temperature was not the same for every location, which will unavoidably cause some "noise".

Which model can we expect to work here?
You can find some weird "solutions" for this on the internet, like at http://webinquiry.or g/examples/temps/ where a certain Philip Molebash recommends to use a mess of three linear pieces (Fig. 4.103). Sorry, but this hurts my mathematical feelings!

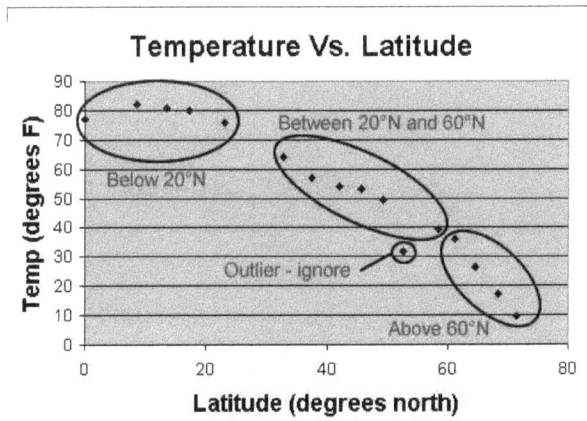

Fig. 4.103. Do NOT do this! Don't even think about it!

If we look at the actual data, we might think that a parabola (Fig. 4.104) fits well, or a cosh function (hyperbolic cosine, also known as "chain line") (Fig. 4.105), both centered around 0°N. Many people would be happy with this result. But, of course the extrapolation makes no sense since they both go down infinitely!

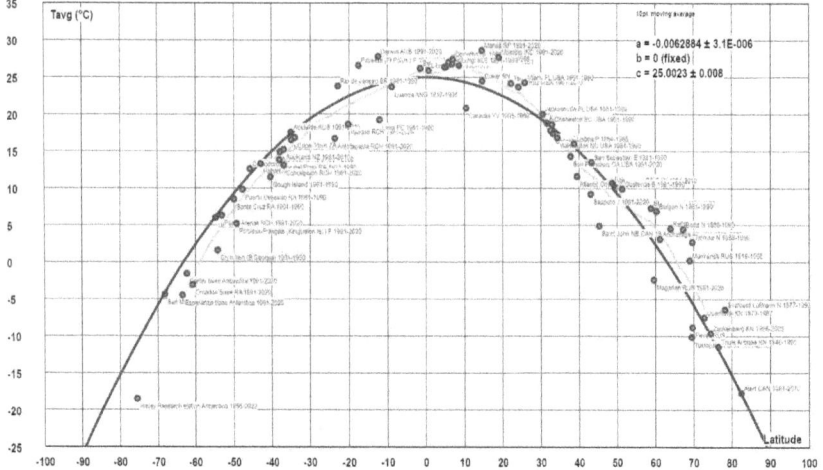

Fig. 4.104. A parabola fitted through the data.

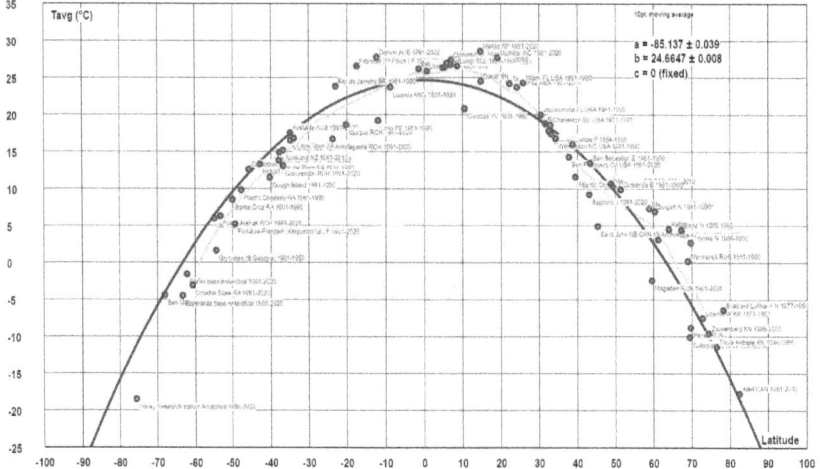

Fig. 4.105. A chain line fitted through the data.

We better look at the underlying physical reality, even a simplified version. If we assume that the Earth is a ball flying at a constant distance from the Sun, and the average temperature is proportional with the average insolation, T_{avg} should be more or less proportional to the cosine of the latitude/2 (= $\sin(2\pi(x - 45)/180)$ if x is the latitude in degrees).

The truth is more complicated but this is fairly close; see *Global Physical Climatology* 2nd ed., by Dennis L. Hartmann, Elsevier (2016). Anyhow, if we walk to the north and we continue over the pole until we did say 110°, we are back at 70° from the Equator. So our idealized model must obligatorily be a periodical function; $f(x)$ has to be the same for $x = 110°$ and $x = 70°$!

Fig. 4.106 shows what we get with a cosine model ($y = T_{avg}(x) = A \cdot \sin(2\pi(x + 45)/180) + m$).

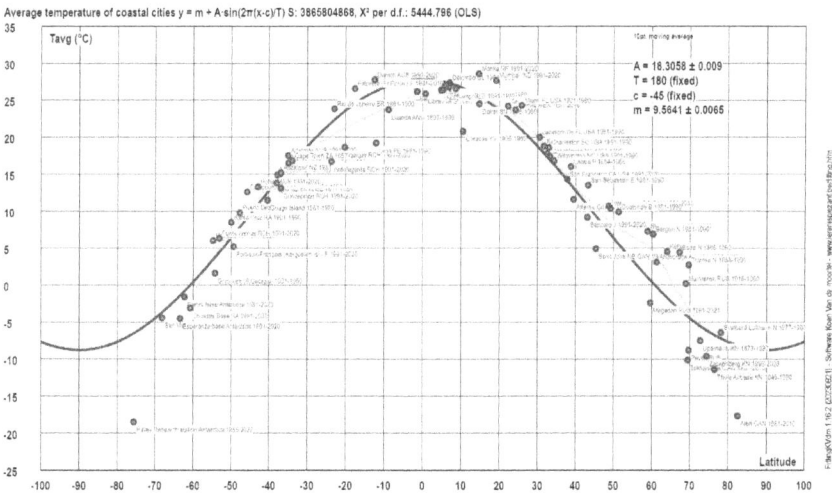

Fig. 4.106. A fitted cosine wave and 10-point moving average through the data.

It tells us that the world's average coastal temperature (parameter m) is 9.56 °C; the shown confidence interval is certainly too small since we assumed too high of a precision for the data.

Now, you might raise the objection that the extreme polar places don't fit in at all, and that the parabola-like curves actually "looked better". Yes... but we are not just "number crunchers"; we don't want things just "to look good". We rather want to obtain some insight. Those "outliers", like Halley (Antarctica) and Alert (Canada), which we took for "coastal cities", maybe they actually aren't really so "coastal". For example, we read "The sea is covered with sea ice for most of the year but the ice pack does move out in the summer, leaving open water", which means the climate is more like a land climate, see (https://en.wikipedia.org/wiki/Alert,_Nunavut)! So they are *supposed* to be outliers in this dataset! Leaving them out will certainly lift the curve a bit.

Interesting things to note are that places like Bergen (Norway) and Oostende (Belgium) lie far above the curve, while Lima (Peru) and Saint John (NB, Canada) lie far below it. That can easily be explained because the first places enjoy a warm ocean current, while the last ones endure a cold current!

FittingKVdm example data file: "Temperature in coastal cities.dta1"

4.2.4. The atmospheric pressure in Karlsruhe

Are there any seasonal cycles in the air pressure, like in the temperature? I obtained a big dataset from Karlsruhe, on the German side of the Rhine, where it forms the border with France, 115 m above sea level. It contains 47208 measurements done every morning from 1 Jan. 1876 to 1 March 2006, except between Nov. 1944 and Sep. 1945.

If you want to find weather cycles related to that of the Sun, the most logical thing to do is not to use the *time* as independent variable but the *ecliptic longitude (or length) of the Sun* (l). Both are strongly correlated of course, since the Sun moves (from our geocentric point of view) a little less than 1° per day, but that speed has small variations that cause – by the way – the differences between a sundial and a steady clock. When the Sun is at longitude 0°, by definition, the northern spring begins. The Sun at 90° defines the beginning of the northern Summer, 180° the fall, and 270° the winter. The longitudes were calculated using astronomical routines from NASA's Jet Propulsion Laboratory. The precision of each value must be better than 0.1° if we can believe that the measurements were done every day around the same time. The pressures (p) were accurate to 0.1 hPa.

The plot of the data was quite surprising to me; apparently, the average pressure almost doesn't change during the year: a fitted sine function has an amplitude of about 1 hPa around the global average of 1001.68566 ± 0.00048 hPa. And the small periodic pattern is very skewed, which is visible when fitted with the "skewed wave" function (see p. 52): the pressure goes up a tiny bit starting in the spring till approximately mid February, and then it descends back about 2 hPa (Fig. 4.107).

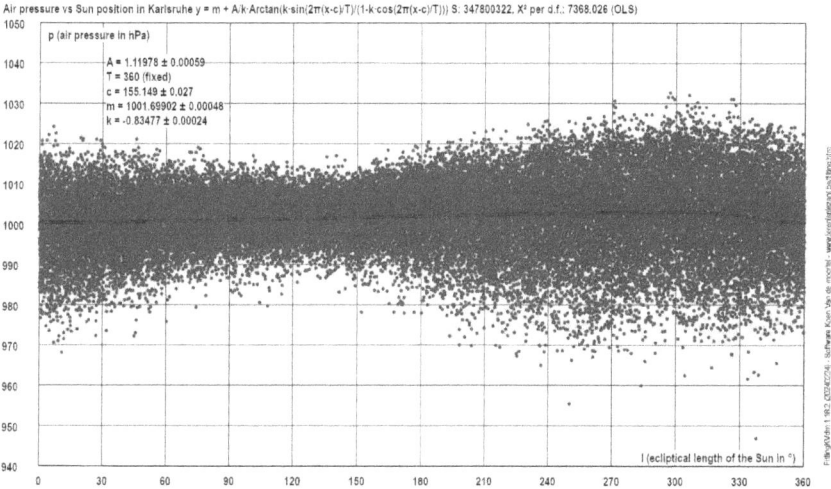

Fig. 4.107. Atmospheric pressure in Karlsruhe versus the position of the Sun (0° = vernal point), and the best fitting skewed wave.

But, an unexpected very clear pattern appeared: the *variability* of the pressure seems to depend a lot on the season! In the hottest part of the summer ($110 < l < 140°$), the pressure was never above 1020 hPa or below 980 hPa, while that is often the case in the winter.

To make this pattern more clearly visible, I calculated the *standard deviation of the pressure*, degree by degree. The errors on those standard deviations are their values divided by the square root of the number of measurements in each interval (see, e.g., www.investopedia.com/terms/s/standard-error.asp).
Now a sine function fits nicely, as you can see in Fig. 4.108!

Air pressure vs Sun position in Karlsruhe y = m + A·sin(2π(x-c)/T) S: 890.9763118, X² per d.f.: 0.623933 (OLS)

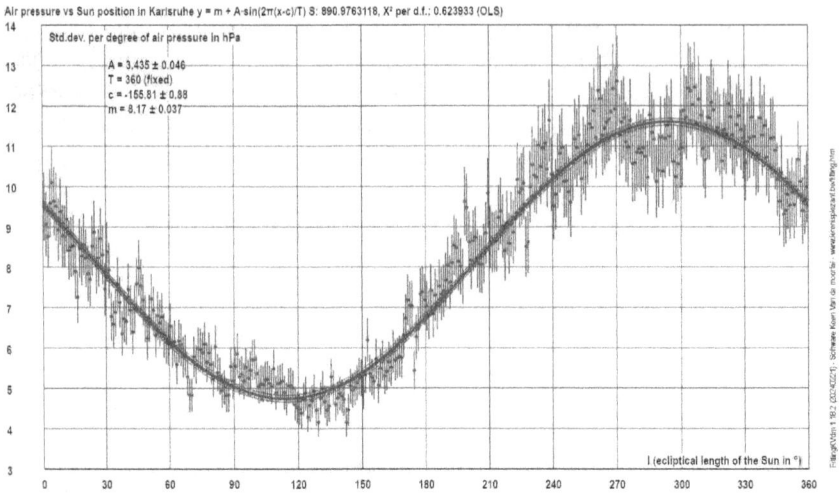

Fig. 4.108. Standard deviations of the pressure in Karlsruhe per degree vs Solar length, and the best fitting sine function.

Are these seasonal variations in agreement with findings in other places? I found at least one article describing the phenomenon, and it is even connected with circannual physiological rhythms of animals – very interesting! Apparently, the pattern becomes especially visible in polar areas.

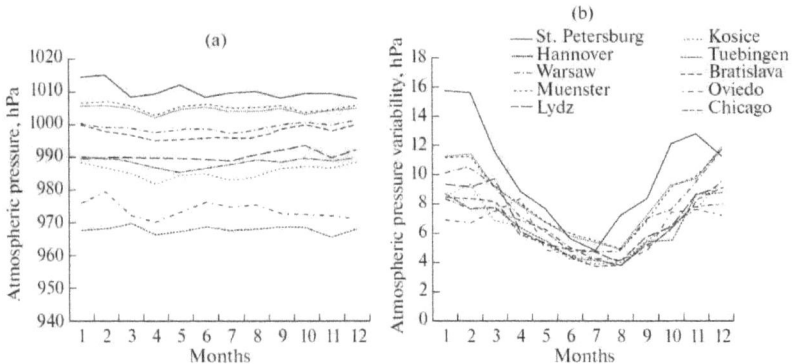

Fig. 4.109. Similar patterns observed in other places [Kuzmenko 2019].

We can also ask ourselves: in these 130 years of observations, is the average pressure stable, or is it going up like the temperature?

To study that question, the longitudes had to be made cumulative, meaning that after the first cycle, 360° had to be added, after the second cycle, 720°, etc. Every 360° is one year (ca. 365.25 days).

The simplest way to detect "any trend" is by fitting a linear relationship: $p = a \cdot l_{cum} + b$ (with OLS since there is no invertibility expected here).

That produced:
$a = -3.5080 \cdot 10^{-5} \pm 0.0033 \cdot 10^{-5}$ hPa/° $= -0.126288 \pm 0.000012$ hPa/year,
$b = 1002.51702 \pm 0.00089$ hPa.

So, there seems to be a tiny *downward* trend going on.
The "predicted" p_{avg} was, at 1 Jan. 1876, 1002.5072 hPa, and at 1 March 2000, 1000.8633 hPa.

Is this "significant"? If you believe the traditional numbers Pearson $r \approx -0.05513$ and Kendall $\tau \approx -0.03626$, you would say no. But those numbers don't say anything about the importance of this indication of climate change.

Can this be extrapolated? No, not unless some physical mechanism is found that explains the trend, and even then one has to be careful.

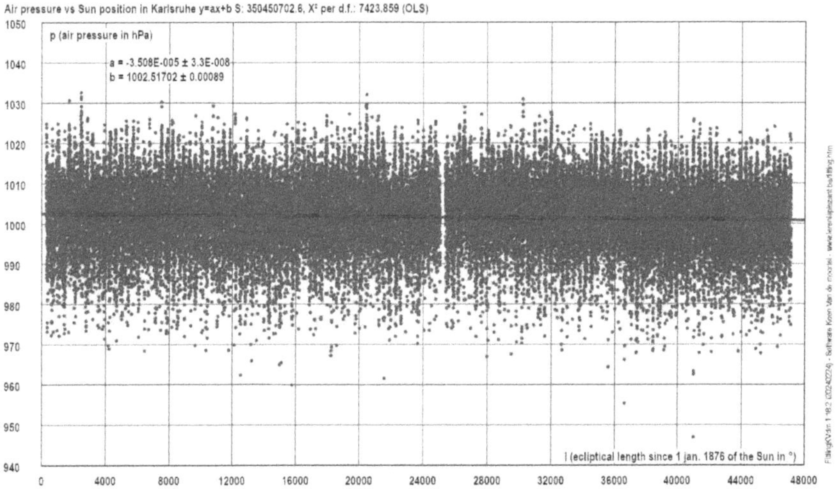

Fig. 4.110. Pressure in Karlsruhe vs. the number of degrees the Sun has moved from our point of view since 1876, and best fitting line.

Reference:
• Kuzmenko, N.V. (2019), "Seasonal Variations in Atmospheric Pressure, Partial Oxygen Density, and Geomagnetic Activity as Additional Synchronizers of Circannual Rhythms", *BIOPHYSICS* **64**, 599–609, https://doi.org/10.1134/S0006350919040080.

• German weather data can be found here:
  www.dwd.de/EN/ourservices/_functions/search/search_Formular.html

FittingKVdm example data files: "Air pressure vs Sun position 0–360 in Karlsruhe.dta1", "Air pressure vs Sun position in Karlsruhe SD per degree.dta1", "Air pressure vs Sun position in Karlsruhe.dta1"

4.2.5. Big rivers, big flow rates?

What is the relationship between the size of a river's drainage basin (A) and the amount of water that flows at its mouth per time unit (Q)?
Let's imagine the simplest situation: the river catches water from an area where the precipitation rate is the same everywhere. In that case, the amount of water caught is directly proportional to the size (area), so the discharge would be proportional too: $Q = aA$, parameter a reflecting the average amount of precipitation per area.

But when water has to flow over a long distance, a part of it will evaporate or just disappear in the soil on the way. The amount of evaporation is of course proportional with the temperature and humidity in the area, but also with the total water surface, or in other words, with the average width of the river. The width at a certain point is somewhat proportional with the distance from the source, so the average width is proportional with the total length. But it's also inversely proportional with the height gradient: steeply descending mountain rivers are usually narrower than rivers through plains, where they might even become wide swamps.
It's complicated of course, but anyhow, it can be expected that longer rivers with wide basins lose a bigger percentage of their water before it arrives in the ocean, compared to short rivers. So it looks plausible that $Q = aA^b$, a power function with $b < 1$. Anyhow, the model function should start in the origin since no basin is equal to no water.

Let's have a look at the reality! Some data can be found, for example, at
https://en.wikipedia.org/wiki/List_of_rivers_by_discharge.
In that list, you can see some rivers with a big discharge but a very small
basin because they just connect two lakes, like the Niagara or the Detroit
rivers. They can't be put in this model of course, since they have
another input besides the precipitation in their area.
Also rivers that are frozen most of the year, like the Mackenzie, have a
dynamic that might differ a lot from the others, so I wouldn't include
those either. Some smaller rivers were added too (from Wikipedia) to
have a better distribution.

For the errors on A, 1% was assumed for all rivers, since I suppose in
this era of satellite observation, we more or less know where all the rain
flows to. The errors on Q are certainly bigger, since the flow depends on
varying precipitation etc. I assumed again that all rivers had the same
variation, say 10%. Maybe it's bigger, but as long as all the Q values
have the same relative error, the parameters a and b will stay the same;
only the "worst cases" will be different.

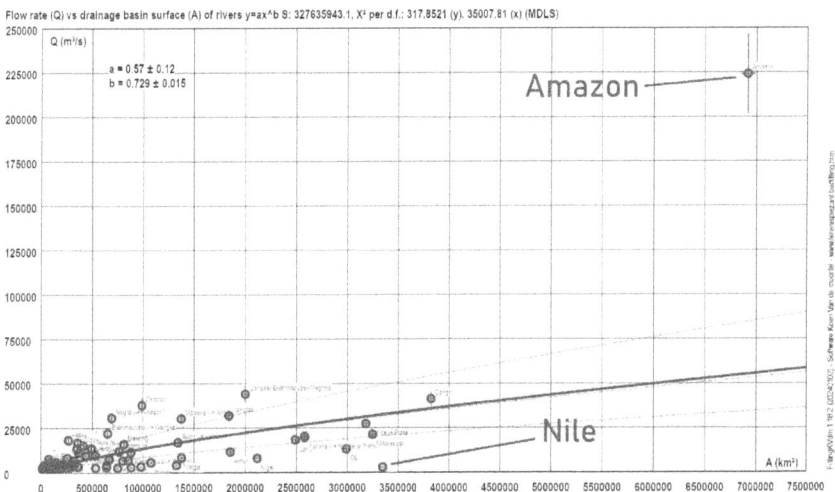

Fig. 4.111. Flow rates vs. basin surfaces of 84 rivers, and best fitting power function.

As you can see in the scatterplot (Fig. 4.111), the model makes sense,
but as expected, there is a lot of heteroskedasticity in the data. And that
is exactly what is interesting: with one glimpse on the graph, you can see
which rivers flow through a wet or a dry area! The Amazon is by far the
biggest water transporter, not only absolutely but also relatively in com-

parison to its basin area. On the other hand, look at the Nile, also trans-
porting a lot of water, but most of it comes from the area of its sources in
Central Africa, and almost nothing is added while it traverses the Sahara
desert. On the contrary: a lot of its water is used for irrigation before it
arrives in the Mediterranean Sea.

A competing model function could be the logarithm shifted through the
origin (Fig. 4.112). This curve is a bit above the first one; the Congo
river switched from "just above" to "just below" the curve.

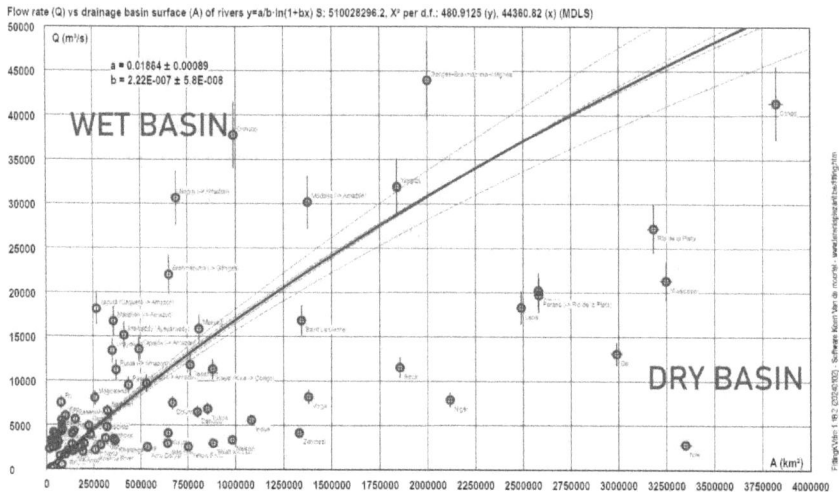

Fig. 4.112. Same data (Amazon included but not shown), and the best fitting logarithmic function through the origin.

Remarks:
- Doing the fit with (weighted) OLS results in much lower curves, since the absolute errors on the high Q values are much bigger than those on the lower Q values!
- Due to the complexity of the Q vs. A relationship, and all the other influences that cause the heteroskedasticity, it is very important that you use many random, unbiased data from all kinds of climates if you want to obtain a realistic idea of what "the average river" does.

FittingKVdm example data file: "Rivers.dta1"

4.2.6. Predicting the tides

Some coastal places have huge variations in sea levels because of the tides. The biggest and most spectacular can be seen in the Bay of Fundy, between Nova Scotia and New Brunswick, Canada: 16 meters difference between ebb and flow. Shippers need predictions to know when it's safe to leave or enter their harbors of course, not only of the level but also the strength of the possibly dangerous currents. That's why websites like the following exist:
https://l-36.com/tide_week.php?location=Grand%20Manan%20Channel%20(Bay%20of%20Fundy%20Entrance),%20New%20Brunswick%20Current.

To have a modeling example, I extracted some data from there: x = time (t) in hours starting 24 August 2020; and y = speed (v, in knots) of the tidal current at the entrance of the Bay of Fundy.

The appropriate model here is the sum of two sines, one with a period of 12 h (reflecting the influence of the Sun) and another with a period of approximately 12 h 25 min ≈ 12.41667 h (from the Moon).

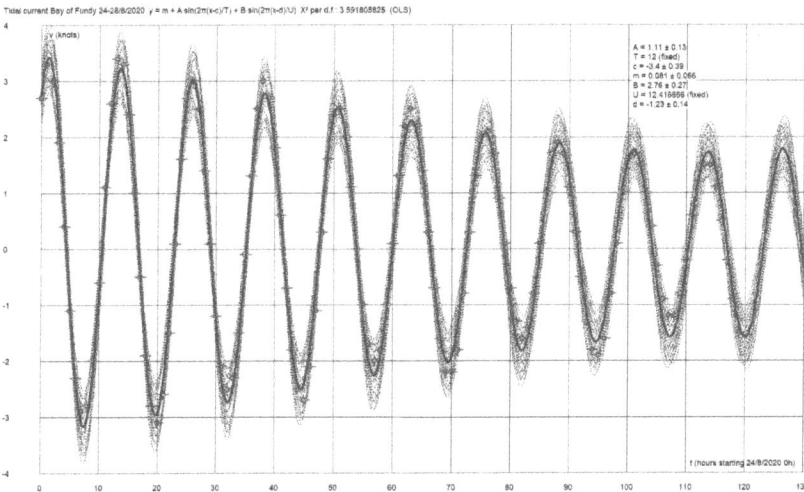

Fig. 4.113. Tidal currents predicted with a simple model.

The fit is good but not perfect, why? Well, the exact movement of the Moon is much more complicated than you might think! Models developed by NASA to calculate its position over long time spans use sums of dozens of sine functions with different periods!

FittingKVdm example data file: "Tidal_current_Fundy.dta1"

4.2.7. The population of Nigeria

Living beings that have enough space and food and no predators, tend to reproduce at a rate that is proportional with the already existing population. And that results in an exponential growth $N = N_0 a^t$ (N_0 being the population when $t = 0$) (see p. 130).

A typical example can be seen in the population of Nigeria. Data for this country can be found at

www.statista.com/statistics/1122838/population-of-nigeria.

The variables here are as follows: x = the time in years starting from 1950 (not the year itself, for reasons of precision: a small imprecision in the growth factor causes a big error in y if the exponent is 2000 rather than 50) and y = the population. The error on x was set to 0.5 since there was no information about the exact time of the year the population was counted. The error on y was set to 1, which is probably too low, but the error on the time compensates for that, since most probably the given numbers were reached exactly somewhere in the middle of each year ± 0.5.

The exponential model seems to fit quite well (Fig. 4.114), and it detects a growth factor $a = 1.02537$, which means that the average growth in 1950–2022 was 2.537%.

Often, the population growth becomes smaller when education levels grow, and then the growth becomes more like a logistic curve, but in this case, the contrary is true. If we fit only the data from 1950 to 2000 (Fig. 4.115), the growth rate was only 2.445%. The points from 2005 to 2022 (in red) are even *above* that curve.

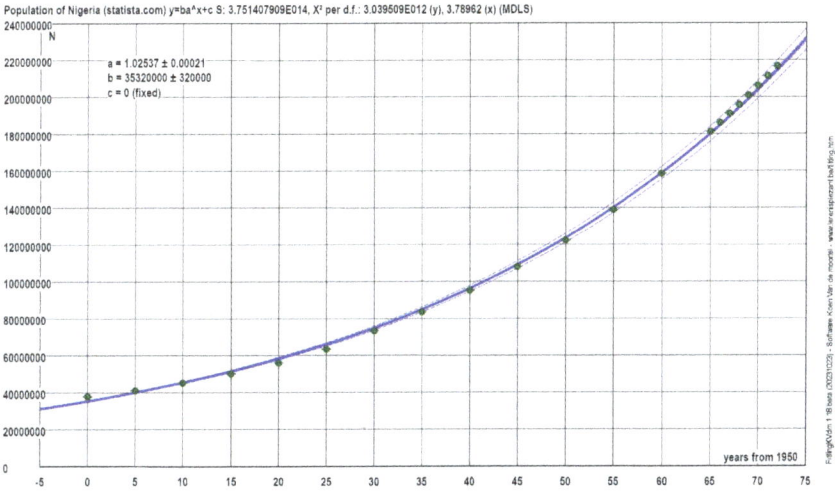

Population of Nigeria (statista.com) y=ba^x+c S: 3.751407909E014, X² per d.f.: 3.039509E012 (y), 3.78962 (x) (MDLS)

a = 1.02537 ± 0.00021
b = 35320000 ± 320000
c = 0 (fixed)

Fig. 4.114. Population of Nigeria 1950–2022.

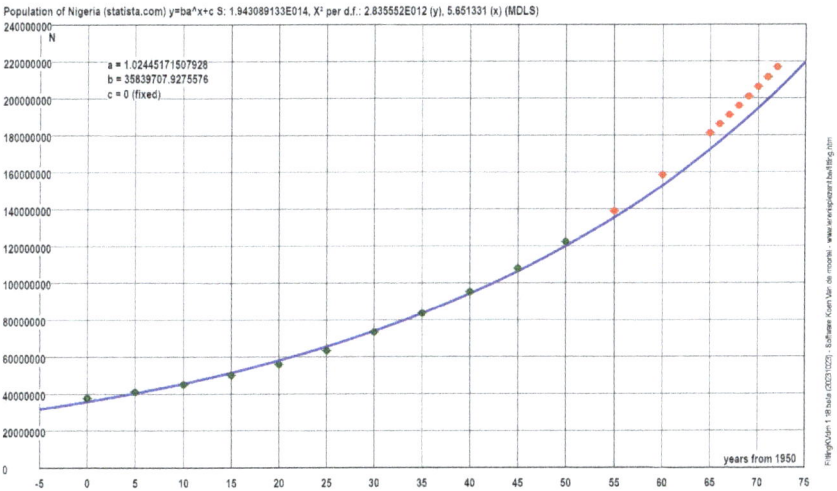

Population of Nigeria (statista.com) y=ba^x+c S: 1.943089133E014, X² per d.f.: 2.835552E012 (y), 5.651331 (x) (MDLS)

a = 1.02445171507928
b = 35839707.9275576
c = 0 (fixed)

Fig. 4.115. The population in the last years grows faster than could be calculated from the 1950–2000 data.

FittingKVdm example data file: "Population_Nigeria1950-2022.dta1".

4.2.8. Driving time versus distance

How long (t) do you expect to drive from location A to B, if you know the distance (d) between them? At first sight, you might expect a linear relationship between t and d, with zero intercept, since $t = 1/v_{avg} \cdot d$. But, is the average speed (v_{avg}) a constant? Most probably not, since the longer the distance, the more you drive on highways, which reduces the time! So, a power function with an exponent smaller than 1 seems plausible.

Luckily, we have navigation software these days, so we can easily check this. I collected a dataset with variables as follows: x = distance from my home to the target location nr. i, by road, in km, and y = average driving time according to Google Maps, starting 8am on a working day, in minutes. For the errors on the distances $\sigma_{x,i}$, half of the least significant digit was used, and $\sigma_{y,i}$ = half of the difference between the longest and shortest given time.

Of course, we should use MDLS, not OLS, since the questions "How much time do we need to drive 200 km?" and "How far can we get in 3 hours?" should be answerable using the same formula!

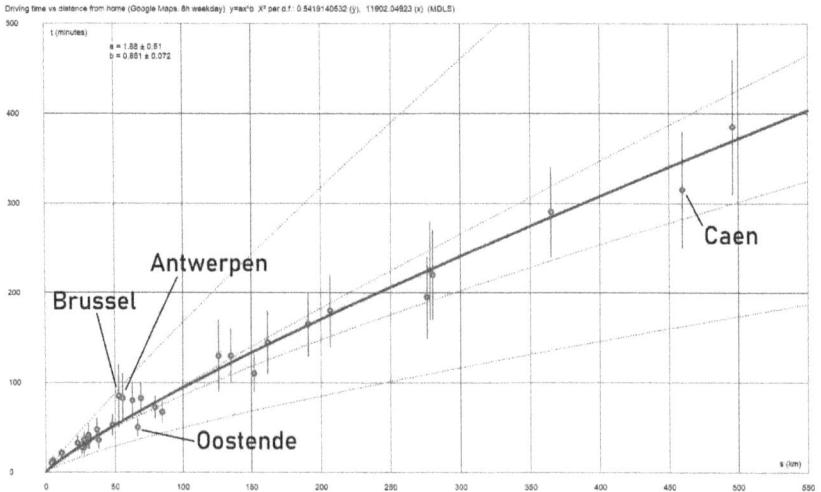

Fig. 4.116. Expected driving times and distances from my home to some places, according to Google Maps, and best fitting power function.

As expected, the exponent is a bit, but significantly, less than one: 0.851 ± 0.072.

What more can we learn from studying this graph? Points above the model curve correspond to places that are harder to reach. For example, Antwerpen and Brussel are infamous for their horrible traffic jams, while Oostende is a bit further for me, but it's quicker to get there most of the time. Driving to Caen goes smoothly since there aren't many busy agglomerations to be passed.

Also, you could apply some geometry to find out where I live approximately...

Remark: The power function model might fail for distances that are too far since the average speed will probably go to a limit value. A rational function might suit better but those are very unstable (see p. 21).

FittingKVdm example data file: "Driving times vs distance.dta1".

4.3. Life Sciences

4.3.1. Shoe sizes versus height

Are the sizes of people's feet proportional to their heights? During childhood, probably not exactly since not all body parts expand with the same rate, but among adults, it seems plausible, and it's not so difficult to test this hypothesis. I requested my friends on social media their heights and shoe sizes, and apparently people like to share this information... in a few days, I received data from 126 adult men and 242 women (mostly from Belgium and Holland, Feb. 2023). If you have students, you can also gather a dataset quickly; it's a nice classroom experiment.

So, our variables here are x = height in cm of a person and y = (European) shoe size. Since the heights were usually rounded to 1 cm, the "error" on x was set to 0.5. The shoe sizes are mostly whole numbers, so someone's "best fitting" shoe size could also be 0.5 higher or lower.

The hypothesis "$y = ax$" (linear model through the origin) is certainly not perfectly true, as can be seen in the scatterplot (Fig. 4.117).

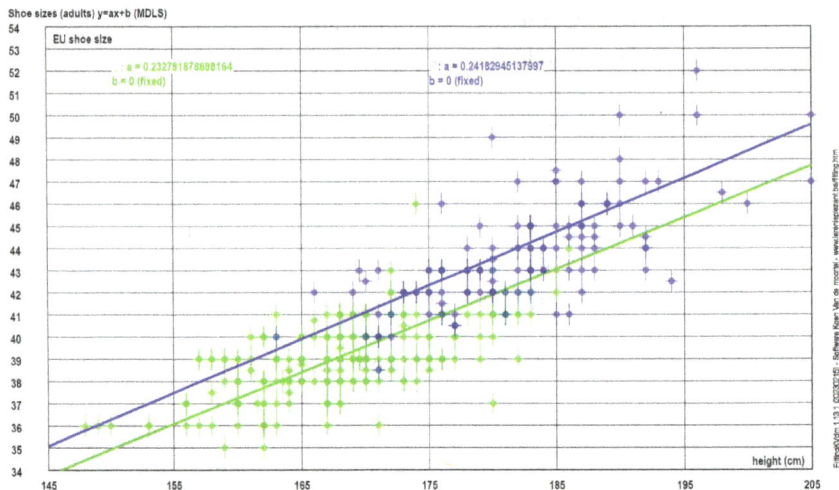

Fig. 4.117. Shoe sizes vs. heights of adult Belgian and Dutch men and women, and best fitting lines through the origin. The darker the dots, the more duplicates.

The proportionality factor a can be fitted quite precisely, but the Kendall τ (see p. 84) and Pearson r values are quite low, which means that other factors play a significant role, or in other words, there is a lot of different body types.

For men: $\quad a = 0.24210 \pm 0.00050,\quad \tau \approx 0.517,\quad r \approx 0.676$
For women: $\quad a = 0.23289 \pm 0.00034,\quad \tau \approx 0.457,\quad r \approx 0.662$

So, if you want to buy shoes for a woman you know is 1.70 m tall, you could estimate that her size will be $0.23289 \cdot 1.70 \approx 39$ or 40, but I wouldn't risk it if the shop doesn't refund, since it could also be 38 or 41.

What would happen if people's feet size and body height would grow at slightly different rates? The exponent of x would not be exactly 1. So, let's see what we get if we fit with a power function model $y = ax^b$.

Using MDLS, which should be used here, since the questions "What shoe size does a person with height x have?" or "How tall is a person with shoe size x?" should be solvable with the same formula, it seems that the exponent is slightly above 1 (1.196 ± 0.020). OLS pulls the curve dramatically flatter ($b = 0.7185 \pm 0.0081$), see Fig. 4.118!

The iteration is very very slow and the uncertainty is quite big, as you can deduce from the dotted lines. Why? Simply because the "correct" curvature is almost impossible to detect from this small cloud far away from the origin. I deliberately scaled the plot so to make this clear.

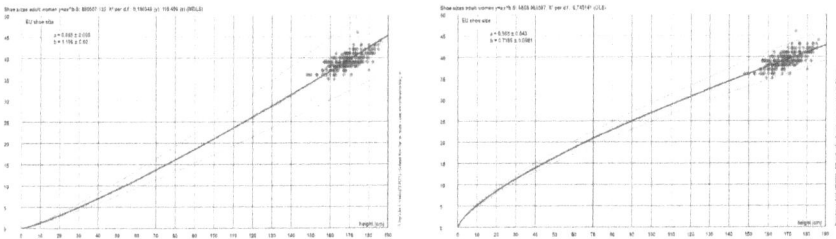

Fig. 4.118. Shoe sizes vs. heights of women, and the best fitting power function using MDLS (left) and OLS (right).

What can we conclude from this? For the men, the results were similar, suggesting there *might* be a small deviation from linearity, but if we would want more certainty, we need more data from exceptionally tall and small people! In order to belong to the same population, they should be normally developed though, not small or tall because of some disease or health problem.

FittingKVdm example data files:
"Shoe sizes adult men.dta1", "Shoe sizes adult women.dta1".

4.3.2. Running records

How do you think the relationship might look between a distance (x) and the time needed to run it (y)?

At first sight, you might expect a *linear* relationship, since the time needed is proportional with the distance, but... a runner is not a car or a train, a runner gets tired... So, when running longer distances, the average speed will be smaller than when sprinting. This sounds like the relationship can be described by a power function $y = ax^b$ with b a bit higher than 1 (the linear case).

To test this hypothesis, I gathered some data: the fastest running times (world records) for all possible distances from 100 to 30000 m that are listed on Wikipedia. You might also consult many books about athletics of course. The precision of these data is most probably very good, like 0.1 m for the distances and 0.01 s for the times.

The model seems to fit quite well (see Fig. 4.119), with an exponent in the expected range: $b = 1.0805442 \pm 0.0000025$. The precision of this parameter is impressive, but the χ^2 values are very big, even though the curve fits very nicely: $\chi_x^2 \approx 961426$ and $\chi_y^2 \approx 3174208$. Why are they so big? Well, the given σ_x and σ_y values are the precisions of the distance and time measurements here: records by single persons. χ^2 would be much lower if the y and σ_y values were averages and standard deviations of a group of random runners, not just records. That would be an interesting experiment to make sports and math teachers collaborate!

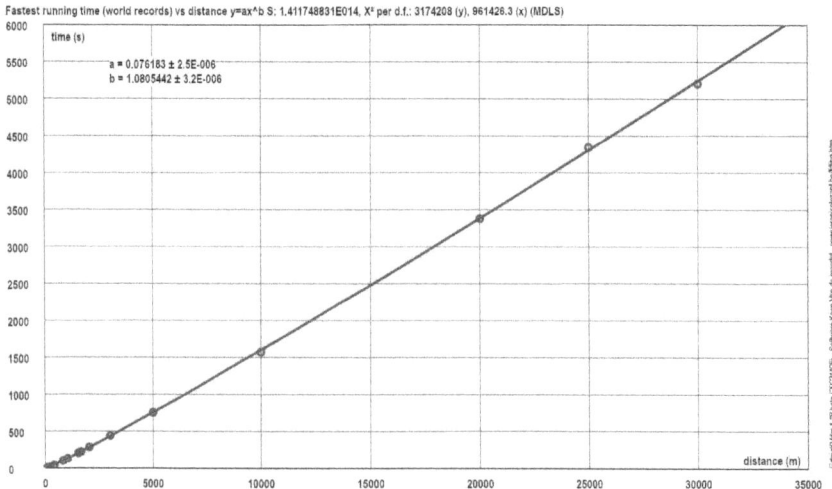

Fig. 4.119. World records of running contests, and the best fitting power function.

You could also collect times from individuals running different distances and find the fit for each of them separately. A high value for parameters *a* and *b* would indicate the person is a good sprinter, while a big *a* and a low *b* probably indicates a better long-distance runner.

FittingKVdm example data file: "Running records.dta1".

4.3.3. Ultra-marathons

I had no idea that so many 90+ people could still run 100km, but apparently they do, and not even that much slower than younger people.
This is reported in an article by Angelika Stöhr *et al.* (2021), "An Analysis of Participation and Performance of 2067 100-km Ultra-Marathons Worldwide", *Int. J. Environ. Res. Public Health*, 18(2), 362 which can be found online at https://doi.org/10.3390/ijerph18020362 or www.mdpi.com/1660-4601/18/2/362.

The variables here are x = center of an age group in years (e.g., for ages [40, 50[: $x = 45 \pm 5$) and y = average running speeds for 100 km ultra-marathons $\pm$ their standard deviations in km/h.

A model describing these data should definitely start from the origin, since newborns don't run marathons, and it should go down after the age of optimal endurance and body control. That age, around 50 (!), seems to come significantly later than the usual age of maximal power (in the thirties).

A candidate is a power function multiplied by an exponential decay (see p. 42). This fits quite well, but it's a pity that we didn't have the raw data. Those might reveal more details about the steepness in the first and the last age bin!

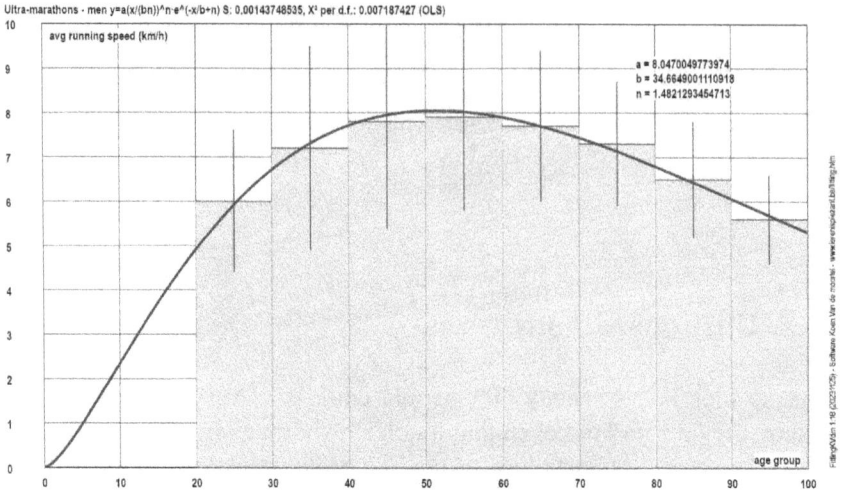

Fig. 4.120. Running speeds in ultra-marathons vs age – men.

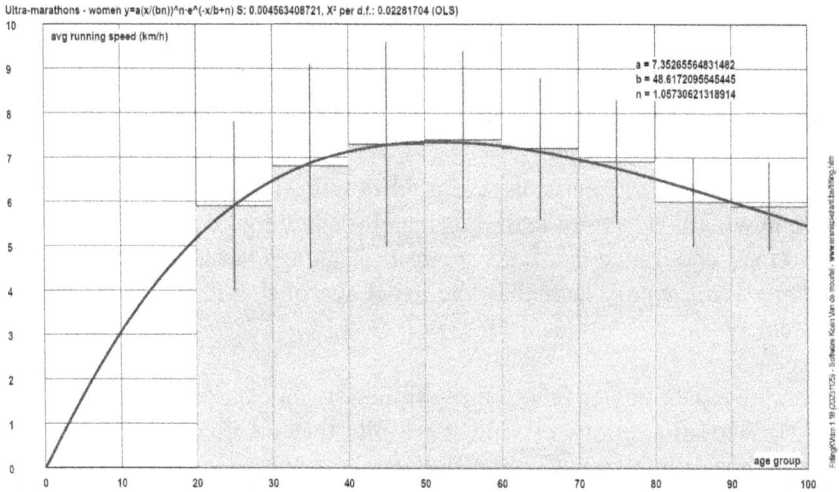

Fig. 4.121. Running speeds in ultra-marathons vs age – women.

FittingKVdm example data files: "Running speeds ultra-marathons men.dta1", "Running speeds ultra-marathons women.dta1".

4.3.4. The body mass index recalculated

The so-called "Body Mass Index" (BMI) is a good example of quantification: its purpose is to put a number on "overweight". As you probably know, it is calculated by taking a person's mass m (in kg) and dividing it by the square of the height h (in meters). Now, this is quite awkward, since the masses of objects with the same shape and similar density distributions are proportional with the *third* power of the height (or any longitudinal dimension).

So I started digging... Why did the inventor, Adolphe Quêtelet (1796–1874), who happens to have lived in the same city as me (Ghent, Belgium), define this index with h^2 in 1832? I wanted to find the original data that he analyzed, the "reference people" to calibrate it. Strangely, there seems to be no trace of them on the internet, and also no other dataset could be found! Thousands of websites offer "ideal mass" tables or calculators, and many use obscure disclosed formulas, clearly not using h^2, some even using a linear relationship!

For me as a physicist, it's hard to believe, but apparently it took almost a century before someone (Fritz Rohrer, a Swiss physician, 1888–1926) came up with the idea to calculate the index with h^3 anyway. This number m/h^3 is then called "**Corpulence Index**" (CI) or "**Ponderal Index**" (PI) [Rohrer 1921]. It took another century until someone (Sultan Babar, SA) found what was to be expected: "It has the advantage that it doesn't need to be adjusted for age after adolescence." [Babar 2015]. In spite of that, the general public still only knows the BMI: today, a search on www.academia.edu on "BMI" produces more than 795,000 results, while only 1,735 articles mention the CI, and 6,428 the PI (June 2021). It took until 2013 before someone like Nick Trefethen (numerical analyst at the University of Oxford, GB) raised his eyebrows and dared to make this remark in *The Economist*: "The body-mass index that you (and the National Health Service) count on to assess obesity is a bizarre measure. We live in a three-dimensional world, yet the BMI is defined as weight divided by height squared. It was invented in the 1840s, before calculators, when a formula had to be very simple to be usable. As a consequence of this ill-founded definition, millions of short people think they are thinner than they are, and millions of tall people think they are fatter.". And then he said, "You might think that the exponent should simply be 3, but *that doesn't match the data at all*. It has been known for a long time that people don't scale in a perfectly linear fashion as they grow. I propose that a better approximation to the actual sizes and shapes of healthy bodies might be given by an exponent of 2.5. So here is the formula I think is worth considering as an alternative to the standard BMI: 'new BMI' = $1.3m/h^{2.5}$'' [Trefethen, 2013, my emphasis].

Now, how could it "not match the data"? I was curious now to inspect some data myself. After a long search, I came in touch with Nir Krakauer (The City College of New York), who was also doing BMI-related modeling, and he was so kind to refer me to his data https://rdrr.io/github/dtkaplan/NIMBIOS/man/nhanesOriginal.html from the American National Health and Nutrition Examination Survey (NHANES).

From this large collection, I extracted the masses and heights of 90 adult men who had a more or less "ideal" body fat percentage: between 11.6 and 13.8%. I'm not a medical doctor, but according to different sources, these percentages seem to be considered good for young adults. The most important point for this selection was to have a more or less homogeneous group with a range of sizes, but with similar densities. Of course, I know other factors like bone density and body type play a role as well, but this is the best I could do.

First, I put the data in the popular math program GeoGebra (version 5 Classic). This calculated the following relationship as "best fitting": $m = 18.8541 \cdot h^{2.1419}$.

Aha, that must have been the reason why Quêtelet decided to round the exponent of h to 2 because the empirical value seems to be 2.1419! Then I realized that this program takes the logarithms of the variables, in order to reduce the regression problem to a linear one. This causes errors, as I illustrated earlier (p. 145).

So I decided to put the data in my own software program (FittingKVdm), which uses an iterative algorithm to estimate the parameters. The measurement errors were assumed to be the least significant digit (0.1 kg and 1 mm). This produced $m = 19.331 \cdot h^{2.1084}$. Now the exponent was even closer to 2! Strange!

GraphPad Prism 9.0.2, a program that seems well designed to me, and also uses iteration, produced an identical result. Their writers also condemn the logarithm habit, by the way.

Still not being happy, I wanted to see the difference between a fit with a fixed exponent of 2 and one with exponent 3. The results were $m = 20.5918 \cdot h^2$ and $m = 11.4640 \cdot h^3$.

The value of 20.5918 is indeed considered to be a good BMI, and 11.4640 is close to the "good" value of 12 for the CI according to Sultan Babar.

Now, a picture is worth a thousand words, so I would like you to take a look on the mass versus height graphs of both fitted curves (don't mind if you can't read the small letters, just look at the dots and the lines).

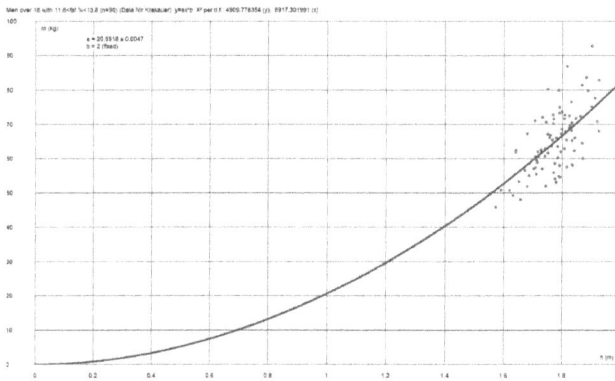

Fig. 4.122. Height vs. mass data and best fitting 2nd degree function.

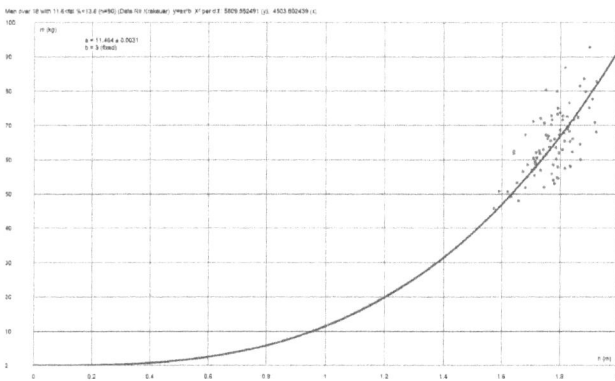

Fig. 4.123. Height vs. mass data and best fitting 3rd degree function.

Which line visually fits the best through the cloud of points? Everyone I asked this question answered the one on the right, obviously!

Now watch this: if the "best fit" for our data, with free moving exponent, $m = 19.331 \cdot h^{2.1084}$, was indeed the best fit, it shouldn't make any difference if we switched the h and m columns and fitted again, should it? The expected outcome of this procedure would be:

$$\Rightarrow h = \left(\frac{m}{19.331}\right)^{\frac{1}{2.1084}} = 0.24534 \cdot m^{0.47429}$$

Now, what was the actual outcome? $h = 0.72877 \cdot m^{0.21394}$ or, inverted $m = 4.38814 \cdot h^{4.67421}$.

I double-checked it using GraphPad... same result.
GeoGebra gave almost the same: $h = 0.7225 \cdot m^{0.2159}$.
This is not just a small difference, like a "rounding error" or so. This is obviously shocking and dramatic!

This example is actually the one that signaled me there might be something wrong with the regression method itself, which brought me to the idea of using the vertical and horizontal distances between the points and the curve ("MDLS"), to make the method symmetrical.

Fitting the same data with MDLS (see Fig. 4.124), yielded
$m = ah^b$, with $a = 10.482 \pm 0.058$, $b = 3.1581 \pm 0.0092$.

Men over 18 with 11.6<fat %<13.8 (n=90) (NHANES) y=ax^b S: 5950565806, X² per d.f.: 5956.567 (y), 4299.013 (x) (MDLS)

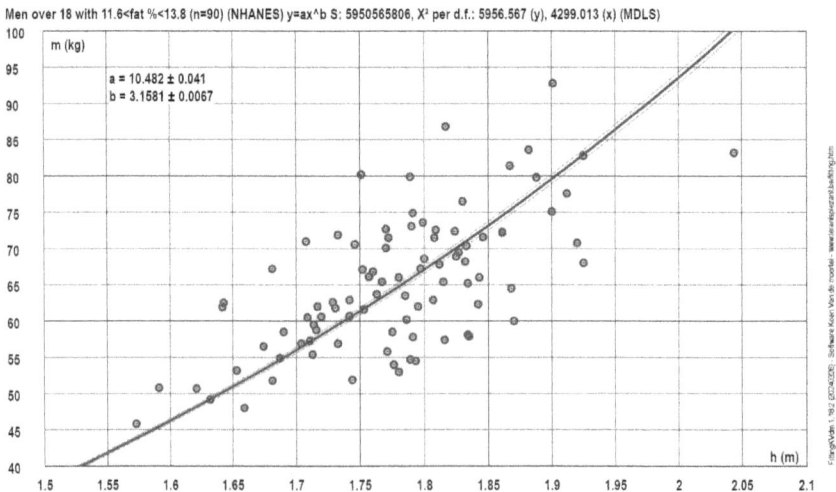

Fig. 4.124. Mass vs. height for "ideal" men, and the best fitting power function with free floating exponent.

That exponent b is a lot closer to 3, as we physicists always expected! And 3.1581 is approximately equal to the geometric mean of 2.1084 and 4.67421, which makes sense.

We can ask ourselves now: was this result only valid for "ideal" people, or also for more voluptuous people?

So, the same analysis was done for men and women (aged 16 and more) with different fat percentages, first with parameters a and b floating, then with a fixed value of $b = 2$, so in that case "a" represents the classic "BMI", and then with b fixed to 3, so in this case "a" represents the "BMI with h^3" aka "CI".

Men		*a* and *b* floating		*b* = 2 fixed	*b* = 3 fixed
fat % (*F*)	*n*	*a*	*b*	*a* = "BMI"	*a* = "BMI3" = CI
[10, 12[	9	9.800	3.192	19.318 ± 0.017	10.921 ± 0.012
[12, 14[	169	10.84	3.093	20.5621 ± 0.0051	11.4527 ± 0.0036
[14, 16[	344	7.695	3.761	21.2382 ± 0.0043	11.9307 ± 0.0029
[16, 18[	423	10.82	3.186	21.4028 ± 0.0033	12.0430 ± 0.0022
[18, 20[	467	10.28	3.337	22.3881 ± 0.0038	12.5053 ± 0.0022
[20, 22[	528	9.587	3.540	23.2707 ± 0.0035	13.0892 ± 0.0022
[22, 24[	562	10.24	3.543	24.5169 ± 0.0030	13.9356 ± 0.0022
[24, 26[	806	11.81	3.366	25.8724 ± 0.0029	14.5751 ± 0.0022
[26, 28[	960	12.47	3.326	26.6912 ± 0.0022	15.0336 ± 0.0019
[28, 30[	943	12.10	3.480	28.0654 ± 0.0027	15.8937 ± 0.0019
[30, 32[	837	11.80	3.600	29.3692 ± 0.0035	16.6075 ± 0.0021
[32, 34[	721	12.15	3.628	30.8169 ± 0.0031	17.4054 ± 0.0024
[34, 36[	501	10.31	4.016	32.9015 ± 0.0041	18.4909 ± 0.0030
[36, 38[	351	10.55	4.070	34.6159 ± 0.0046	19.5014 ± 0.0035
[38, 40[	224	9.109	4.485	38.3387 ± 0.0063	21.5662 ± 0.0041
[40, 42[	118	8.993	4.567	39.8371 ± 0.0084	22.4331 ± 0.0067
[42, 44[	51	10.89	4.419	44.184 ± 0.014	24.8267 ± 0.0099
[44, 46[	15	35.67	2.355	43.407 ± 0.024	25.308 ± 0.021
[46, 48[	9	10.52	4.735	46.445 ± 0.032	27.236 ± 0.024
[48, 50[	1			43.711 ± 0.054	24.836 ± 0.049

When a and b are left free to be adjusted, the fitting is not very stable. We can presume that this is because many other factors besides the body fat percentage play a role, like the body type, muscle weight, etc. and the groups are also too small. The measurement points form clouds rather than precise lines. Anyhow, the exponents are almost always more than 3. The average is even 3.668, and the standard deviation 0.579. With fixed exponents, we see a very nice correlation of a versus the fat percentage. The "goodness-of-fit" (indicated by S and the χ^2 per degree of freedom but also by the estimated errors on the fitted parameters) was always better with $b = 3$ than with $b = 2$.

We can make the same remarks for the women. The average b is here even more, 3.859, and the standard deviation 0.556.

Women		*a* and *b* floating		*b* = 2 fixed	*b* = 3 fixed
fat % (*F*)	*n*	*a*	*b*	*a* = "BMI"	*a* = "BMI3" = CI
[16, 18[	2			17.209 ± 0.031	10.927 ± 0.023
[18, 20[	13	6.472	4.187	18.000 ± 0.021	11.302 ± 0.015
[20, 22[	32	7.763	3.807	18.834 ± 0.011	11.5395 ± 0.0076
[22, 24[	51	9.110	3.469	18.869 ± 0.092	11.5032 ± 0.0072
[24, 26[	124	9.542	3.479	19.8013 ± 0.0077	12.0853 ± 0.0054
[26, 28[	201	9.814	3.461	20.4095 ± 0.0055	12.3724 ± 0.0037
[28, 30[	275	11.08	3.314	21.1685 ± 0.0054	12.9417 ± 0.0040
[30, 32[	383	13.82	2.914	21.6030 ± 0.0039	13.2474 ± 0.0027
[32, 34[	459	11.48	3.389	23.1021 ± 0.0046	13.9777 ± 0.0030
[34, 36[	562	9.557	3.878	24.5112 ± 0.0042	14.8582 ± 0.0028
[36, 38[	720	11.03	3.707	25.5553 ± 0.0031	15.6460 ± 0.0028
[38, 40[	738	12.49	3.570	26.9707 ± 0.0030	16.5106 ± 0.0025
[40, 42[	892	15.12	3.316	28.7631 ± 0.0030	17.6573 ± 0.0022
[42, 44[	896	11.49	4.014	30.6793 ± 0.0030	18.8671 ± 0.0023
[44, 46[	756	14.76	3.606	32.4786 ± 0.0032	19.8652 ± 0.0029
[46, 48[	583	12.14	4.156	34.6952 ± 0.0039	21.3284 ± 0.0034
[48, 50[	404	12.94	4.199	38.0436 ± 0.0049	23.2991 ± 0.0038
[50, 52[	255	11.62	4.590	41.1029 ± 0.0074	25.2133 ± 0.0057
[52, 54[	99	10.22	5.061	46.046 ± 0.011	28.013 ± 0.011
[54, 56[	26	10.84	5.096	47.842 ± 0.027	29.441 ± 0.020
[56, 8[	4	20.06	3.977	52.369 ± 0.048	32.142 ± 0.036

If the body mass is better correlated with h^3 than with h^2, as it was found, we would also expect that the body fat percentage (F) is better correlated with the CI than with the classic BMI. Fig. 4.125 and 4.126 show BMI and CI vs. F for men (from the table above). The relationship is clearly not linear, but it seems to follow an exponential pattern. That can only be a pragmatical approximation in a limited domain, since it can definitely not be extrapolated!

Men, age>=16 y=ba^x+c S: 513425.427, X² per d.f.: 11953.54 (y), 1.794317 (x) (MDLS)

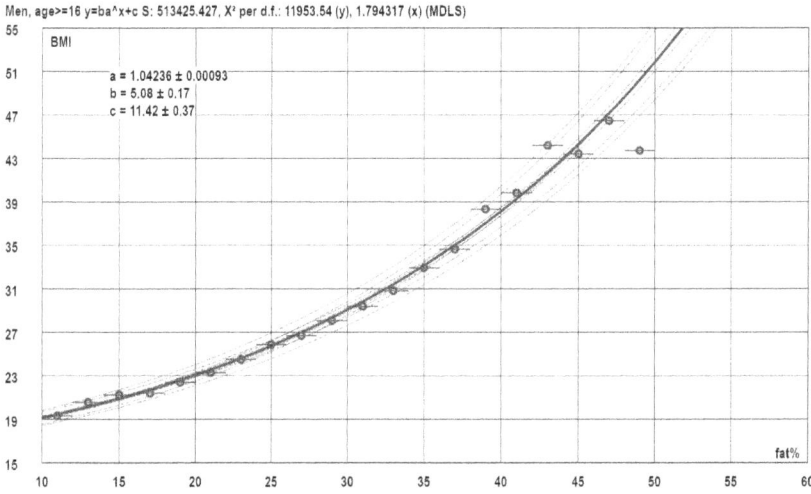

Fig. 4.125. BMI vs. body fat % and best fitting exponential function (men).

Men, age>=16 y=ba^x+c S: 194991.0998, X² per d.f.: 6344.76 (y), 1.595892 (x) (MDLS)

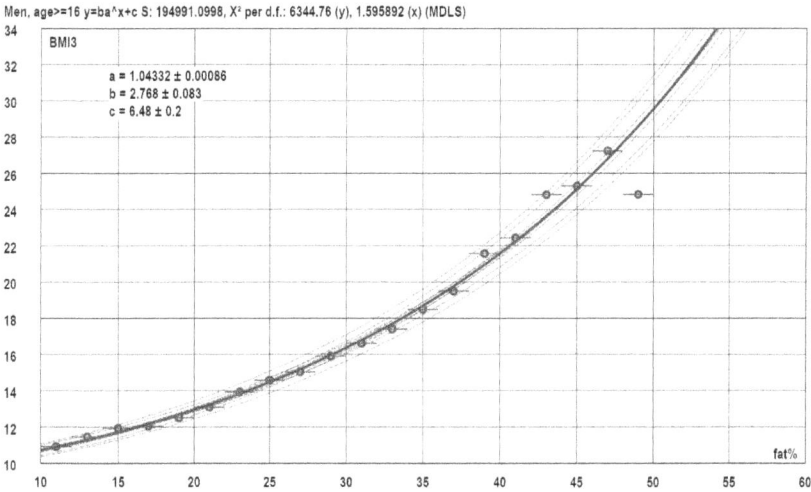

Fig. 4.126. CI vs. body fat % and best fitting exponential function (men).

The curve through the data points is the best fitting exponential function $B = ba^F + c$, with a, b, and c fitted parameters and $B =$ "classic BMI" (Fig. 4.125) or the "CI" (Fig. 4.126). The graphs are similar for men and women, for BMI and CI, but with different parameters of course. Those are listed in the following:

	men aged 16 or more ($n = 8039$)		women aged 16 or more ($n = 7475$)	
	B = classic BMI	B = BMI3 = CI	B = classic BMI	B = BMI3 = CI
a	1.0424 ± 0.0011	1.0433 ± 0.0011	1.0671 ± 0.0019	1.0687 ± 0.0019
b	5.08 ± 0.26	2.77 ± 0.14	0.97 ± 0.081	0.564 ± 0.050
c	11.42 ± 0.37	6.48 ± 0.21	14.7 ± 0.31	9.15 ± 0.16
χ_x^2	1.79432	1.5959	0.581528	0.444813
χ_y^2	11953.5	6344.76	3654.17	2086.1

By comparing the χ^2 values (in both x and y directions), we see that the **body fat percentage (F) is better correlated with the CI than with the BMI**, at least if we use an exponential model.

We see that the Corpulence Index can be estimated from the body fat % (F), using:
"CI" $\approx 2.77 \cdot 1.0433^F + 6.43$ (men)
"CI" $\approx 0.564 \cdot 1.0687^F + 9.15$ (women)

Actually, an exponential function without baseline fits quite well too, and much quicker than if we leave the c parameter free. And for the most precise regression, I used the raw data (n = 8039 men, 7475 women). The errors on the y values had to be calculated with the derivative now:

In case $y = m/h^2$,
$$\frac{\partial y}{\partial m} = \frac{1}{h^2} \qquad \frac{\partial y}{\partial h} = -\frac{3m}{h^3}$$

and in case $y = mh^3$,
$$\frac{\partial y}{\partial m} = \frac{1}{h^3} \qquad \frac{\partial y}{\partial h} = -\frac{4m}{h^4}$$
so
$$\sigma_y = \sqrt{\left(\frac{\partial y}{\partial m}\right)^2 \sigma_m^2 + \left(\frac{\partial y}{\partial h}\right)^2 \sigma_h^2}$$

This is how the raw data look: quite "cloudy", which means that not only the body fat is responsible for one's BMI or CI.

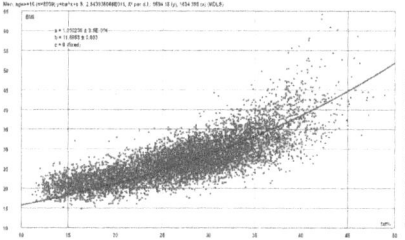

Fig. 4.127. BMI vs. fat% and exponential fit without baseline (men).

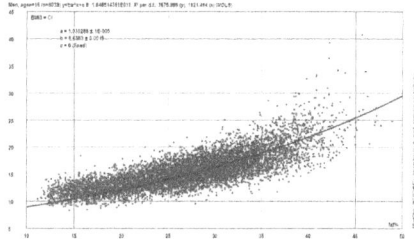

Fig. 4.128. CI vs. fat% and exponential fit without baseline (men).

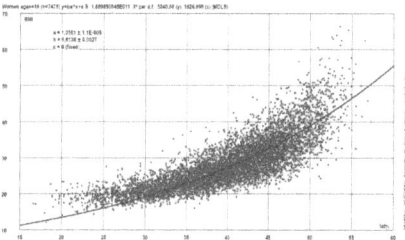

Fig. 4.129. BMI vs. fat% and exponential fit without baseline (women).

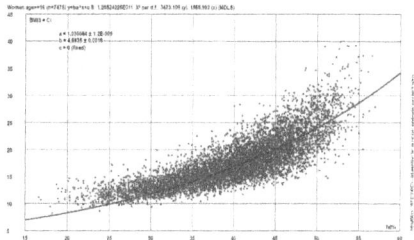

Fig. 4.130. CI vs. fat% and exponential fit without baseline (women).

The model is clearly not perfect, but for men and women, using the CI instead of the BMI produces a better fit: S is much smaller.

From this fitting we get the following Corpulence Index estimation from the body fat% (F):

"CI" $\approx 6.6383 \cdot 1.020288^{F}$ (men)

"CI" $\approx 4.0835 \cdot 1.036064^{F}$ (women)

As you can see in the graphs, the approximation is not very good, especially for $F < 25$ and $F > 55$, but it's up to the biologists to find a better curve with theoretical support.

We can conclude that there is no reason anymore to use the classic BMI. The "BMI" with h^3 in the denominator, also known as the Corpulence Index, should logically be a better estimator for all kinds of health issues. Maybe the theoretical exponent of h should even be bigger than 3, as suggested by the empiric evidence.

Remark: I showed you two versions of the "fat vs. CI" graphs: one with only 20 data points representing averages, and the other with thousands of raw data points. Both show "the truth", the same data. In the first, it's easier to see the fit of the model function, but that might also give the illusion that you can calculate y from x quite exactly and vice versa. In the second version, it's harder to see the pattern, but it gives a more realistic idea of the precision of the relationship. So they both have their purpose.

References:

- Babar, Sultan (March 2015), "Evaluating the Performance of 4 Indices in Determining Adiposity", *Clinical Journal of Sport Medicine, Lippincott Williams & Wilkins*, 25 (2): 183.

- Rohrer, Fritz (1921), "Der Index der Körperfülle als Maß des Ernährungszustandes", *Münchner Med. WSCHR.* 68: 580–582.

- Trefethen, Nick (2013): on his own website: https://people.maths.ox.ac.uk/trefethen/bmi.html.

FittingKVdm example data files:
"adult men 11.6-13.8 percent fat - h vs m.dta1", "adult men 11.6-13.8 percent fat - m vs h.dta1", "BMI2 vs fat Men 16 and older DIRECT DATA.dta1", "BMI2 vs fat Men 16 and older.dta1", "BMI2 vs fat Women 16 and older DIRECT DATA.dta1", "BMI2 vs fat Women 16 and older.dta1", "BMI3 vs fat Men 16 and older DIRECT DATA.dta1", "BMI3 vs fat Men 16 and older.dta1", "BMI3 vs fat Women 16 and older DIRECT DATA.dta11", "BMI3 vs fat Women 16 and older.dta1".

4.3.5. The heartbeat of land mammals

Newborns have a faster pulse than adults, you may have noticed. Apparently, this rule can be generalized for all mammals (and other classes of animals): the heartbeat of a mouse is much faster than that of an elephant. It's a matter of resonance frequency related to the size of a pulsating or a vibrating object; big clocks also have a lower resonance frequency than small ones.

Now, a scientist asks himself: is there a way we can predict how fast the heart of a 5 kg animal "should" beat? Or, a mammal with a heartbeat of 200 pulses per minute, how small can we expect that to be? Are humans "normal" members in this series of animals?

To my surprise, it was not easy to find data about this phenomenon. Everybody seems to copy the same short list over and over in school books... But I found some in
Jacopo P. Mortola (2015), "The heart rate - breathing rate relationship in aquatic mammals: A comparative analysis with terrestrial species", *Current Zoology* 61 (4): 569–577.
Online: https://academic.oup.com/cz/article/61/4/569/1803113.

The variables here are x = mass of a (terrestrial) mammal (m, in kg) and y = its average heart beat when resting (in pulses per minute). The uncertainties on the values were not known, so we assumed them to be half of the last significant digit. They seem to come from different origins since some masses have three significant digits, some only one.

What could be the connecting formula? If all mammals had the same shape and differed only in size, it would be purely a matter of scaling, represented by a power function.
Anyway, land mammals also have a lot in common, so it's still worth trying to fit a power function ($y = ax^b$) through the data.

Since there is no reason here to prefer the mass or the pulse as "independent" variable, we should definitely use MDLS.
FittingKVdm produces:
$a = 171.52 \pm 0.72$
$b = -0.21353 \pm 0.00065$

The χ^2 values in both directions (x: 316247, y: 16183) are very high.

This means that other factors besides the body mass have an influence (very plausible), or that the confidence intervals of the data were estimated too small (certainly the variability of the heartbeats must be greater than the least significant digit). To figure that out, it might be interesting to distinguish, for example, between predators and vegetarian animals, or to group a lot of data from more similar animals, e.g., all kinds of rats and mice, or different primates, and also to collect many measurements from the same species. Well, that's just a first thought of me as a non-biologist...

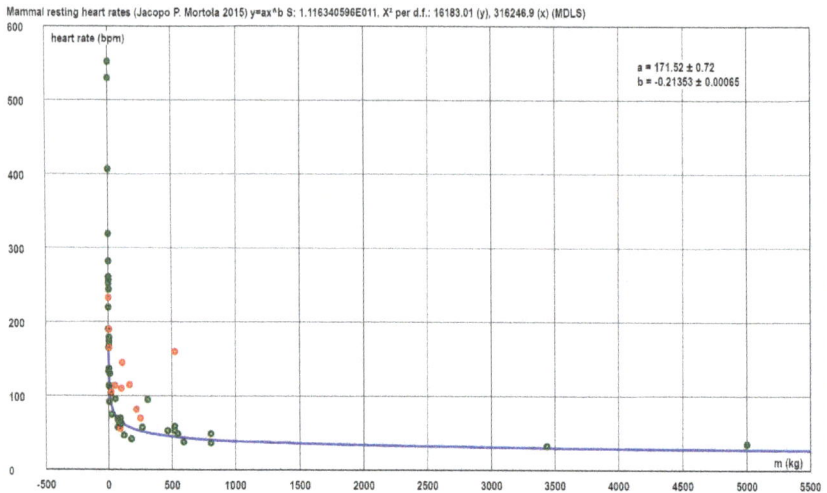

Fig. 4.131. A power function fitted through the data (MDLS method).

The attentive reader might notice that some of the dots are in red, meaning they were set "inactive" for the fitting, why? We should always be prudent to categorize measurements as "outliers", but in this case, some measurements were taken from *sedated* animals, and it seems quite obvious that might have a significant influence, see, for example, the giraffe ($m = 525$kg) high above the curve, so I switched them off. Maybe it would be better for the comparison, and easier, to measure sedated animals only. How does one measure the pulse of a non-sedated wild tiger anyway? With implanted sensors? It's not written in the article.

You might wonder wether it would make a big difference if we used OLS instead of MDLS.

Well, apparently it does... quite dramatically, as you see in Fig. 4.132.

How could it possibly be that the curve is so far from most of the points? I must admit, it took me some time to find the culprit.

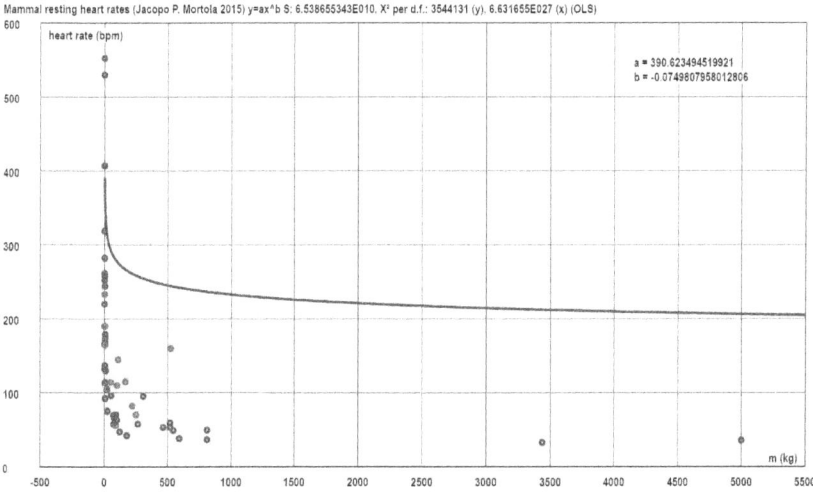

Fig. 4.132. A power function fitted with (weighted) OLS.

The first two points are from very small animals, so the error on the mass is also very small. That gives them a big weight in the fitting, which they should have, because it's the truth.

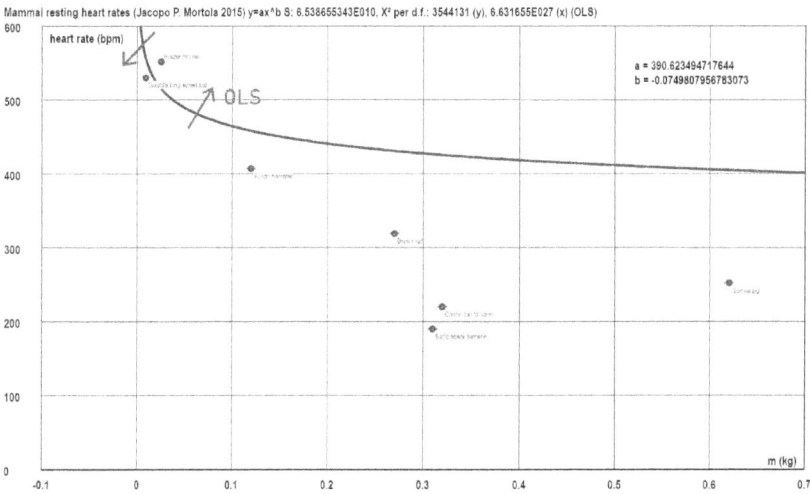

Fig. 4.133. OLS pulling the curve in the wrong direction.

Now, this first couple of points happens to be "atypical" (the second point has a bigger x and yet a bigger y), so the curve is pulled upward enormously since OLS tries to minimize the *vertical* distances between the points and the curve only. MDLS, however, knows how to deal with this!

If the "errors" (and hence the weights) of the masses are all set equal, OLS produces almost the same graph as MDLS, but that is *ad hoc* cheating, like making two sign errors in a calculation to get a "good" result!

Now, the attentive reader may have noticed that the first animal is a *bat*, which is naturally lighter than walking mammals of the same size, so it might be a good idea to leave it out of the dataset. We can rightfully consider it as an "outlier". Without the bat, even with the correct weights, OLS also produces a much better looking fit (but still not symmetry-proof if you switch x and y)!

Data:			x > 0, y > 0 !!!		
Description:		Mammal resting heart rates (Jacopo P. Mortola 20			
x-legend:		m (kg)			
y-legend:		heart rate (bpm)			

x	σ_x	y	σ_y	Active	Label
▸ 0.009	0.0005	530	0.5	True	Gould's long-eared bat
0.025	0.0005	552	0.5	True	House mouse
0.12	0.005	407	0.5	True	Syrian hamster
0.27	0.005	319	0.5	True	Brown rat
0.31	0.005	190	0.5	True	Saddleback tamarin
0.32	0.005	220	0.5	True	Cotton top tamarin
0.62	0.005	252	0.5	True	Guinea pig
1	0.5	282	0.5	True	European polecat
1	0.5	165	0.5	True	Capuchin monkey
1	0.5	133	0.5	True	Fennec
1	0.5	261	0.5	True	Mink
1.04	0.005	233	0.5	False	Talapoin monkey (sedated)
1.5	0.05	257	0.5	True	Steppe polecat
3.05	0.005	244	0.5	True	European rabbit
3.58	0.005	137	0.5	True	Rhesus macaque
4	0.5	113	0.5	True	Cat domestic
4.8	0.05	114	0.5	True	Dik-dik
5	0.5	190	0.5	False	Meerkat (sedated)

Fig. 4.134. Part of the data, entered in FittingKVdm version 1.6.

The moral of the story is one single inconspicuous "wrong" point can seriously influence the fitted parameters but much less if you use MDLS. MDLS is a much more "robust" algorithm since it pulls the curve in two directions.

FittingKVdm example data file: "Heart beat land mammals - Mortola 2015.dta1".

4.3.6. Heart rate recovery

After an intensive physical exercise, your heart rate will go back to normal after some time. This time is used as an indication for the health condition of the person; the shorter the time, the better. Now, how should you define that time exactly? Some people compare the heart rate one minute after an exercise with the rate during the exercise, others say it might be better to check after 10 seconds.

A discussion about this topic can be found, for example, in an article by Yordi J. van de Vegte, Pim van der Harst and Niek Verweij (2018). "Heart Rate Recovery 10 Seconds After Cessation of Exercise Predicts Death", *Journal of the American Heart Association.* 7:e008341, online: www.ahajournals.org/doi/full/10.1161/JAHA.117.008341.

This discussion can be avoided by making a *model* of how the heart beat evolves after an exercise. This process of "going back to normal" might very possibly be described using an exponential function $y = ba^x + c$, like many other "decay" phenomena. That would produce one simple parameter that can be used to compare people.

I collected some sample data for you from a test person, but you can do the experiment yourself too. I used a very inexpensive (ca. 20€) sports watch with a chest band: Kalenji Onrythm 110. This is a very basic device, so I had to write down the values of x (time in seconds starting from the end of a physical exercise) and y (heart rate in beats per minute) manually. More sophisticated devices have data loggers, etc.

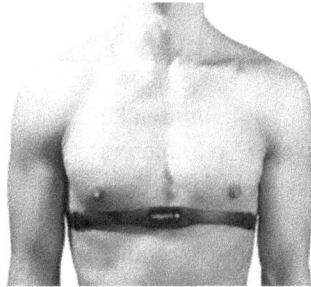

Fig. 4.135. The measurement tool.

You can see the results in Fig. 4.136. Apparently, there can easily be some "ups and downs" in the measurements, maybe due to bad contacts of the sensor, or unknown physiological or psychological events, etc. Anyway, *regression* reduces these kinds of errors maximally, so it should give better results than individual measurements, e.g., after 10 or 60 s! Unfortunately, because of this "noise" around the horizontal asymptote, MDLS can't be used.

Heart rate after exercise KVdm 20221008 y=ba^x+c S: 2729416.579, X² per d.f.: 2.785119 (y), 1.077893E008 (x) (OLS)

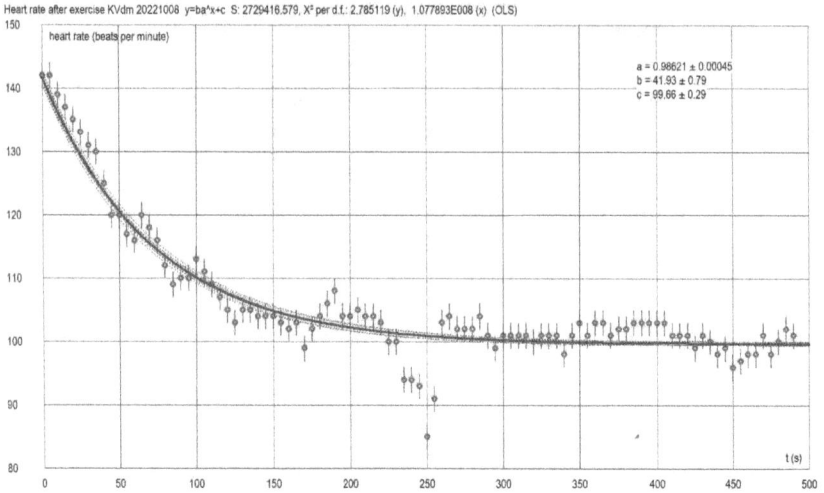

Fig. 4.136. Heart beat measured after an exercise, and the best fitting exponential function.

As an indicator for the heart rate recovery you could take several parameters (Fig. 4.137):

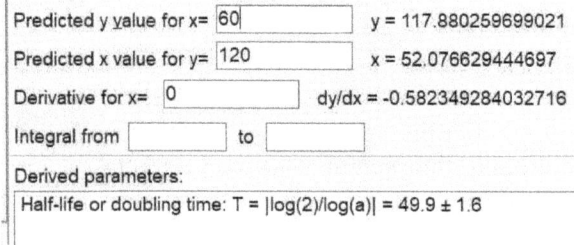

Predicted y value for x= 60	y = 117.880259699021
Predicted x value for y= 120	x = 52.076629444697
Derivative for x= 0	dy/dx = -0.582349284032716
Integral from ___ to ___	
Derived parameters:	
Half-life or doubling time: T = |log(2)/log(a)| = 49.9 ± 1.6	

Fig. 4.137. Screenshot from "FittingKVdm".

- the starting value minus the predicted value after 60 seconds ($x = 60$) (or whatever time span you want),
- the slope of the tangent line (derivative) at $t = 0$ (here -0.582),
- the time span needed to go back half the way to normal (here 49.9 s),
- the parameter a (here 0.98621), which says how much the rate drops every second (here $1 - 0.98621 = 0.01379 = 1.379\%$).

All of them are equivalent. Use whichever you find convenient.

The asymptotic value (c) is probably also a useful indicator for the person's condition, I suppose.

FittingKVdm example data file: "Heart rate after exercise.dta1".

4.3.7. How much energy does a person or an animal consume?

Since there are more problems in the Western world due to obesity than due to lack of food, quite a bit of attention is paid to the "calories" (or better, the joules) we consume. Everything we buy is required to state how much energy it contains so that fanatics can neatly calculate how much they can eat before night falls. But, we can ask ourselves a few questions about this:

(1) **How do we actually know how many joules we get from a bag of chips or a pint of beer?**
(2) **How much do we need and how do we measure it?**

The answers turn out to be not simple at all!

A simplistic answer to question (1) can be obtained with a **"calorimetric" home experiment**: simply put the chips on fire under a pot of water and measure how much it heats up! If m is the mass of the water and c is the specific heat (4186 J/kg/K), then the following is the heat energy absorbed by the water:

$$Q = c \cdot m \cdot \left(T_{end} - T_{begin} \right)$$

This is done effectively, as it is demonstrated in this video by British science teacher Hazel Lindsey:
https://www.youtube.com/watch?v=dZB7kzKUZlc.

By dividing this Q by the mass of the burned chips, you can find out how many joules are produced per kg or per 100 g. That pint of beer is a bit more difficult: you would first have to be able to separate all the water from the flammable dry matter and the alcohol... Okay, a good chemist can do that.

But,... there are still some issues to resolve:
(1) Heat is clearly also lost to the air and the heating of the pot itself. Okay, you can take the latter into account by also calculating the Q of the jar if you know the specific heat of the glass. Measuring the losses, or limiting them to a minimum, is somewhat more difficult because you cannot really isolate the test properly, there is no combustion without oxygen supply!
(2) Who says that all possible combustion energy will also be absorbed by our body? After all, you could do the same experiment with a piece of wood, and I don't think we would be able to jump very far by chewing

on wood.

According to the first consideration, the measured Q is therefore too small, and according to the second, too large.

To answer the first question, chemists have studies tons of reactions by now and the energy they can produce, so the theoretical (maximal) energy in a food substance can be calculated with an acceptable precision.

For the second question, the American chemist Wilbur Atwater (1844–1907) estimated how much energy is actually absorbed, by also measuring how much we excrete in solid, liquid, and gaseous form. This appears to be on the order of 85%. This way we can still estimate our consumption based on what we eat but never with a better accuracy than about 5%.

Of course, we can also try to *directly* measure the consumption of our body (or that of any animal). But that is also quite complicated!

Another calorimetric method that is used consists of placing the person or animal in an enclosed space and measuring the temperature inside and outside. If the thermal conduction coefficient (λ) of the wall material is known, you can determine how much heat flows from inside to outside per unit time using Fourier's law:

$$\Phi = \lambda \cdot d \cdot \left(T_{inside} - T_{outside}\right)$$

where Φ is the heat flux (in Watts per m²), d is the wall thickness, and T is the temperature. To know the creature's consumed power, we simply need to multiply the flux by the total wall area. This seems simple, but the practice is somewhat different:

- The person or animal also has to breathe. So you still have to take heat loss through air exchange into account.
- The creature will also sweat, and if it is an animal, other substances may be excreted during the measurement...
- The temperature difference across the wall is usually very small, so that very precise equipment is required (a grid of thermocouples), and the entire setup becomes extremely sensitive to temperature fluctuations in the environment; a door opening or the heating turning on is enough to ruin the measurement...

The first problem can still be solved neatly: calibrate the room by taking a measurement with a known heat source, such as a resistor through which a current flows, and measure the temperature difference. φ is then

known, and you can immediately calculate the corrected proportionality coefficient that replaces λd.
The rest is less simple. The details would lead us too far here.

For your information, for an adult person sitting quietly on a chair, this results in a consumption of approximately 104 W, but during intense sporting activities, this can be almost 10 times as much. Also, children use up more than older people. A summary table can be found at
www.engineeringtoolbox.com/met-metabolic-rate-d_733.html.

Another method, which is more precise, consists of **measuring the CO_2 produced**. To know how much of that CO_2 comes from the combustion of food, the animal must be injected with a little "special" water, in which the hydrogen atoms have been replaced by the isotope deuterium (^2H) and the oxygen atoms by the isotope ^{18}O (in the literature: "double labeled water"). These isotopes can then be traced, for example, in the exhaled air. There is still a lot to consider, but that's the general idea.

For enthusiasts, the American physiologist John R.B. Lighton wrote a detailed practical book on this theme: *Measuring Metabolic Rates - A Manual for Scientists* (Oxford University Press 2008).

A question that can be investigated using *regression* is as follows: "**How is the body mass related to an animal's metabolism?**".

Big animals need more energy than small ones, especially if they are warm-blooded, but how much more? The heat radiation (Q) of a body is proportional to its surface. And that is proportional to the square of its size (x), while the volume and the mass (m) are proportional to the third power of the size. So, we would expect something like $Q \sim m^{2/3}$. But, apparently, the basal metabolic rate (M) is not exactly proportional with the radiation, since the Swiss biologist Max Kleiber (1893–1976) found evidence suggesting the truth looks more like $M \sim m^{3/4}$. His original article "Body size and metabolism", *Hilgardia* 6:315–53, 1932) can be found at (including data!)
https://hilgardia.ucanr.edu/fileaccess.cfm?article=152052&p=VOWQRB

A power function fit through these data produces an exponent of 0.743 ± 0.12, which confirms Kleiber's theory nicely. The values of χ_x^2 and χ_y^2 are near 1, indicating that the assumed errors on m and M were realistic.

animal	m (kg)	M (kcal/day)	
ring dove	0.15	19.5	
rat ♀	0.173	20.2	
rat ♂	0.226	25.5	
pigeon	0.3	30.8	
hen	1.96	106	Kleiber's data
dog ♀	11.6	443	
dog ♂	15.5	525	For the regression, precisions of ± 10% on
sheep	45.6	1219.9	m and M were assumed.
woman	56.5	1349	
man	64.1	1632	
cow	388	6421	
steer 1	342	6255	
steer 2	679	8274	

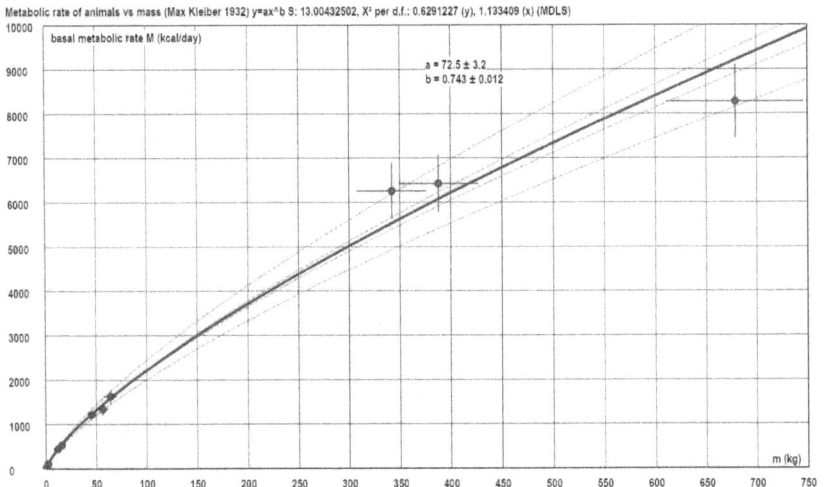

Metabolic rate of animals vs mass (Max Kleiber 1932) y=ax^b S: 13.00432502, X² per d.f.: 0.6291227 (y), 1.133409 (x) (MDLS)

a = 72.5 ± 3.2
b = 0.743 ± 0.012

Fig. 4.138. Max Kleiber's data and best fitting power function.

If you want to know more, the following is interesting to read:
Niklas, K.J. & Kutschera, U. (2015), "Kleiber's Law, How the Fire of Life ignited debate, fueled theory, and neglected plants as model organisms", *Plant Signal Behav.* 10(7):e1036216,
www.ncbi.nlm.nih.gov/pmc/articles/PMC4622013/.

FittingKVdm example data file: "Animal_metabolism_vs_mass_Kleiber_1932.dta1".

4.3.8. Height distributions

Regression can be used to visualize the "ideal" curve in a distribution. The example data to illustrate this come from NHANES (see also p. 294); the *x* variable here is height (in cm, grouped in bins per 2 cm) and *y* the number of subjects in each bin. The error on each *y* value is its square root but also 1 if $y = 0$. The sample consists of 415 boys aged 15.

Fig. 4.139 shows what we get if we fit a Gaussian curve through those data.

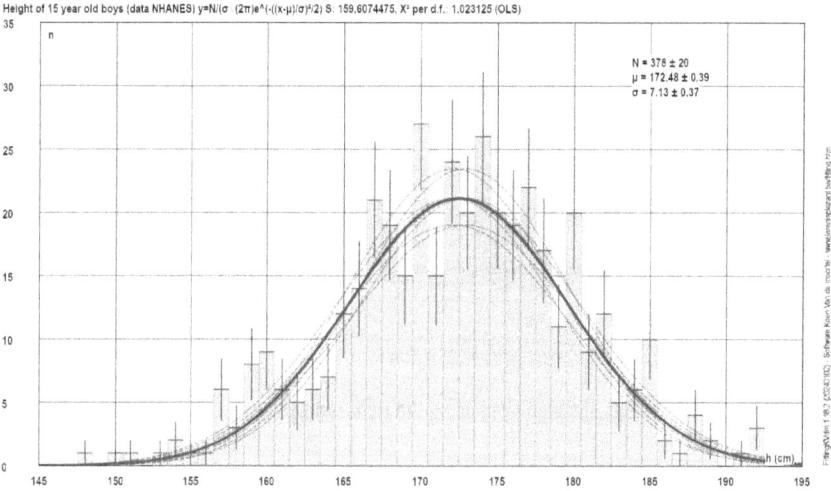

Height of 15 year old boys (data NHANES) y=N/(σ·(2π)e^(-((x-μ)/σ)²/2) S: 159.6074475, X² per d.f.: 1.023125 (OLS)

$$N = 378 \pm 20$$
$$\mu = 172.48 \pm 0.39$$
$$\sigma = 7.13 \pm 0.37$$

Fig. 4.139. Heights histogram and best fitting Gauss curve.

This curve doesn't "confirm" that the distribution is normal; it means only that *if* the heights are normally distributed, the most probable peak value $\mu = 172.48 \pm 0.39$ (the weighted sample average) and the weighted standard deviation of the sample $\sigma = 7.13 \pm 0.37$. The fit does look good though, $\chi_y^2 \approx 1.0$, so most error flags intersect the curve, as it should be.

An interesting other "power" of regression is that it can reveal hidden subgroups in a mixed dataset. For example, if we mix 12 and 18 year old boys together, it's still possible to find the characteristics of both groups, assuming they are both normally distributed, which is probable in this case.

Very important here is the right parameterization of the model function! Don't express the total function as $f(x) = f_{\mu 1}(x) + f_{\mu 2}(x)$ (being distributions around averages μ_1 and μ_2) but, for example, as $f(x) = f_{\mu 1}(x) + f_{\mu 1 + \delta}(x)$ or something like that, see p. 125. In the second form, you can keep δ locked above a specified minimal value, while in the first form, the two distributions might just merge together and converge to the same μ value.

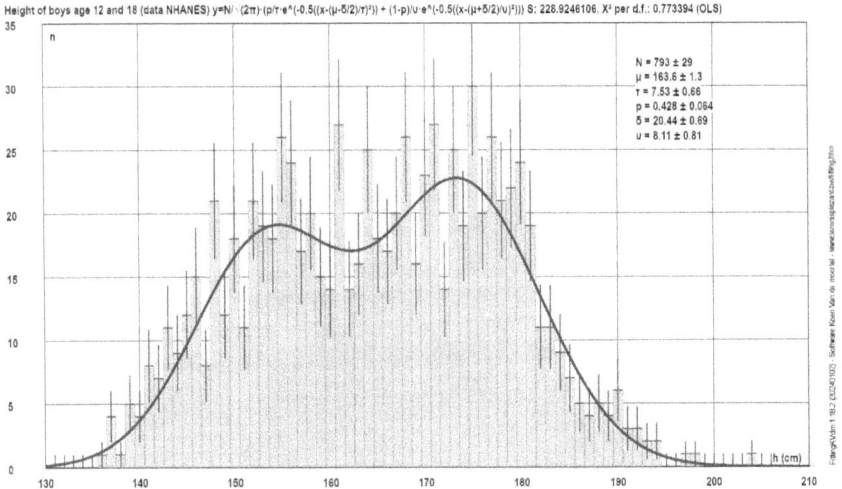

Height of boys age 12 and 18 (data NHANES) y=N/·(2π)·(p/τ·e^(-0.5((x-(μ-δ/2)/τ)²)) + (1-p)/u·e^(-0.5((x-(μ+δ/2)/u)²))) S: 228.9246106, X² per d.f.: 0.773394 (OLS)

N = 793 ± 29
μ = 163.6 ± 1.3
τ = 7.53 ± 0.66
p = 0.426 ± 0.064
δ = 20.44 ± 0.89
u = 8.11 ± 0.81

Fig. 4.140. Mix of two groups, distinguished using the right regression.

Fig. 4.140 shows the height histogram of 845 boys (415 of age 12, 430 of age 18), divided in bins of 1 cm. Even from these mixed data, you can derive that boys grow on average $\delta = 20.44 \pm 0.69$ cm between the age of 12 and 18.

Remark about this assumed "normality" of the distribution: if you *want* to see an asymmetry, you *can* see it! Just use another model, like the skewed peak I proposed on p. 38. Of course, as you can see in Fig. 4.141, this fits a little bit better since it has more parameters.

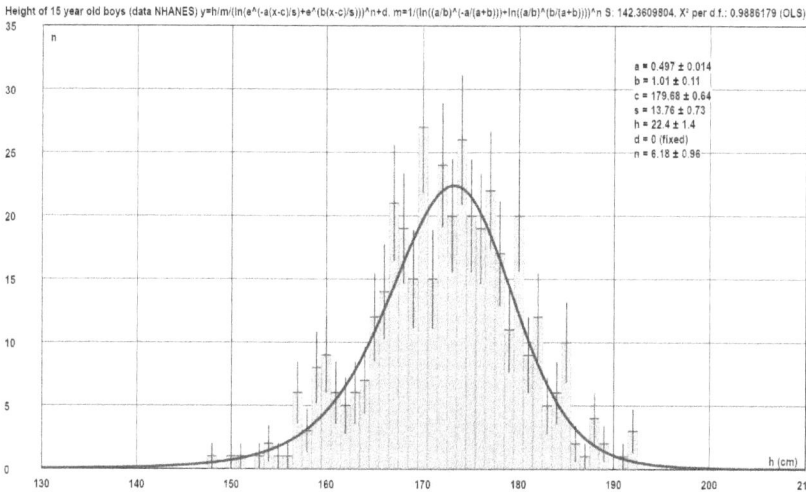

Fig. 4.141. Again the heights of 15 year old boys, but now with a skewed peak fitted.

FittingKVdm example data files: "Height_boys_15.dta1", "Height_boys_12_18.dta1".

4.3.9. Life expectancy versus healthcare spending

Will you live longer if you spend more on healthcare? That's a question that keeps many people busy: health care specialists, economists, insurance companies, etc.

And here you can find some interesting data, per country and per year: https://ourworldindata.org/us-life-expectancy-low.

If we look at Fig. 4.142, it's quite obvious that the health care expenses do matter. But the differences between countries are significant. For example, the highlighted USA is a bad performer: a horribly expensive healthcare system and yet a life expectancy way below the average of the industrialized countries. I suppose most people have an idea about the unhealthy lifestyle of Americans, which suggests an explanation for these observations...

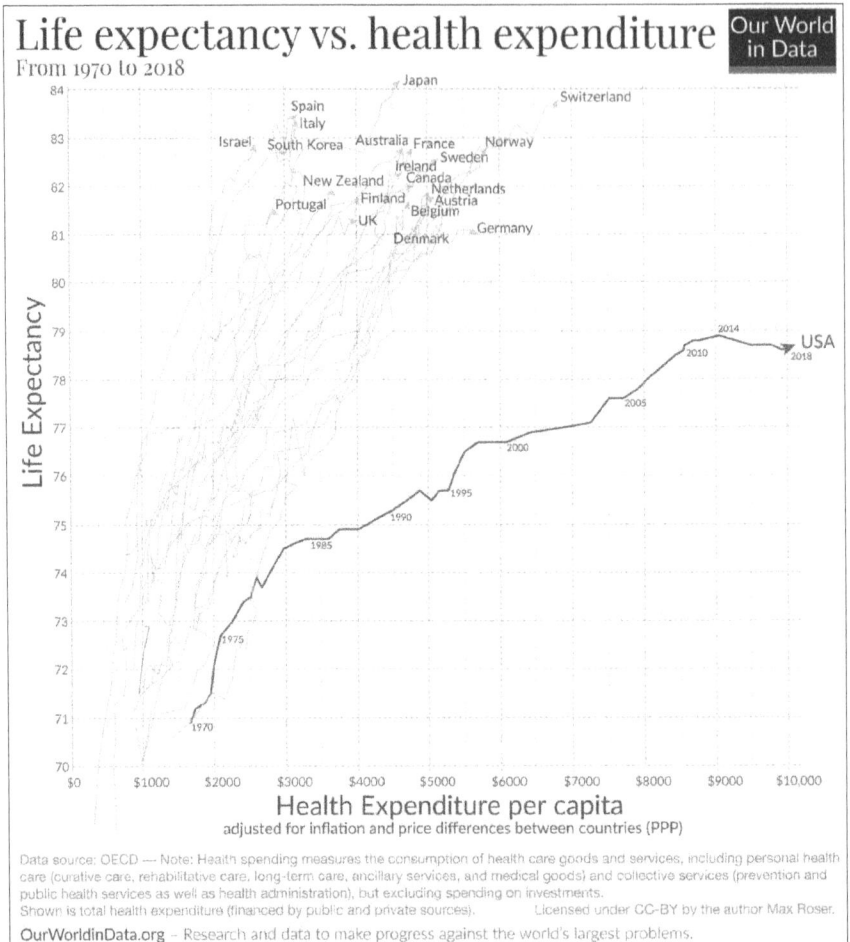

Fig. 4.142. *Source:* ourworldindata.org/us-life-expectancy)-low.

What kind of function could model this relationship?

Linear is out of the question; that would be way too simplistic. Any higher polynomial makes no sense too. Most probably it should be ascending all the time. Important questions to answer are as follows:

(1) Is there an upper limit to the life expectancy? I suppose there is, unless you consider "keeping someone in the freezer to revitalize him when a cure for his disease has been found" as a valid solution.

(2) Should the curve start at the origin? Definitely not, of course. People who lived a nice life without ever spending anything on medicines, they might be rare in "developed" countries, but they do exist in this world.

So, that limits our choices to rational, exponential, logistic and other sigmoid curves.

For your education, I selected data from one country: **Japan**. The variables here are x = health expenditure per capita (converted to 2010 intl. $) and y = life expectancy (years), 1970–2015.

Let's compare four models.

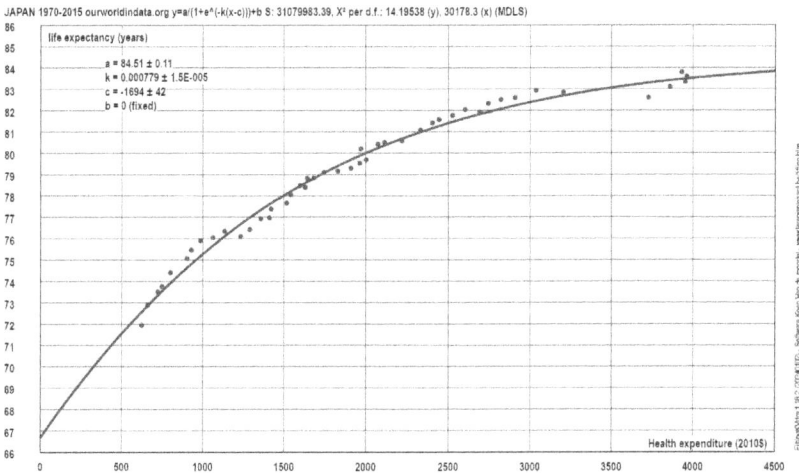

Fig. 4.143. Model 1 – logistic with baseline 0.

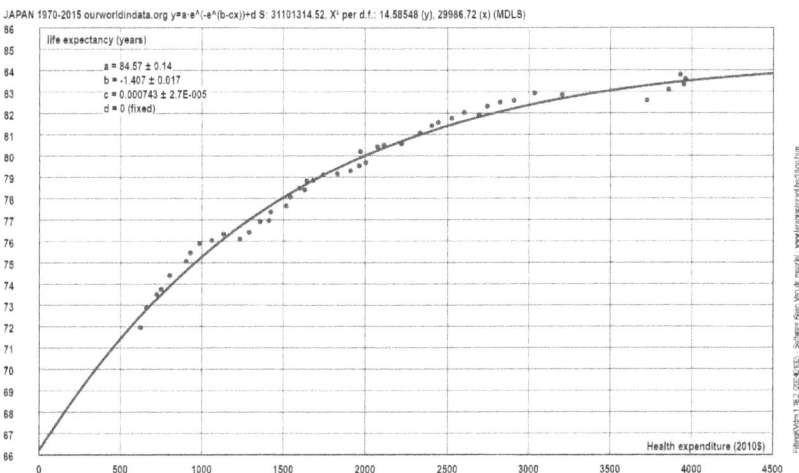

Fig. 4.144. Model 2 – Gompertz with baseline 0.

JAPAN 1970-2015 ourworldindata.org y=a(1-e^(-(x-c)/b)) S: 1.054405064E010, X² per d.f.: 18.44201 (y), 1.462456E007 (x) (MDLS)

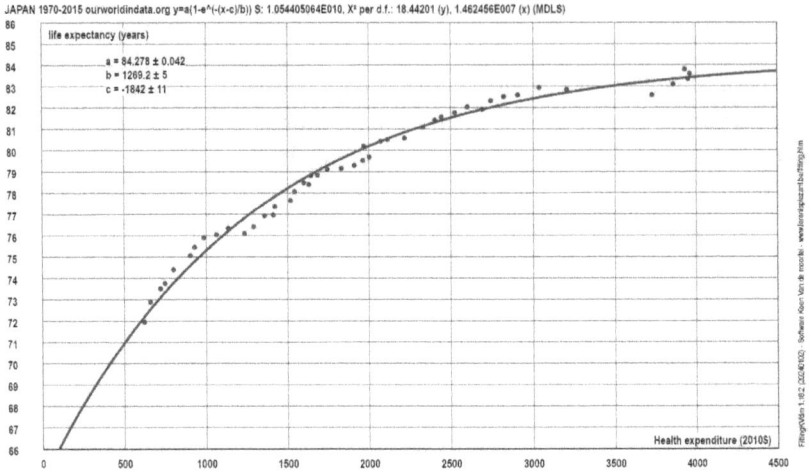

Fig. 4.145. Model 3 – plateau-exponential.

JAPAN 1970-2015 ourworldindata.org y=a-b/(x+c) S: 37603995.61, X² per d.f.: 22.1388 (y), 29815.8 (x) (MDLS)

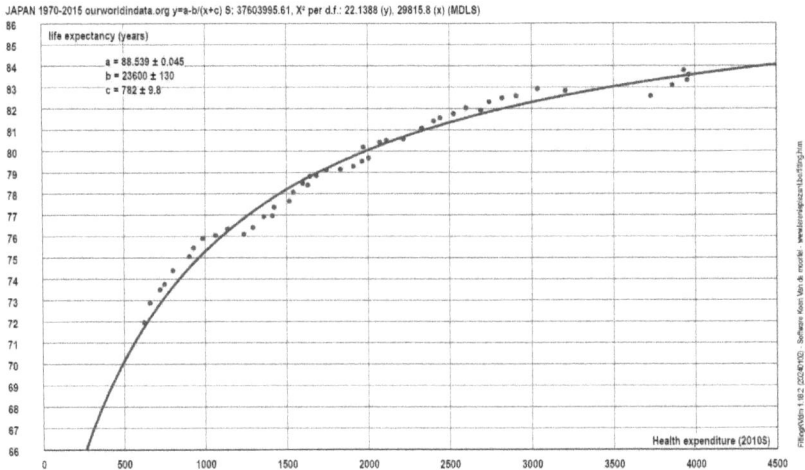

Fig. 4.146. Model 4 – rational.

All four models have three free parameters, so we can look at the S (total sum of squares) values to compare the goodness-of-fit. We are especially interested in their predictions for the age limit (the asymptotic y value for x going to infinity) and the age without any medical expenses ($f(0)$).

Model	*S*	*f(0)*	limit (HA)
Logistic (baseline 0), see p. 28, 34	$31.1 \cdot 10^9$	67	84.5
Gompertz growth (baseline 0), see p. 30	$31.1 \cdot 10^9$	66	84.6
Exponential, parameterization (4), see p. 24	$10.5 \cdot 10^9$	65	84.3
Rational, homographic parameterization (3), see p. 19	$37.6 \cdot 10^9$	58	88.5

One model clearly fits better (lowest *S*): the exponential, but does that mean it is the "best"? The one that gives the most credit to health care, is the rational: the promise for the longest life (88.5 years), and the shortest life if you don't go to the doctor (58.4 years), but it fits the worst!

Since there is no detailed theoretical explanation for the phenomenon we observe, except the obvious "yes, the doctor can help, but not forever", it's impossible to decide which model is the best. So what's the use of it? Well, at least, you might use one of them (maybe the best fitting) to compare different countries or groups with different diets as an indication of lifestyle quality: a high *f(0)* and a high limit are good signs, and also a steep curve, since that means high quality care for less money.

FittingKVdm example data files: "Life exp vs health spending CH.dta1", "Life exp vs health spending JAPAN.dta1", "Life exp vs health spending USA.dta1".

4.3.10. The chance to be alive or dead

One of the favorite questions of insurance companies is as follows: "What is the probability that a person is alive or dead at age x?". Not a very simple question, since a lot of influences play a role, like the health condition, diet, finances, social situation, etc. And Congo in the 16th century will most probably differ from Switzerland in the 21th century. But okay, you might ask the question for the average population of a given region in a given time.

How do we measure this? This is a tricky one! You might suggest to collect the life spans of a number of deceased people and study their distribution. This will certainly give you insights, but the sample is asymmetrical in time, if you want to know what the current probabilities are, at this moment in time. Mmm... another thing you could do, and this is a simple home or classroom experiment; collect this information from all the people you know or knew who were born alive: how old would each person be today (x), and is he/she alive ($y = 1$) or dead ($y = 0$)?

Try to be as precise as possible: you can estimate the age in years by using "1 month $\approx$ 1/12 of a year", but you can use a function that every spreadsheet program has, to calculate the number of days between two dates, and then divide that by 365.25 to get the age in years. You should collect a random sample with all kinds of ages, more or less evenly distributed!

Then, you could put them in bins like: aged 0–4.999..., 5–9.999..., 10–14.999..., etc. and calculate the percentage of people alive from each bin. The first values will be almost 100%, and the last ones will inevitably be 0%. Then you could connect the dots to see the pattern. Two disadvantages of this procedure are (1) you need quite a lot of people to get a more or less smooth pattern, and (2) you waste a lot of information if you have to divide the ages in such wide bins (you can make them narrower but then you will have many empty ones and the percentages will go up and down too often).
Instead of connecting the dots, you might also fit a simple function through the percentages. This will act like a global "smoothing filter".

Which function is suitable? Certainly not a linear one! It should start at 1 (100%) and slowly descend toward 0. That sounds like a sigmoid shape! The most commonly used is the "logistic" function (with fixed upper and lower limits of 1 and 0):

$$y = \frac{1}{1 - e^{-k(x-c)}} \qquad k < 0$$

Actually, it is often "mis-used" in a strange way, but it works: instead of fitting the curve to the percentages, the regression is done directly with the 0 and 1 values. Strange, because those values have no "measurement uncertainty", as opposed to the average percentages, that have a standard deviation, and... the curve cannot possibly "fit" with the dots. Unfortunately it's not possible to use MDLS in this case, since *all* the y values have asymptotic values.

The data I collected are the ages of 156 people I know/knew personally, at July 1, 2023, accurate to the day. To do the regression, I had to enter some artificial fake value like 0.001 as "measurement error" on y (small enough to make the error flags invisible). The actual values don't matter for the fitting: as long as they are the same, all the points get the same weight.

You can see in Fig. 4.147 that the curve follows more or less the 15 point-moving average line, which is nice of course. The χ^2 value doesn't mean anything here, because we just invented the y errors.

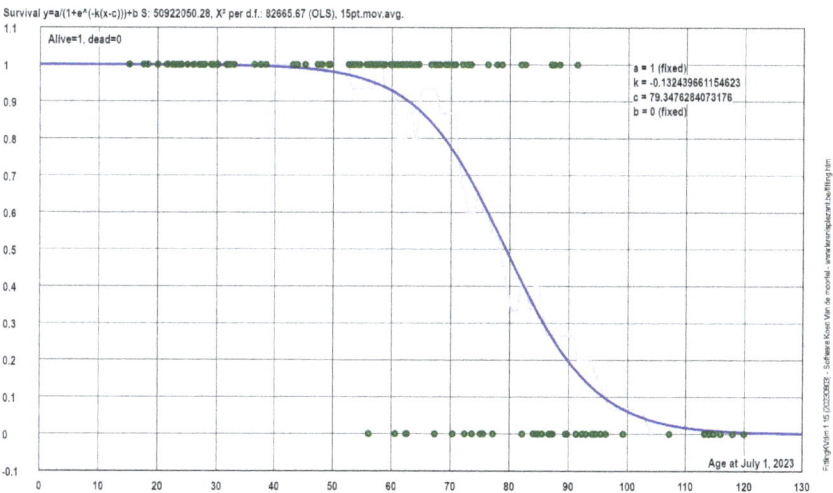

Fig. 4.147. "Dead or alive" data and the best fitting logistic function (dark blue). Light curve = 15pt. moving average.

It is usually *assumed* that $f(x)$ will give you "the probability of being alive after x years", but there is no "law" whatsoever dictating this. The curve is just an "idealized" moving average.

From this fitting, we can estimate that half of the people die before the age of 79.3 years. This is actually astonishing because we did not enter a single person's lifespan in the program!

There are other models like "Weibull decay" (with upper limit $a = 1$) that also fit well, as you can see in Fig. 4.148!

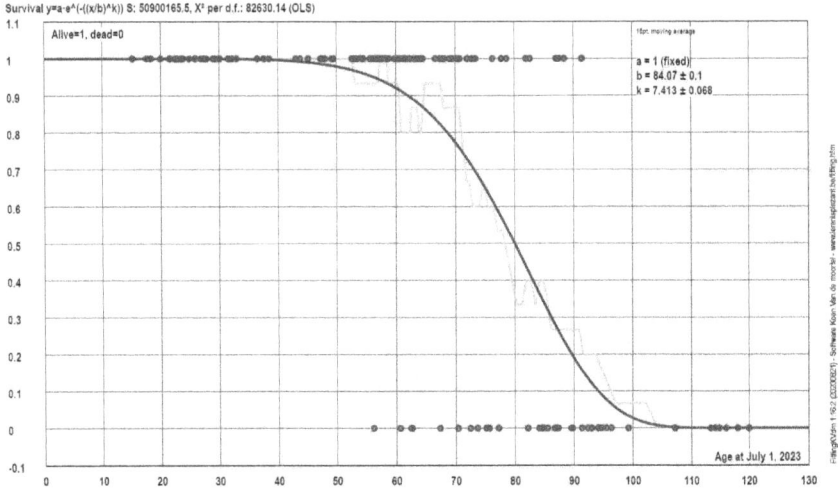

Fig. 4.148. "Dead or alive" data, the best fitting Weibull decay function and 15pt. moving average (lighter curve).

Or, the "Transition" model (see p. 32, 32, with $a = -0.5$ and $m = 0.5$), also looks plausible, as you see in Fig. 4.149.

And also the Gompertz model (with lower and upper limits fixed to 0 and 1) might work, see Fig. 4.150.

Survival y=a(x-b)/((x-b)^k+c^k)^(1/k)+m S: 50900794.29, X² per d.f.: 83171.23 (OLS), 15pt.mov.avg.

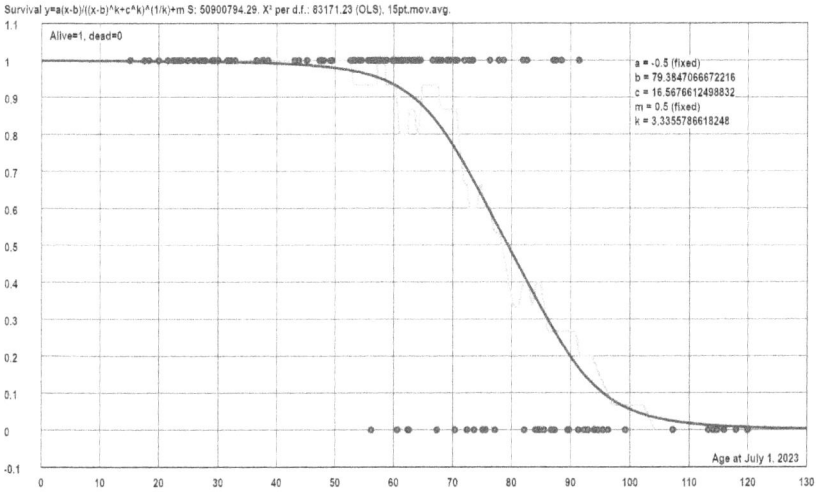

Fig. 4.149. "Dead or alive" data, the best fitting "transition" function and 15pt. moving average (lighter curve).

Survival y=a·e^(-e^(b-cx))+d S: 51051148.09, X² per d.f.: 82875.24 (OLS)

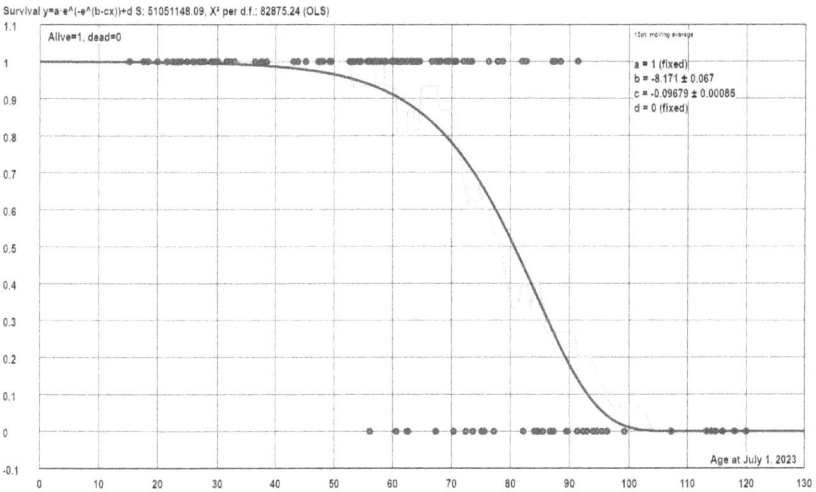

Fig. 4.150. "Dead or alive" data, the best fitting "Gompertz" function and 15pt. moving average (lighter curve).

All four models cross the 50% line around the age of 79–81. It's quite impossible to distinguish which one is the "best fit" because we have only two possible y values! We could compare their numbers of "correct predictions" ("$f(x_i) < 0.5$ and person nr. i *was* actually dead" is a good prediction), but that wouldn't make much difference, certainly not to find out which model is usually best for this kind of data! For that, we would need tons of data.

Actually, we could use χ^2 here for comparison, not their meaningless absolute values but their differences.

Model	χ^2
Logistic	82666
Weibull	82630
Transition	83171
Gompertz	82875

The differences are minimal, but Weibull comes out slightly better. Besides that, this model also has at least one theoretical advantage: it's the only one that has $f(0) = 1$!

FittingKVdm example data file: "Survival_friends-family.dta1".

Fortunately, it is possible to find large datasets related to life expectancy, if you look around on your national statistics website. For Belgium, I found one at
https://statbel.fgov.be/nl/themas/bevolking/sterfte-en-levensverwachting/sterftetafels-en-levensverwachting#figures.

The variables here are x = age (years) and y = the average number of men that will survive until that age (Belgium, 2020). No standard deviations were available, so the "errors" on x and y were simply set to 0.5 and 1, which will cause the χ^2 values to be unrealistically high, but the influence on the parameters is most probably very low since the number in each age bin is high anyway, so the "noise" is low.

Again, we can try the same models as above, e.g., "Logistic" (with fixed $a = 1000000$ and $b = 0$) and "Transition" (with fixed central value $m = 500000$ and "amplitude" $a = 500000$). But they don't really follow the pattern. The Weibull decay function fits reasonably, as you can see in Fig. 4.151.

The Gompertz model though (Fig. 4.152), seems to fit almost miraculously. χ^2 is a lot less: $6.2 \cdot 10^7$ instead of $3.0 \cdot 10^8$.

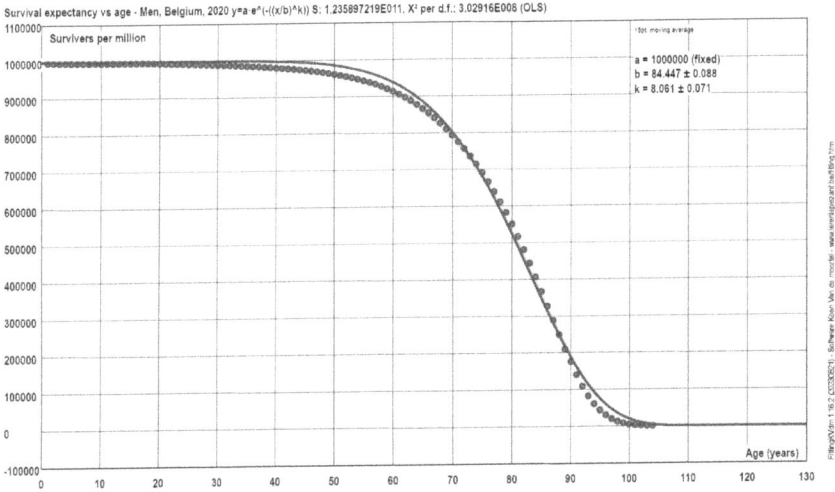

Fig. 4.151. Belgian men survival data, the best fitting Weibull function and 15pt. moving average (lighter curve).

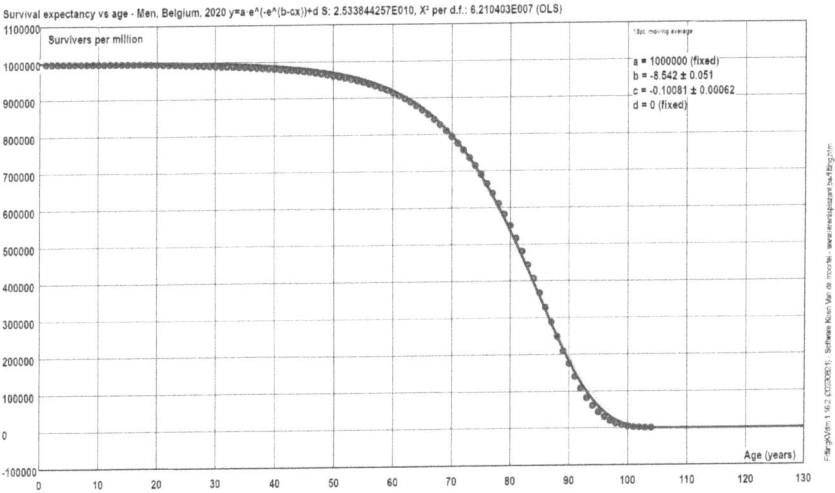

Fig. 4.152. Belgian men survival data, the best fitting Gompertz function and 15pt. moving average (lighter curve).

Does this "prove" anything about the "mechanisms" how people die? I guess not. But for demographers and insurance companies, it can be a nice tool to make estimations.

Also, medical researchers might do this regression for different groups (men/women, smokers/non-smokers, vegetarians/carnivores, rich vs.

poor people, different genetic groups, etc.) to gain insight about how these factors influence one's life expectancy.

Remark: For those who are still tempted to try *polynomials* for approximating such special data patterns...

See what happens in Fig. 4.153: as long as you are in the "inside" area of the *x* range, the fit seems not so bad, but once you try to extrapolate, the absurdness becomes obvious!

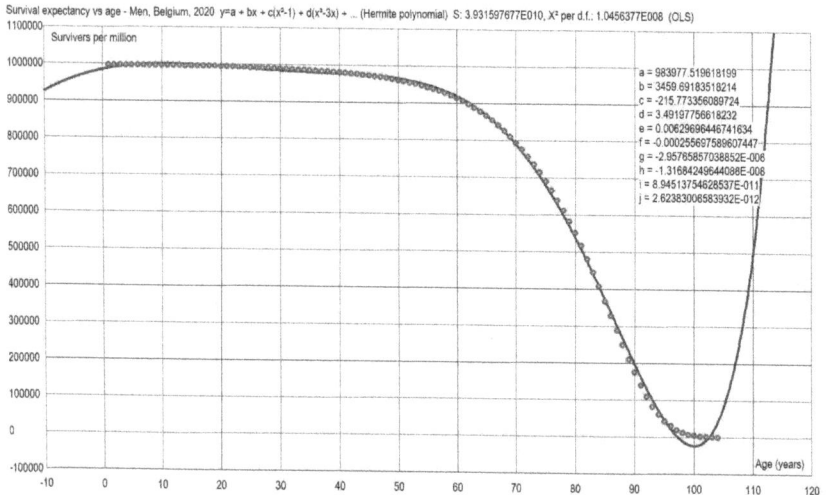

Survival expectancy vs age - Men, Belgium, 2020 y=a + bx + c(x²-1) + d(x³-3x) + ... (Hermite polynomial) S: 3.931597677E010, X² per d.f.: 1.0456377E008 (OLS)

a = 983977.519618199
b = 3459.69183518214
c = -215.773356089724
d = 3.49197756618232
e = 0.00629696446741634
f = -0.000255697589607447
g = -2.95765857038852E-006
h = -1.31684249644088E-008
i = 8.94513754628537E-011
j = 2.62383006583932E-012

Fig. 4.153. Belgian men survival data and the best fitting 9th degree polynomial. DON'T do this!

FittingKVdm example data file: "Survival chances Men Belgium 2020.dta1".

4.3.11. Measuring the effect of fertilization

4.3.11.1 Potassium and cotton

Farmers use fertilizers to obtain a bigger crop production (at least in quantity; whether it is also in quality is another matter), for example, potassium to get more cotton.

Some data can be found in an article by Morteza Mozaffari (University of Arkansas, USA) (June 2018), "Cotton and Soil Responses to Annual Potassium Fertilization Rate", *Agricultural Sciences*, 9(6). You can read it at www.scirp.org/journal/paperinformation.aspx?paperid=85771.

I put these data into FittingKVdm; the variables here are x = cumulative (3-year) fertilization with potassium (K) rate (in kg/ha) and y = total 3-year yield (kg/ha) of cotton on this field. The $\sigma_{x,i}$ and $\sigma_{y,i}$ values were estimated here only from the measurement precision, not taking into account the yearly variability, so the estimation of the precision the parameters might be too optimistic.

Which model function could describe the relationship?
One thing is for sure, $f(0) > 0$, since plants grow also when you add nothing to the soil. Probably there is a limit, and adding more fertilizer won't add any more profit. Even more probable is that there is a maximum since using way too much might damage the crops. But to detect that, we would need more data, so let's limit ourselves to some simple models with a horizontal asymptote that start above the origin: the homographic (see p. 19) and the logistic (with baseline zero, see p. 28, 34). The logistic might make the most sense if extrapolated on the negative side since that one goes to zero. But to check that out, we would have to see what happens if we *extract* potassium from the soil...

Both models give very similar predictions (a maximal 3 year yield around 12000 kg/ha and about 8800 without fertilizer) and the sum of squares is almost the same:

Model	$f(0)$	Limit	S
Homographic	8757	12097	$5.2 \cdot 10^{10}$
Logistic	8821	11704	$5.5 \cdot 10^{10}$

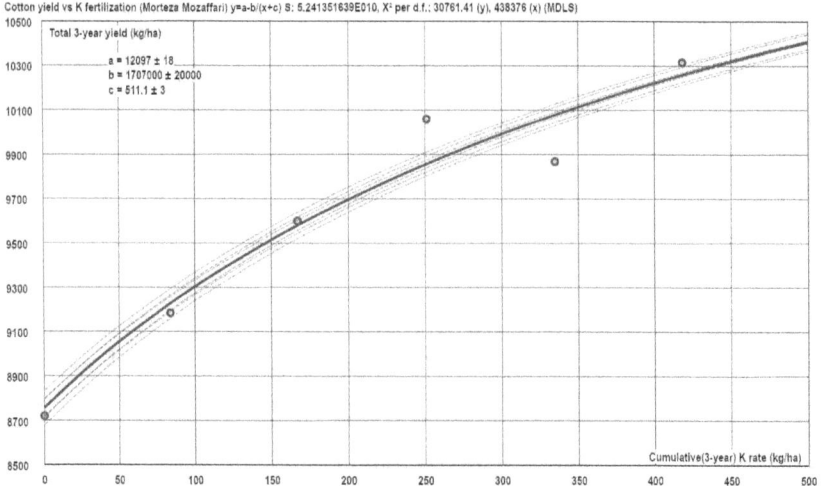

Fig. 4.154. Cotton yield vs. K usage, and the best fitting homographic function.

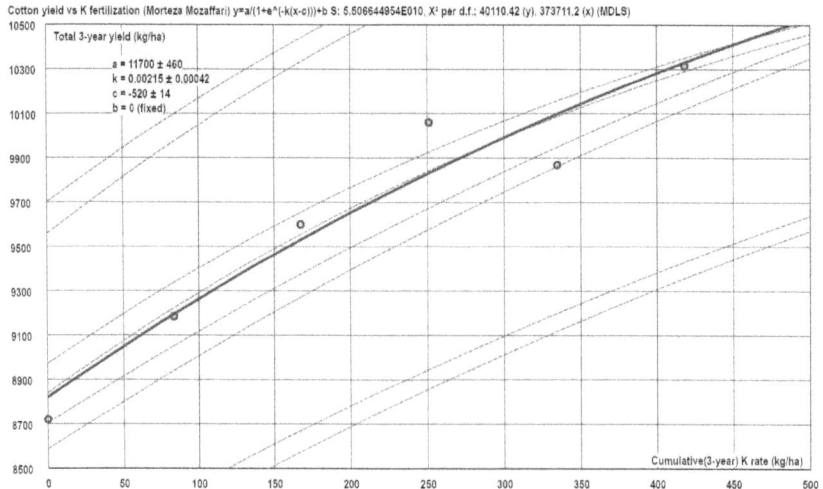

Fig. 4.155. Cotton yield vs K usage, and the best fitting logistic function with baseline 0.

But, as you can see from the dotted lines in the graph, the parameters in the logistic model are much more sensitive to small changes in the data. In other words, the homographic model is much more robust here!

Note 1: The convergence is very slow here because the points are too much aligned. It would have been better to do at least one more measurement with more fertilizer.

Note 2: I also tried an exponential model but that didn't fit at all.

With more complex models (more than three parameters), the risk for overfitting is way too big since we only have six data points.

4.3.11.2 Nitrogen and cereals

How much nitrogen fertilizer is used in different countries, and how does this compare with their cereal yield? That information can be found at
https://ourworldindata.org/grapher/cereal-crop-yield-vs-fertilizer-application.

I extracted data from 148 countries. The numbers differ every year, just like the weather, so I calculated the averages and standard deviations of x = applied amount of nitrogen fertilizer (kg/ha) and y = cereal yield (tonnes/ha).

The same models can be applied, but with these data the scatterplot looks more "cloudy" of course, since the soil, the climate, the techniques, the seed quality, diseases, etc. also influence the yield, besides the fertilization. That's why OLS had to be used here: there is no simple invertible causal relationship between x and y, like in the previous example where only one quantity was varied.

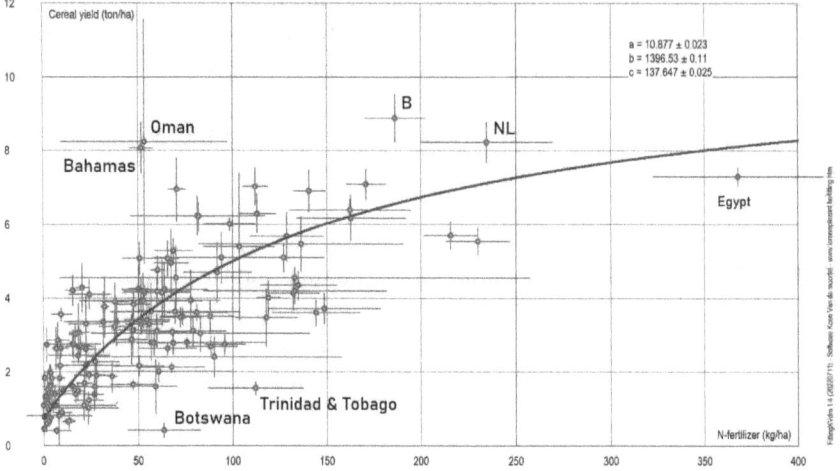

Fig. 4.156. Nitrogen fertilizer used in different countries, and their yield. Curve = best fitting homographic function.

Conclusions that can be made are as follows: points above the curve represent countries that are doing well (Belgium, the Netherlands, but especially Oman and the Bahamas since they need much less fertilizer). Egypt has good results but uses way too much fertilizer which doesn't look very efficient. The lowest points are countries that perform very badly, e.g., Botswana or Trinidad and Tobago. So, this model can be a starting point to investigate how things can be improved.

FittingKVdm example data files: "Cotton yield vs K conc Morteza Mozaffari.dta1", "Cereal-crop-yield-vs-fertilizer-per-country.dta1".

4.3.12. Tumor growth

How fast does a tumor grow? Can it be predicted how big it will become? Can we detect if a treatment has any good influence?

Anything that has enough space and resources tends to grow exponentially. The space for a tumor, however, is limited because it grows somewhere in a living body. According to people in this field, this kind of growth can be described quite accurately with the Gompertz model function (without baseline):

$$f(x) = a \cdot e^{-e^{b-cx}}$$

Parameter a describes the maximal size the tumor will get and c the speed of the growth.

I checked it out, with data measured in mice, from Constantine Daskalakis (2016), *Tumor Growth Dataset*, TSHS Resources Portal, case 101 (no treatment): www.causeweb.org/tshs/tumor-growth.
The variables here were x = the time in days (I assumed that the measurements were done more or less at the same time of the day, so ± 0.1) and y = the size of the tumor in mm³ (assumed ± 0.1 since that was the least significant digit).

Since the limit value (the finally expected tumor size) was far above the last measurement, MDLS could be used, but carefully, by entering a starting value for parameter a, higher than the measurements (e.g., 1800). Anyway, OLS and MDLS don't differ very much in this case.

The Gompertz model predicts a final size of 2013 ± 17 mm³, see Fig. 4.157.

Out of curiosity, I also tried the logistic model (also starting from zero), see Fig. 4.158. This actually fitted even better (smaller S: $6.06 \cdot 10^6$ instead of $8.19 \cdot 10^6$). But it predicts a smaller final size (1774.6 ± 6.6 mm³).

Which model is the best to depict this phenomenon? The best way to find that out is probably empirical, by repeatedly comparing predicted and observed values.

Mouse tumor growth - C.Daskalakis case 101 y=a·e^(-e^(b-cx))+d S: 818676806.2, X² per d.f.: 403362.6 (y), 132.9584 (x) (MDLS)

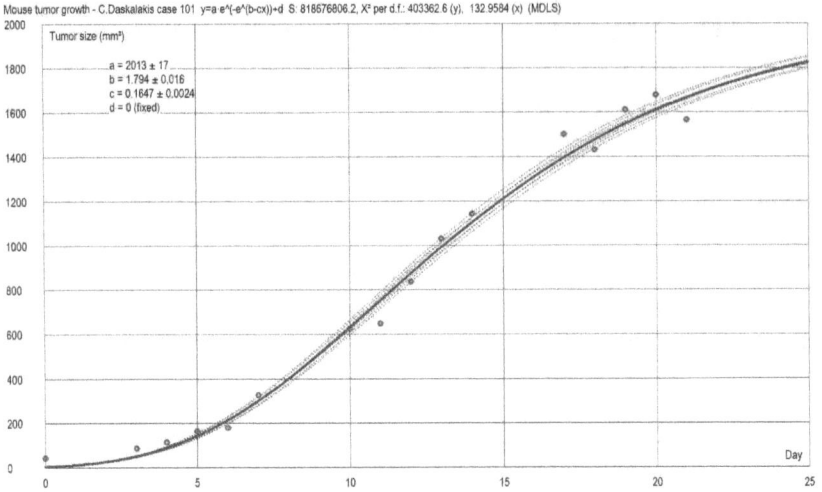

Fig. 4.157. Tumor growth in a mouse, and a Gompertz function describing it.

Mouse tumor growth - C.Daskalakis case 101 y=a/(1+e^(-k(x-c)))+b S: 606524799.6, X² per d.f.: 307300 (y), 90.93702 (x) (MDLS)

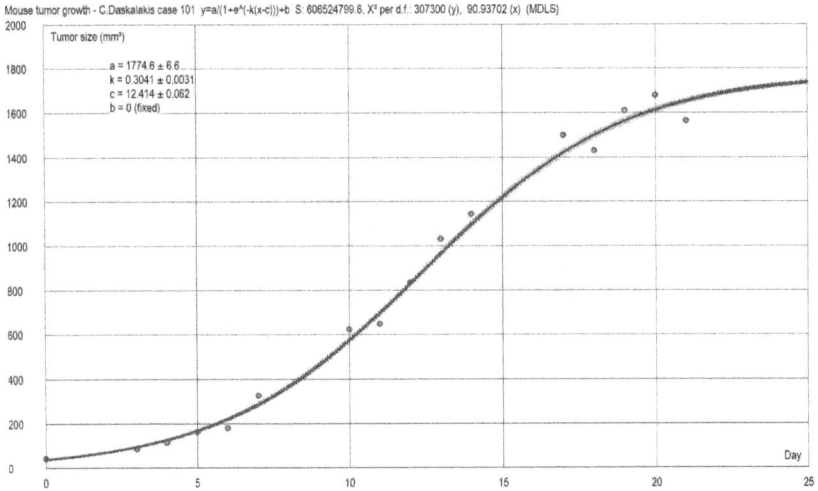

Fig. 4.158. The same data but with the best fitting logistic function.

FittingKVdm example data file: "Tumor growth Daskalakis 101.dta1".

4.3.13. Predicting liver disorders

Medical researchers try to predict health problems from blood tests, for example, liver disorders by measuring the presence of alanine amino-transferase (ALT) and glutamyltransferase (GGT). This is mentioned on the Wikipedia page about "robust regression": https://en.wikipedia.org/wiki/Robust_regression (Dec. 2023) and it refers to the following article:

Fraser, A. *et al.* (Nov. 2009), "Alanine aminotransferase, gammagluta-myltransferase (GGT) and all-cause mortality: results from a population-based Danish twins study alanine aminotransferase, GGT and mortality in elderly twins.", *Liver Int.* 29(10):1494-9, which can be read at www.ncbi.nlm.nih.gov/pmc/articles/PMC3633107/.

The point that the Wikipedia authors want to make is that there is a correlation between the two markers (ALT and GGT) and that some "robust" regression method (called MM estimation, but without further explanation, checked 12 Feb. 2024) would be better than OLS. A graph is shown with regression lines for the log(ALT) vs. log(GGT) data, so the expected relationship between ALT and GGT must be a power function. This intrigued me, and I downloaded the data, measured by "The British United Provident Association Limited" (BUPA), from the site https://web.archive.org/web/20171023174701/http://ftp.ics.uci.edu:80/pub/machine-learning-databases/liver-disorders/, and put them in FittingKVdm.

Strangely, no units were mentioned in the dataset, but I assume it must be concentrations, some quantity per liter. All the numbers were rounded to integers; I assumed errors of ± 1 by lack of given confidence intervals.

First, I did the fitting using OLS, with x = ALT, y = GGT. This yielded $y = 7.687 \cdot x^{0.4019}$.

The Wikipedia page mentions an exponent of 0.420 instead of 0.4019, which is probably because they took the logarithms. The "robust" method gave 0.373.

The data look very scattered, see Fig. 4.159, Kendall τ is only 0.40!

BUPA liver disorders y=ax^b S: 91444.35618, X² per d.f.: 269.7474 (OLS)

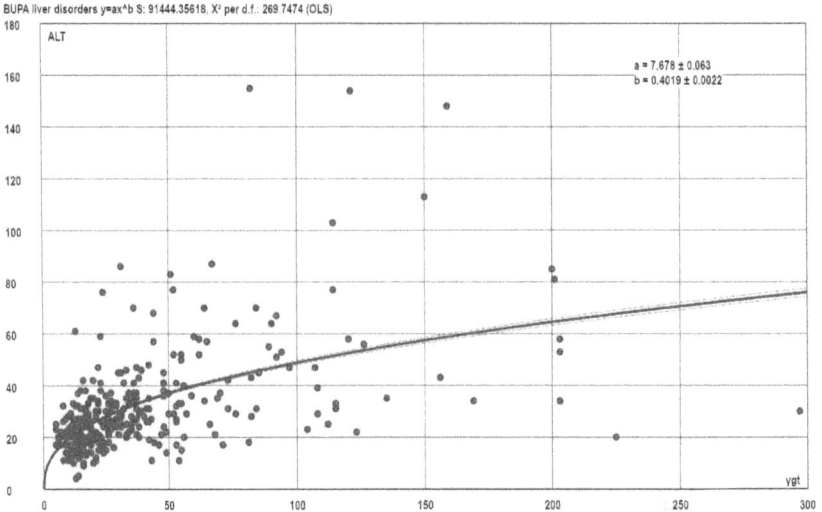

Fig. 4.159. ALT vs. GGT marker concentrations, and the best fitting power function using OLS.

By switching the variables, we would expect an exponent of $1/0.4019 \approx 2.488$.
But... fitting gives $x = 2.352 \cdot y^{0.8262}$. This huge difference was to be expected since the correlation is so bad.

BUPA liver disorders y=ax^b S: 391898.8934, X² per d.f.: 1156.044 (OLS)

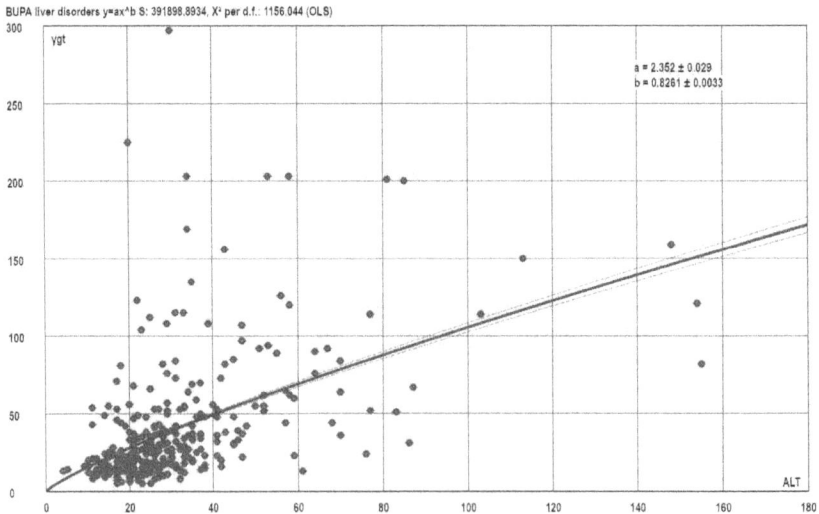

Fig. 4.160. The same data but now the OLS fitting is done with switched variables.

You can clearly see the fitted curves in the previous plots are not each other's mirror images, contrary to what one would logically think, but you know already this is due to the asymmetry of OLS.

MDLS produces $y = 0.970 \cdot x^{1.1395}$, very different from the OLS fitting, and now if you switch the variables, you will get just the same but mirrored!

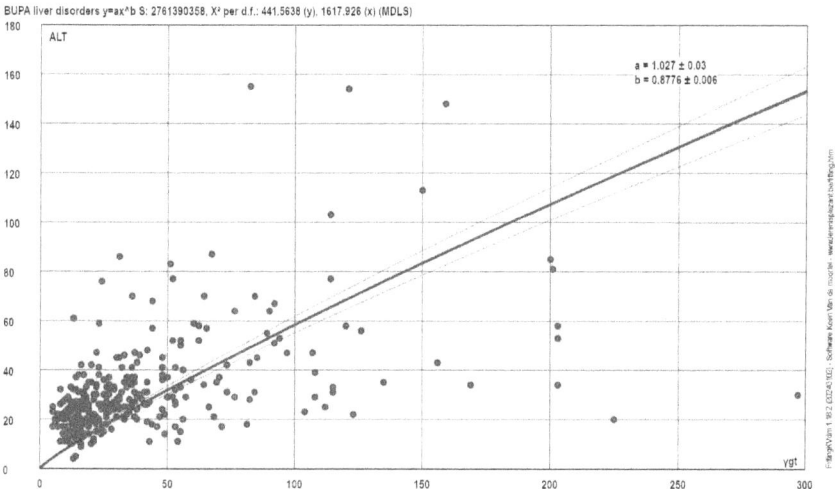

Fig. 4.161. The same data but now fitted with MDLS.

I suppose the reason why they calculated this correlation was to see if one of both markers (GGT or ALT) would be sufficient to use as a predictor. My conclusion would be no! The relationship between the two is clearly too fuzzy!

FittingKVdm example data files: "BUPA ALT vs GGT.dta1", "BUPA GGT vs ALT.dta1".

4.3.14. Death by microwaves

Everyone knows you should not put your hand in a working microwave oven, but what happens if you are exposed to regular small doses of microwave radiation with the same frequency as in that oven (2.45 GHz)?

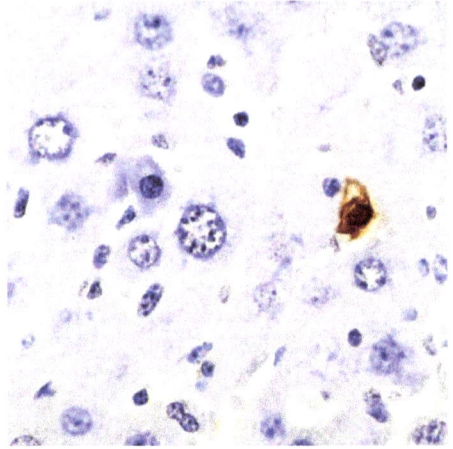

Fig. 4.162. Tunel stained cell in a mouse liver. (Image: Wikimedia commons, by Laboratory of Experimental Pathology, Division of Intramural Research, NIEHS).

This is not a simple home or classroom experiment; it requires a lot of patience and complex procedures with high-tech equipment. A group of Turkish researchers put six groups of eight rats to the test. The first was a control group, not exposed to any electromagnetic field, and the others were exposed to increasing strengths of continuous microwaves for 1 hour per day during 45 days. Then the unfortunate animals were sacrificed for science...

One of the measurements that were made was counting the number of dead cells in slices of bone tissue, using the so-called "Terminal deoxynucleotidyl transferase dUTP nick end labeling" method (short: TUNEL) that detects DNA fragmentation from aptosis (programmed cell death).

Their data:

(fortunately with error margins, except for the "zero" field of the control group)

Field strength (V/m)	Tunel count
0 ± 0.001?? (control group)	2.62 ± 1.4
0.65 ± 0.11	3.00 ± 1.85
1.95 ± 0.25	17.00 ± 3.46
4.97 ± 0.32	18.5 ± 2.44
10.26 ± 0.52	23.5 ± 2.44
15.4 ± 0.6	34.5 ± 4.03

Obviously, the radiation had an adverse effect on the health of the animals: the higher the radiation, the more damage was visible, even if the field strength was much lower than that of a microwave oven (250 V/m or more) and a significant thermal effect could be excluded.

Can we mathematically model what happens here? The pattern is not very simple, but two things are sure:
(1)   The model function should start above the origin since cell aptosis occurs all the time, even in healthy animals, and
(2)   I suppose there has to be an upper limit, namely, when all the cells are dead.

So, that excludes a linear model! A homographic function might be tried.

As you can see in the graph (Fig. 4.163), the fit is far from perfect, but we risk overfitting if we use models with more parameters. To improve the model, more data would be needed, but that would require a lot of expensive work in this case. And the most important goal was achieved: to find the level of radiation that is harmful (significantly different from the control group): about 2 V/m.

Since WiFi signals are often in the same frequency range (2.4 GHz) and stronger than 2 V/m if you are close to the modem in your house or to your mobile phone (5–15 V/m easily), this result should make us a bit concerned about possible adverse effects on humans, especially since most people nowadays are continuously exposed. According to the European Commission, everything below 61 V/m is "safe", and on the Environmental Protection Agency website, we still read the following: "There is no scientific evidence that exposure to low levels of electromagnetic fields (EMF) of any frequency causes damage to human health." [EPA 2024]. Maybe this site needs to be updated?

Cell aptosis vs MW radiation in rats (Karadayi et al. 2024) y=a-b/(x+c) S: 351.6797733, X² per d.f.: 2.954205 (y), 143.0543 (x) (MDLS)

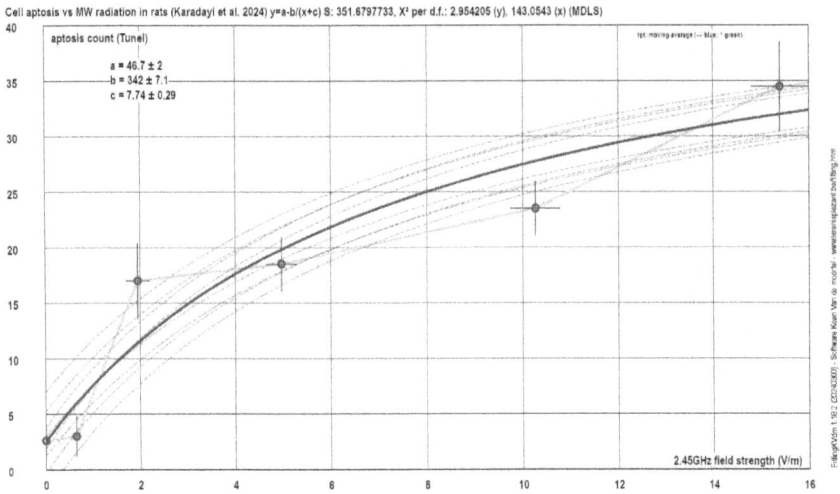

Fig. 4.163. Dead bone cells in rats versus the received microwave strength, and best fitting homographic function.

References:

- Karadayi, A. *et al.* (Mar 2024), "Does Microwave Exposure at Different Doses in the Pre/Postnatal Period Affect Growing Rat Bone Development?", *Physiol Res.* 11; 73(1):157-172. See: www.biomed.cas.cz/physiolres/pdf/73/73_157.pdf

- Konduru, Jahnavi: *TUNEL Staining: The method of choice for measuring cell death*, www.assaygenie.com/blog/tunel-staining.

- EPA (consulted 21 March 2024): *Wi-Fi, Smart Meters & Your Health* www.epa.ie/environment-and-you/radiation/emf/emf-and-your-health/wifi-and-your-health/.

FittingKVdm example data file: "Cell aptosis vs microwave field strength with rats (Karadayi 2024).dta1".

4.3.15. The Canadian lynx population cycle

Foxes eat rabbits, so if there are many rabbits, the foxes will have enough to eat and grow an offspring successfully. But then, many rabbits will be eaten and eventually the foxes will starve, and that in turn will give the rabbits a break to get their population rising again, etc. This way, cycles in predator and prey populations are caused in many biotopes.

Such a phenomenon was observed by the fur trading Hudson company in their McKenzie River district from 1821 to 1934. The number of Canadian lynx pelts they traded was clearly going up and down, and that can be seen as an indicator for the total lynx population.
Their data can be found in this article by Charles Elton and Mary Nicholson (1942), "The Ten-Year Cycle in Numbers of the Lynx in Canada.", *Journal of Animal Ecology*, 11(2), pp. 215–44. JSTOR. It can be downloaded at https://doi.org/10.2307/1358 or https://jxshix.people.wm.edu/2009-harbin-course/classic/Elton-1942-J-Anim-Ecol.pdf.

Since they are counts from a sample, we can use the square root of each number as an estimation of the uncertainty.

A periodicity of almost 10 years is clearly visible in the scatterplot, but it doesn't look like a simple wave. Adding just one harmonic (a sine with half the period, see p. 48) produces a somewhat better fit (Fig. 4.164), but the model does not explain the varying peak heights. Maybe there is actually also a cycle of almost 40 years.

Also the "periodic peaks" model (see p. 49) fits reasonably (Fig. 4.165). The difference is that the peaks are symmetrical and the curve is a bit flatter between the peaks.

Both models find a period of 9.626 ± 0.03 years and are able to predict the minima and maxima quite well. The second one "sees" higher peaks, which is more realistic, and it reveals the baseline easier: on average, 106 ± 27 pelts were sold in the worst years. That figure is less easy to deduce from the first model.

Canadian lynx pelts traded (MacKenzie River, Hudson company records) y = m + A·sin(2π(x-c)/T) + B·sin(4π(x-d)/T) + C·sin(6π(x-e)/T)+ D·sin(8π(x-f)/T) S: 226626.2488, X² per d.f.: 524.6

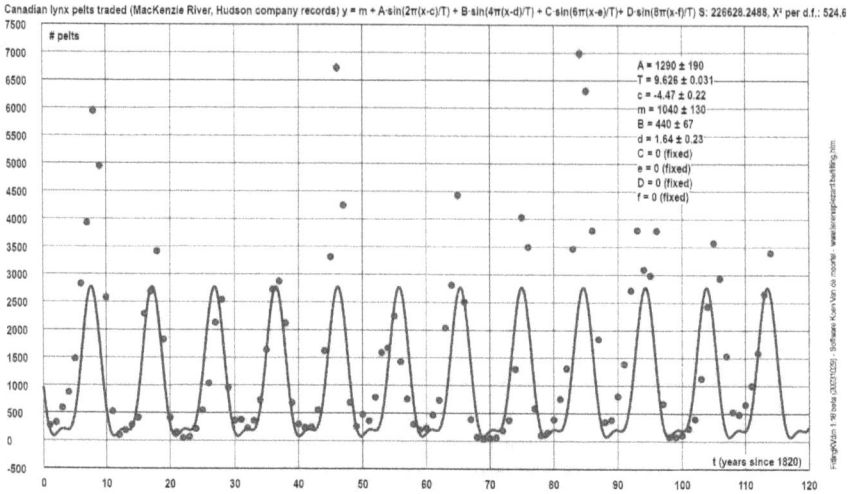

Fig. 4.164. Number of Canadian lynxes traded, and the best fitting sine function with one harmonic.

Canadian lynx pelts traded (MacKenzie River, Hudson company records) y = A·(k(k+1)/(sin²(π(x-c)/T)+k)-k)+m S: 223851.8008, X² per d.f.: 513.4216 (OLS)

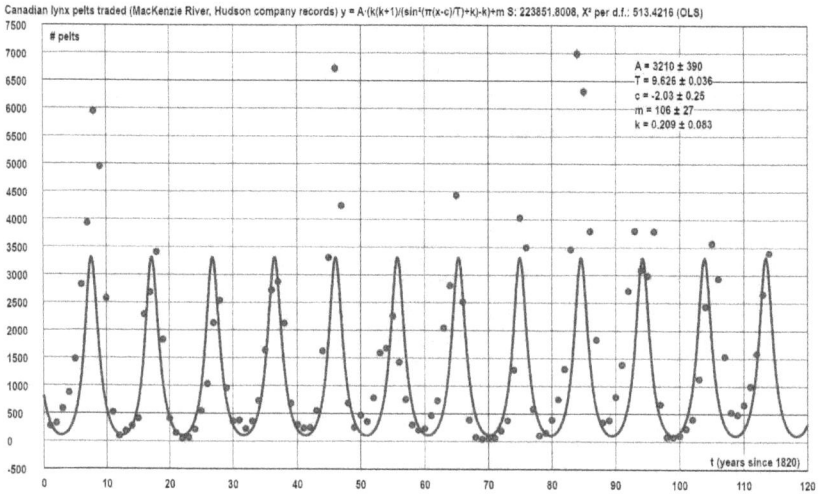

Fig. 4.165. The same data but with a "periodic peaks" function fitted.

FittingKVdm example data file: "Lynx.dta1".

4.4. Psychology, etc.

4.4.1. The points of the Eurovision Song Contest

I am not a fan of the annual Eurovision Song Contest, but the 2022 edition inspired me to do a bit of statistics. Why? The points given by the jury (J) and those from the viewers (televoting, T) were often *very* different (*Source*: https://eurovisionworld.com/eurovision/2022):

	UA	GB	E	S	SRB	I	MD	GR	P	NL	NL	PLST	EST	LUS	AZ	AZH	CHO	RO	B	AM	SMF	CZ	ISL	F	D
J	192	283	231	258	87	158	14	158	171	36	129	46	43	35	123	103	78	12	59	40	12	33	10	9	0
T	439	183	228	180	225	110	239	57	36	146	42	105	98	93	2	3	0	53	5	21	26	5	10	8	6

To check if there was any correlation at all, I calculated the Kendall τ with www.gigacalculator.com/calculators/correlation-coefficient-calculator.php. This produced: $\tau = 0.278$, so yes, there is still a weak positive correlation ($p = 0.26$).

So, there should be a way to estimate J from T and vice versa, and the simplest way to do this is by linear regression. Now, do we take J as the "independent" variable and T as the "dependent", or vice versa? It shouldn't matter, should it? Let's see, using OLS...
Choosing J as the independent, we get
$T \approx 0.6016J + 36.96$, or, by simple rearranging,
$J \approx 1.662T - 61.436$.

But if we do the regression with T as the independent variable, OLS produces
$J \approx 0.3667T + 58.77$, which is dramatically different!

With MDLS (with $\sigma_x = \sigma_y = 0.5$), we get: $T \approx 1.420J - 27.2$.

And by doing the regression with the switched variables
$J \approx 0.7067T + 19.2$, which is, of course, the same as what we get from

rearranging T vs. J, except for a small rounding error, because of symmetry in the algorithm.

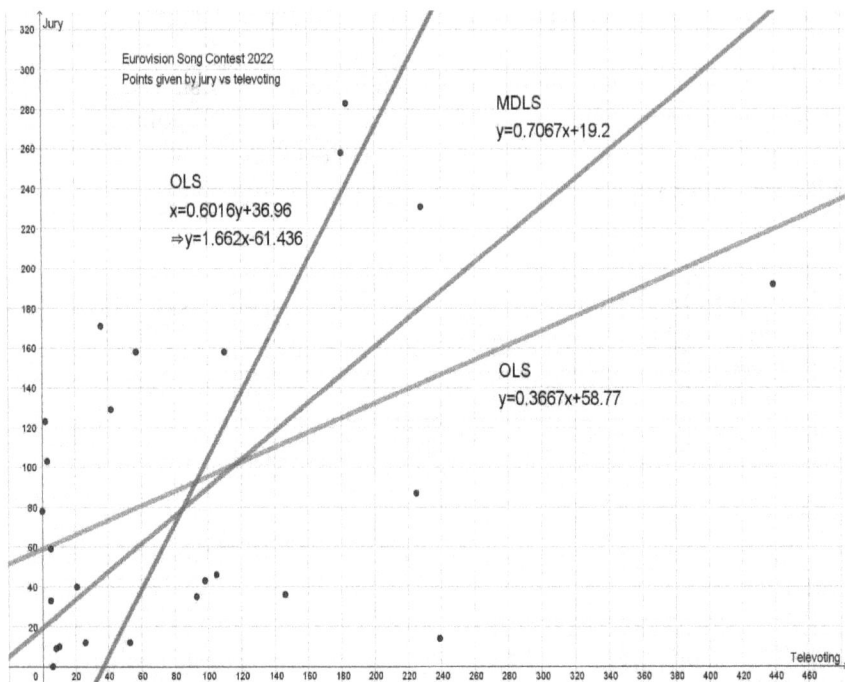

Fig. 4.166. Jury vs. televoting points, Eurovision Song Contest 2022, and "best" fitting lines.

This illustrates again the advantage of using MDLS, especially with very dispersed data like these!

FittingKVdm example data files:
"Eurovision Song Contest 2022 jury vs tele.dta1",
"Eurovision Song Contest 2022 tele vs jury.dta1".

4.4.2. Ages of spouses

Can we predict the age of the wife from the age of the husband (or vice versa) in a traditional marriage or partnership? This is an easy-to-do statistical test in a classroom; just ask the students the ages of their father and mother when they started their relationship!

The data I used for this example, are from C. Marsh (1988): *OPCS study in Great Britain*, quoted in the already mentioned *A Handbook of Small Data Sets* (D.J. Hand *et al.*). The variables here are x = age of husband and y = age of wife (years) from 170 random married British couples in 1980. Assume $\sigma_x = \sigma_y = 0.5$ because the ages were rounded to integers.

This is a typical example where you can expect a more or less linear relationship because in most cases there is not very much age difference between spouses.

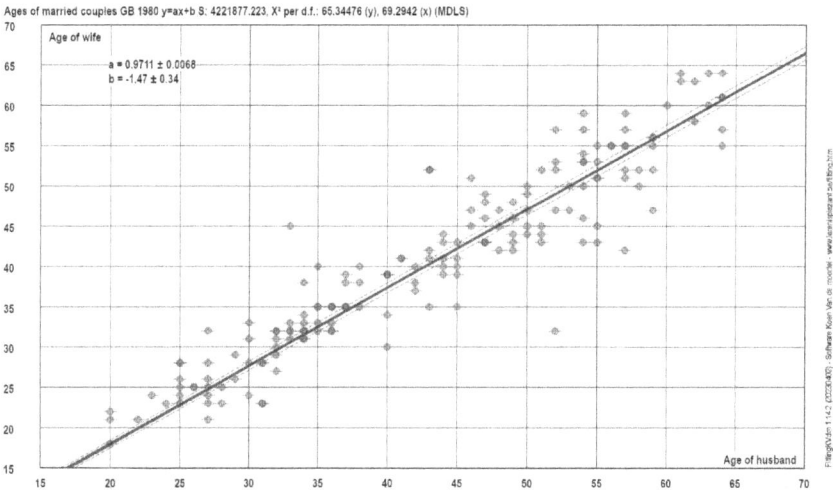

Fig. 4.167. Age of wife vs age of husband, and best fitting line (MDLS method). Darker dots = multiple occurrences of the same ages.

MDLS produced $y = ax + b$ with $a = 0.9711 \pm 0.0068$ and $b = -1.47 \pm 0.34$ (see Fig. 4.167).
The data are less chaotic than in the previous example, so the difference with OLS is not as big, but still... OLS yields $y \approx 0.9112x - 1.5740$.

FittingKVdm example data file: "Ages of married couples.dta1".

4.4.3. Examination scores vs. completion times

If a student completes an exam in a short time, that can be a sign that he knows his stuff very well, or that he is a quick problem solver, both predicting a high score, or... it might be that he just gave up quickly and will have a low score. If he needs a long time, it might mean he is slow, but also that he is thorough and he wants to check everything twice...

So, is there a pattern? Some researchers investigated it (I. Basak *et al.*, 1992), and again, the data are given as an example in D.J. Hand's *A Handbook of Small Data Sets*. If you are teaching, you can collect your own data of course. The variables here are x = time to complete an exam (in seconds) and y = score (max = 75).

The results... well, what shall we say? This is a typical example of a cloud of points in which you can recognize nothing or whatever you want (a line, a power function, an exponential function, etc.). MDLS and OLS give very different results since there is no clear pattern.

Kendall $\tau = -0.069$, meaning there is a very small downward trend in the data: longer times mean lower scores, but take that with a large pile of salt, and forget about making any other conclusion.

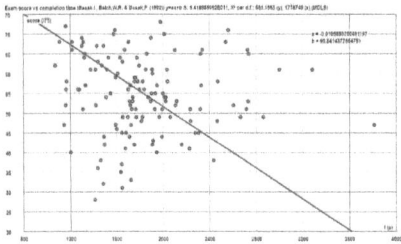

Fig. 4.168. Examination scores vs. times and best fitting line.

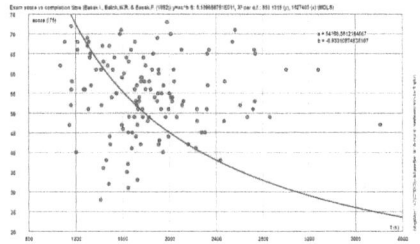

Fig. 4.169. Examination scores vs. times and best fitting power function.

The power function (right) with a negative exponent makes more sense than the linear fit (left), since the power function reflects the simple fact that the score can never be less than zero.

FittingKVdm example data file: "Examination scores vs completion time.dta1".

4.4.4. How many words are we supposed to know at a certain age?

At birth, we start with a knowledge of zero words, until after many months the first words are spoken. From that moment on, if everything evolves well, things seem to go quicker and quicker, until children can have a normal conversation. But most people keep on learning new words during the rest of their lifetimes.

Now, answering the question is really not easy! With young children, you can still literally count the words they know, but once your vocabulary is in the thousands, people diversify, some know words like "spectroscopy" or "capacitance" and others might know "shuttlecock" or "gambit". You might know some words "half", like "I know it has something to do with... but...". That makes it extremely difficult to measure someone's vocabulary; you can't torture people walking through the whole dictionary. So, the existing tests are based upon taking samples and extrapolating. For example, https://my.vocabularysize.com.

If we want to model the "normal" vocabulary evolution, we have to know also if there is a limit, like "the vocabulary of an average adult". Maybe we can keep storing more and more in our memory, but maybe we forget more than we learn after a certain age?

I found some interesting data to analyze, from children between the age of 0 and 6, in an article by Madorah Elizabeth Smith (May 1926): "An investigation of the development of the sentence and the extent of vocabulary in young children", *Iowa Child Welfare Research Station*, vol. III nr. 5, p. 52. It can be found online at https://pure.mpg.de/rest/items/item_2385505_3/component/file_2464108/content.

So our variables are x = age in years (0–6) and y = average number of words observed in the vocabulary of children of that age. The error on x is assumed to be half a month, and on y, the square root.

This seems to be a tricky data set to model though. Typical models for this, like "Weibull growth", "Logistic", or "Transition", predict impossibly low limit values. Only "Gompertz growth" (starting from 0), as shown in Fig. 4.170, produces something that looks reasonable, but it also predicts a limit of only 2616.7 words. So, this model might be used as an estimation for age 0–6, but not much further!

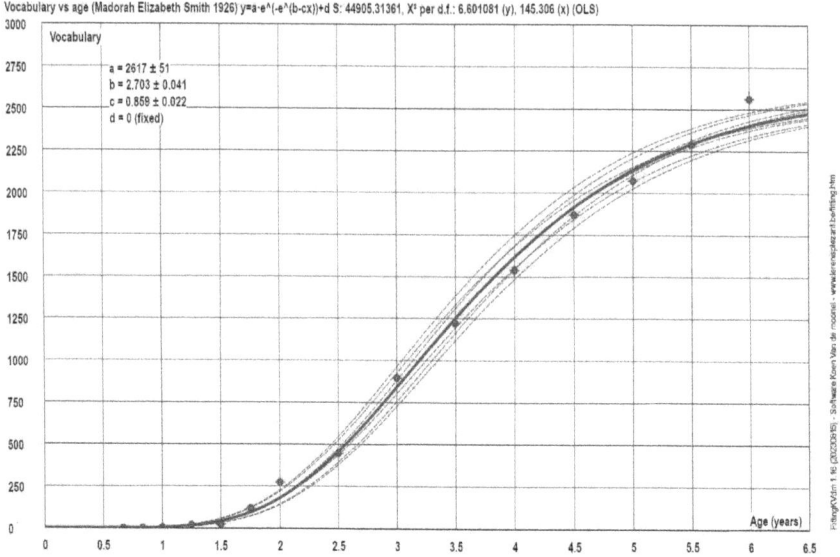

Fig. 4.170. Vocabulary vs. age (M.E. Smith 1926) and the best fitting "Gompertz growth" function.

Remark: Such investigations are not only useful to judge if a child is doing well or not but also to compare the difficulty of different languages in a more or less objective way!

FittingKVdm example data file: "Vocabulary vs age (Smith 1926).dta1".

4.4.5. Does money make us happy?

Does money make you happy? That is a question I often heard!

To find out, you need to gather data. The money is the easiest information to get, just ask people about their income. But how do you measure the "happiness"?

Since it is a subjective quantity, the best thing to do is ask the subjects directly: "How happy do you feel today/the last month or year?" or "Are you satisfied about your life in the last year?". If they answer with a rating between 0 and 10 (the so-called **Cantril ladder scale**), you have a number you can work with.

This would be an interesting assignment for students. You can expect a lot of "noise" in the data though! People might lie about their income, or they might have a good income now but fear to lose it; or they might feel temporarily more or less happy not related to money but to love or health issues for example.

Another way to get an answer is to look at large groups of people, entire countries, for example. You could take a look at the "World Happiness Report", for example, published every year at

https://worldpopulationreview.com/country-rankings/happiest-countries-in-the-world.

Now, before using any data, you should always look carefully *how they were collected*!

On this site, they say, "To determine the world's happiest country, researchers analyzed comprehensive Gallup polling data from 149 countries for the past three years, specifically monitoring performance in six particular categories: gross domestic product per capita, social support, healthy life expectancy, freedom to make your own life choices, generosity of the general population, and perceptions of internal and external corruption levels.".

Oops! Apparently they already *assumed* that money makes you happier, since they made some kind of weighted score in which the GDP is one of the numbers they include! Also from the other factors, they *assume* they play a role in your happiness. They might do that indeed, but it's a very biased way of doing research!

By the way, the GDP is not really the best way to estimate people's richness, since it is an average, not a median.

Better sources can be found here:
https://ourworldindata.org/happiness-and-life-satisfaction
(by Esteban Ortiz-Ospina and Max Roser, first published in 2013) and
https://ourworldindata.org/grapher/daily-median-income   (originating
from the World Bank Poverty and Inequality Platform (2022)).

From these, I extracted data from 142 countries, so our variables are $x =$
median level of income or consumption per day (in international-$ at
2017 prices) in 2019 and $y =$ "Life satisfaction" in the Cantril Ladder
scale. For some countries, one of the previous years was used if 2019
was not available. Both variables were independently measured here, no
bias! The errors on x and y were set equal since no information was
available about the variance of both.

The simplest statistic that tells us something about our question is
Kendall τ (see p. 84). That was 0.64149, so it definitely suggests there is
a positive correlation!

Of course, we like to see a more specific relationship, a "model". Now,
since there is no "physical law" telling us what to expect of the relationship between money and happiness, we have to experiment with models
that make sense. A requirement is that they need to have a *horizontal
asymptote* $y = 10$, since the scale has a maximum of 10. So, a linear
model is no option. Should it go *through the origin*? That's a trickier
question! Probably people with a very low income who are still reasonably happy exist, but how happy can you be if you have literally
nothing? Let's have a look at the data and a few models...

First, the homographic (see p. 19):)
$$y = 10 - \frac{b}{x + c}$$
At first sight, it seems to fit reasonably (Fig. 4.171). It predicts you
should earn 84.5 $/day to have an 8/10 happiness score, and 214.8 $/day
for 9/10. Okay... but to have 4.3/10 with absolutely nothing in your
pocket seems unrealistic.

Let's try models that start from the origin. The simplest is a homographic relationship that is forced through the origin:
$$y = \frac{10x}{x + b}$$

That predicts a happiness of 8 if you earn 45.1 \$/day, and 9 with 101.4 \$/day. Hmm... Clearly, as you can see in the scatterplot (Fig. 4.172), the one free parameter cannot bend the curve enough to adapt to the measurements. The residuals are not well distributed: in the lower half, the curve is way below the dots, and in the higher income half, its way above the dots. We might call that "**underfitting**".

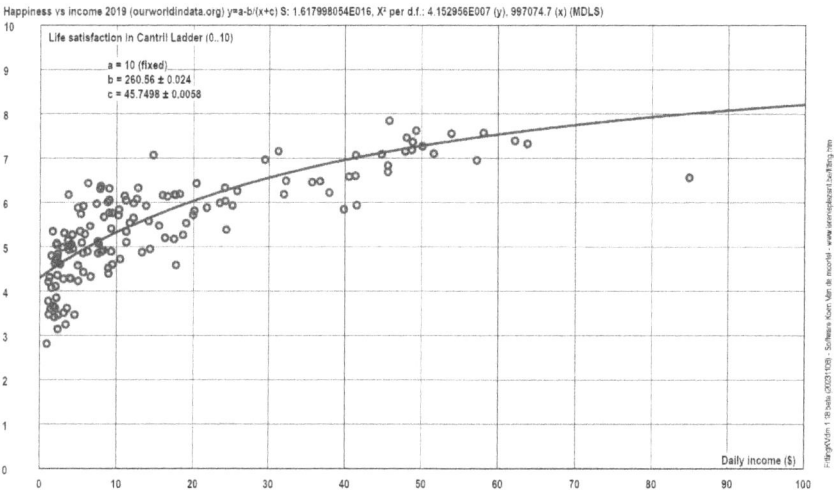

Fig. 4.171. Happiness vs. income data and a homographic model function, not through the origin.

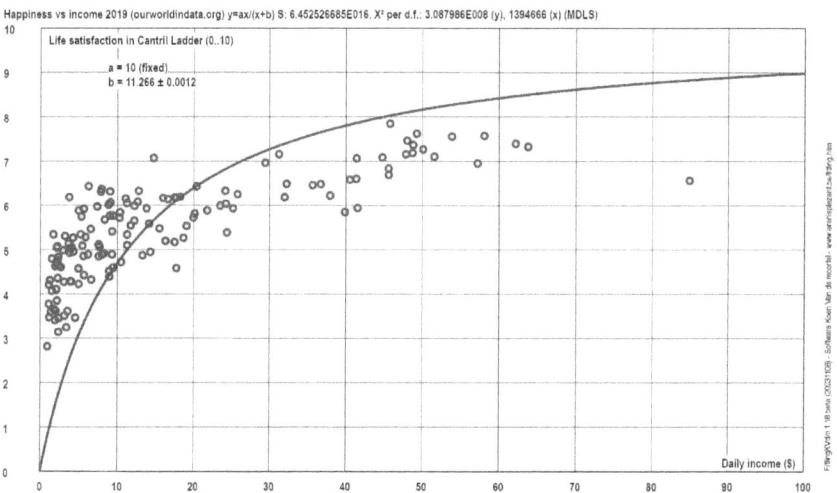

Fig. 4.172. The same data with a homographic function through the origin; clearly not fitting well.

You might think that a similar looking, exponential model, might do the job:

$$y = 10 \cdot \left(1 - e^{-\frac{x}{b}} \right)$$

But it also doesn't have enough flexibility.

Now, remember, with one additional parameter we can make this more "bendable". It was called the "*Weibull growth*" model (see p. 33):

$$y = 10 \cdot \left(1 - e^{-\left(\frac{x}{b}\right)^k} \right)$$

This really adapts well to the lowest income data, with $k \approx 0.275$ (instead of 1, which was the simple exponential model), and it kind of confirms what people know intuitively (Fig. 4.173). The curve is very steep near the start, meaning that the first dollars will be the most effective to increase your happiness; the difference between being rich and very rich is much less important.

Note: To get happiness 8, you need 116.6 \$/day, but OLS predicts 192.9 \$! OLS flattens the curve too much! To get happiness 9 would require 428.7 \$/day, and using OLS even 881!

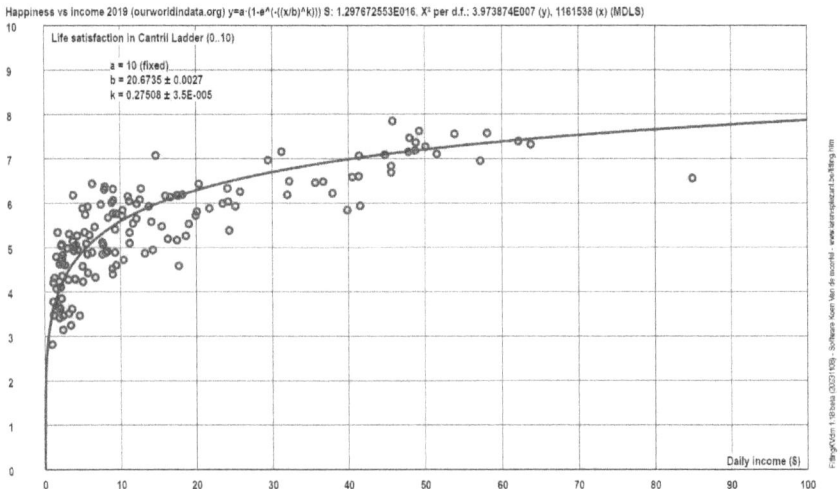

Fig. 4.173. The same data with the Weibull growth model.

By the way, you might wonder what country that rightmost dot might be? Well... it's the Arabs! With all their money they still seem relatively unhappy in all the models! Maybe we should try a model with a peak?

If you are curious about other countries, have a look at Fig. 4.174.
Finland seems to be the happiest country! Japan and South Korea are quite unhappy in spite of being among the wealthier countries. Costa Rica, on the other hand, with a modest income, produces quite happy people! The saddest country was South Sudan, as far as we know... No data from North Korea were available.

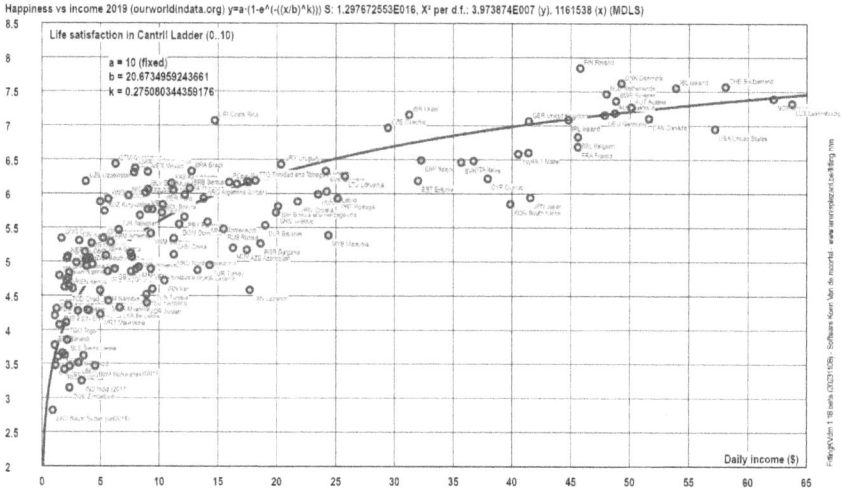

Happiness vs income 2019 (ourworldindata.org) y=a·(1-e^(-((x/b)^k))) S: 1.297672553E016, X² per d.f.: 3.973874E007 (y). 1161538 (x) (MDLS)

Fig. 4.174. The same graph as the previous but with the United Arab Republic (rightmost point) left out (not for the fitting).

FittingKVdm example data file: "Happiness vs income.dta1".

4.4.6. The evolution of smoking habits

Once upon a time – if you were born before 1970 you will remember this – it was supposed to be "cool" if you smoked cigarettes. Even medical doctors appeared in commercials to recommend "their" brand. Some time after it became clear that it was not so healthy, and lobbyists from the industry lost their power and credibility, governments started campaigns to discourage smoking.

Can we measure the effects? Example data, from adolescents in Germany, can be found at
www.destatis.de/EN/Themes/Society-Environment/Sustainable-Develop
ment-Indicators/Publications/Downloads/data-relating-indicator-report-2
021.pdf?__blob=publicationFile.
The variables here are x = the time (year) and y = the percentage of adolescents smoking. Since we didn't know at what time in the year the survey was done, or when exactly people started or stopped smoking, σ_x was set to 0.5, so x can be interpreted as "the middle of the year ± 0.5 year". Looking at the precision of the given percentages, we can guess σ_y to be 0.05, or safer 0.1. Anyway, if all the errors are the same, the absolute value will not influence the parameters, only the parameter confidence interval estimations.

This is a typical example that might be modeled with a *logistic function*: a change in behavior caused by an external stimulus. MDLS can be used since the y values seem to be between the horizontal asymptotes.

According to this model, the starting percentage was about 27.1, the pivoting point was around 2007, and the persistent group seems to stabilize around 4.33% of the population (Fig. 4.175).

Another function that might describe the data well, is the one I refer to as "transition" function (p. 32, 32). That has an extra parameter (k) to bend the curvature of the curve more or less straight (Fig. 4.176).

Both models "see" approximately the same inflection point (0.36 years apart), but the second "thinks" the percentage went from 29.6 to 5.27. The χ^2 values of both models are very comparable. In order to be able to decide which one is best, we would need some more data, especially from before 1995 and after 2020. Anyway, there is no "natural law" that dictates what the evolution of smoking behavior should look like since many influences continue to play a role, like price policies, group pressure, social disapproval, availability of other drugs, etc.

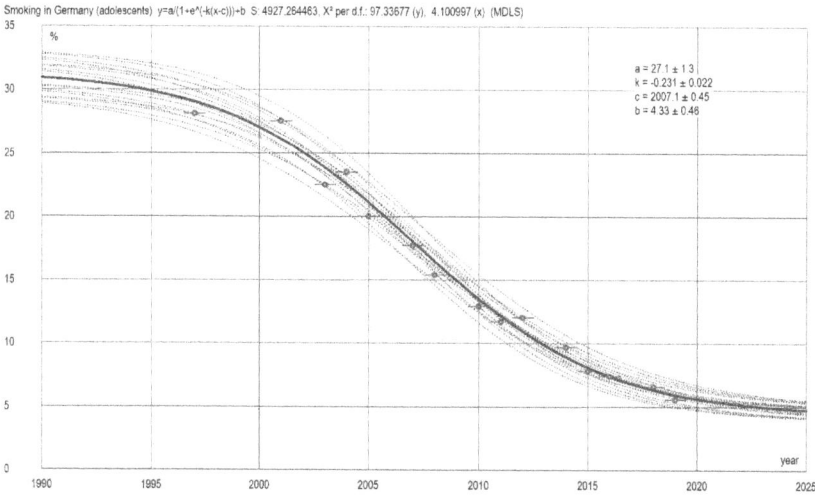

Smoking in Germany (adolescents) y=a/(1+e^(-k(x-c)))+b S: 4927.264463, X² per d.f.: 97.33677 (y), 4.100997 (x) (MDLS)

a = 27.1 ± 1.3
k = -0.231 ± 0.022
c = 2007.1 ± 0.45
b = 4.33 ± 0.46

Fig. 4.175. Percentage of smoking adolescents in Germany, and the best fitting logistic function.

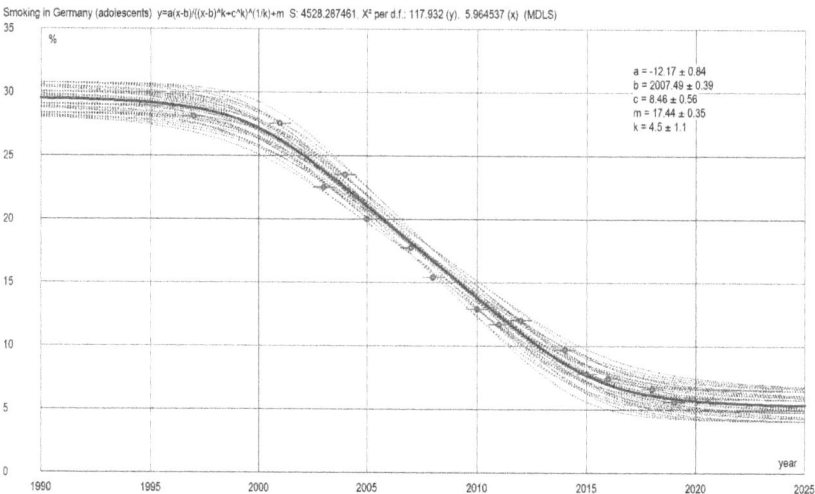

Smoking in Germany (adolescents) y=a(x-b)/((x-b)^k+c^k)^(1/k)+m S: 4528.287461, X² per d.f.: 117.932 (y), 5.964537 (x) (MDLS)

a = -12.17 ± 0.84
b = 2007.49 ± 0.39
c = 8.46 ± 0.56
m = 17.44 ± 0.35
k = 4.5 ± 1.1

Fig. 4.176. Same data, but with the fitted "transition" function.

FittingKVdm example data file: "Smoking - adolescents Germany.dta1".

4.4.7. The vocabulary of a writer

Let's do a thought experiment, a nice one to feed to students: suppose we take a random book, run through the first 100 words, and count the number of different words. Obviously, there will be less than 100 and more than 0. Now we count the number of *different* words in the first 200 words. Will there be twice as many? Probably less because some words are repeated. We can continue like this, and now we want to try to predict what will happen. If we call x the number of words, and y the number of different words, what will approximately be the mathematical relationship $y = f(x)$?

For those who want to try this at home, a reassurance: you don't have to count thousands of words one by one yourself! There is (free) software for this, "QUITA" (Quantitative Text Analyzer 1.1.9.0), available at https://korpus.cz/quitaup. Of course, it doesn't work completely flaw-lessly because it actually counts "tokens" (sequences of characters), so, for example, the word "bat" might appear in multiple meanings (the animal or the stick) and still be counted as one word. *Idem dito* for other homonyms like "bright", which can mean "smart" or "full of light", etc. And on the other hand, singular and plural, or conjugations of the same word will be wrongly counted as different. So there is a certain margin of error on the y values, but hey, manual work would be too time-consuming. "Disruptions" can also occur on the x values, albeit less. Think of abbreviations or Roman numerals, for example.

Before starting, it is a very useful exercise to consider what that function f should look like!

One thing must be certain, $f(0) = 0$, because if there are no words, there are no different ones either. Also certain, it must be a strictly ascending function, because new words can only be *added* on the next page, not lost!
The simplest function that already satisfies these conditions is a linear one that passes through the origin: $f(x) = ax$. An objection to this, however, is that a writer will use fewer and fewer new words after many pages. The slope will therefore have to become less and less. Aha, now hopefully you or your students will come up with the image of the square root function, or something similar: a power function $f(x) = ax^b$ with an exponent between 0 and 1.
Well, that turns out to be quite correct empirically, and the phenomenon

has even been given a name: Heap's law (can be found at
en.wikipedia.org/wiki/Heaps%27_law).

Now, the attentive reader with knowledge of the matter will still make a
critical reflection on this. Such power functions all start infinitely steep
at the origin! So there will always be an x interval where the y value is
greater than x; meaning, more different words than words in total, which
is very strange, actually impossible.

The expected property, that the growth (read: "derivative") of f should
be more or less inversely proportional to x, should actually make us
think of *logarithmic* functions in the first place! Unfortunately, they do
not go through the origin, but if you move them one unit to the left, they
do! And then we get the desired property that the slope at the origin can
never be more than 1, see Fig. 1.12.

With the necessary horizontal and vertical "strain" parameters, a function of this form could therefore be a good candidate to serve as a model
for reality:

$$f(x) = \frac{a}{b}\ln(1 + bx)$$

The b in the denominator is not strictly necessary, but it makes the parameter a more convenient to interpret as the slope at the origin. After
all,

$$f'(x) = \frac{a}{1 + bx} \;\Rightarrow\; f'(0) = a$$

The proof is in the pudding! Let's test our assumptions. The first digi-
tally available book that happened to come across my eyes was the English translation of a book by Guy De Maupassant: *Complete Original
Short Stories*, available at www.gutenberg.org/files/3090/3090-0.txt.
Using QUITA, I did the counts for the first page, the first two, the first
three, etc. up to page 80 (it does require some cutting and pasting). As
an estimate for the error on x, I took 1% and on y, 10%. The exact
values don't matter for the regression (only for the estimation of the
error on the parameters), but it is important that they are proportional to
x and y, to give the points meaningful weights (small error, large
weight). I used MDLS because it meets the logical symmetry require-
ments: x can be calculated from y with the same formula as y from x,
inverted of course.

It's almost miraculous how well the function fits with the data (Fig. 4.177)! Hard to believe that this happens "automatically"; you would almost think that the writer and the translator were consciously counting their words! And I'm betting that it would be noticed if another translator had suddenly taken over starting from a certain page.

The parameter $a = 0.551$, which means that on the first pages 55.1% of the words are used for the first time.

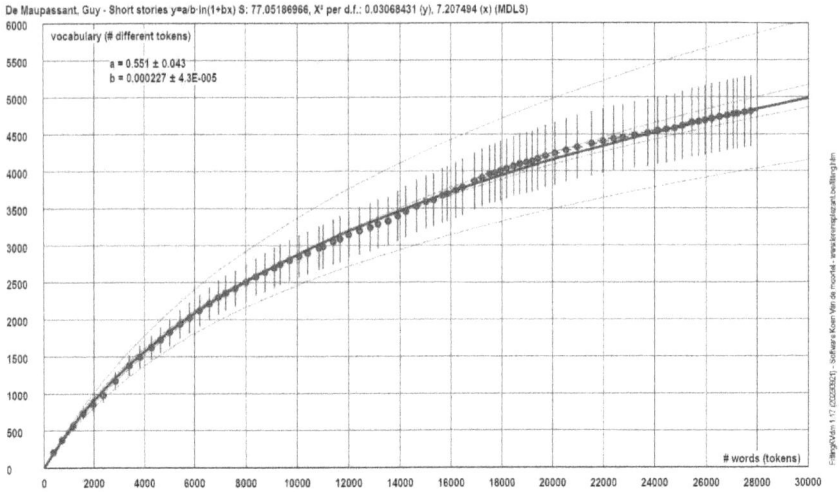

De Maupassant, Guy - Short stories y=a/b·ln(1+bx) S: 77.05186966, X² per d.f.: 0.03068431 (y), 7.207494 (x) (MDLS)

vocabulary (# different tokens)

$a = 0.551 \pm 0.043$
$b = 0.000227 \pm 4.3E\text{-}005$

words (tokens)

Fig. 4.177. Number of different words vs. number of words, counted further and further in a book. Best fitting logarithmic function shifted through the origin.

So, what is this good for? For mathematics teachers, it is a good way to show that there are sometimes applications that we do not immediately think of; for literature researchers, it is a way to quantify the richness of a writer's vocabulary. For example, you can rank James Joyce with his practically unreadable books such as *Ulysses* as an extreme word producer because his curve is far above this one while the *King James Bible* turns out to be extremely "poor", perhaps because they consciously avoided "difficult" words in the translation?

There is one more mathematical detail that we have not addressed: should *f* have a *horizontal asymptote* or not? In other words, will there be a limit to the number of words a writer will use, or will he start inventing neologisms if only the book is thick enough? Can this even be determined empirically? After all, logarithmic functions also increase very slowly over time.

Candidate functions could then be, for example, transformations of exponential or homographic ones that pass through the origin, such as

$$f(x) = \frac{ax}{x+b} \qquad \text{or} \qquad f(x) = a\left(1 - e^{\frac{x}{b}}\right)$$

which have a horizontal asymptote. These two don't seem to fit very well, but there is another one that is often used in growth models: the "Weibull growth" function, similar to the previous one but with an extra parameter to manipulate the curvature:

$$f(x) = a \cdot \left(1 - e^{-\left(\frac{x}{b}\right)^k}\right)$$

This also fits wonderfully at first glance (Fig. 4.178), but it does "predict" a limit of only about 6080 words, and near the origin, it also starts much too steep (derivative about 190), which is impossible.

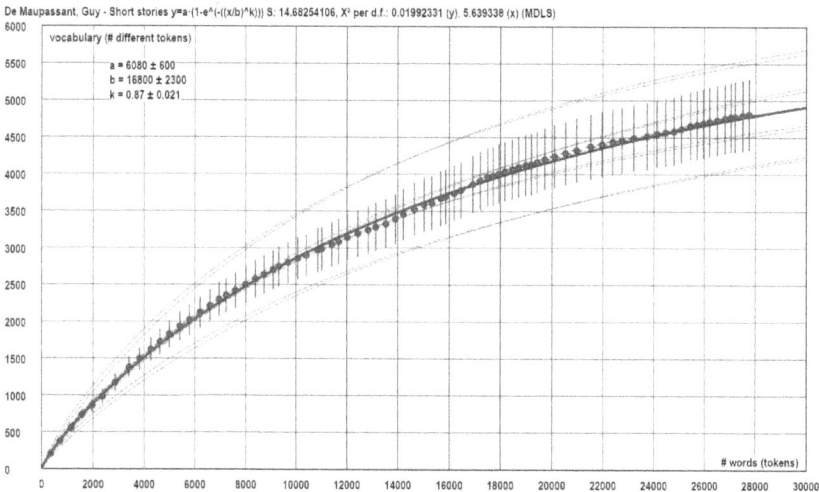

Fig. 4.178. Same data but with "Weibull growth" function fitted.

Moral of the story is don't be fooled by nice-looking pictures, but also look at the details! The mentioned logarithmic function also turns out to be useful in totally different domains! See the case study "How much do we spend on food?" (p. 367).

FittingKVdm example data file:
"Vocabulary vs word count - De Maupassant-Short stories p1-80 & all.dta1".

4.4.8. Measuring psychomotor improvement – throwing pebbles

Coordinating the mind and the body is something we learn especially in the first years of our lives. But of course we are never too old to learn, and by training we can improve the precision of this coordination to get better in sports, playing a musical instrument, driving, or whatever.

How can we measure this? Well, for example, with this easy experiment that is fun to do with students: ask a person to throw a small object to a target (e.g., a line on the floor) and to repeat this a number of times (from the same position). Of course, he/she should try to do this as perfectly as possible. The object should have enough friction on the floor to make it stop quickly, so don't throw with a marble! Each time, a second person measures the distance from the line (y_i for trial number. i); a positive y value means it was too far, negative means it was too close.

My test person tried it with a pebble, a hundred times. The results are shown in Fig. 4.179.

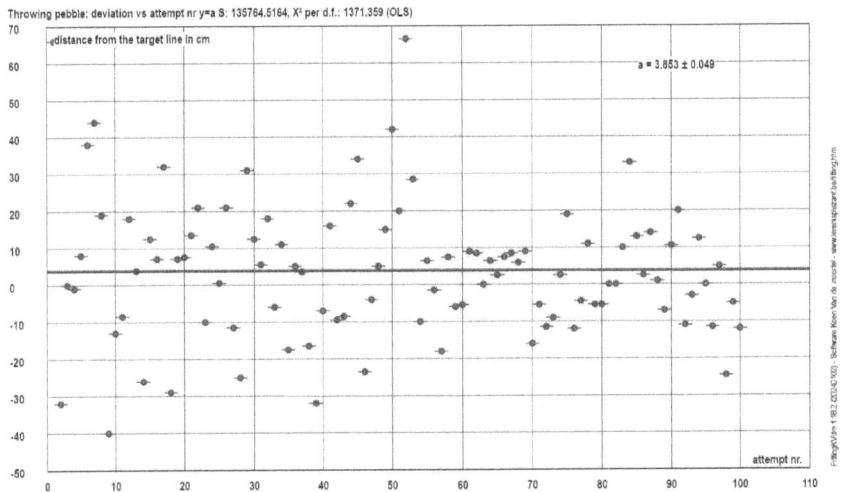

Fig. 4.179. Distances between landed object and target, 100 attempts. Line = average.

The y values (in cm) looked more or less normally distributed with an average value $\mu = 3.853$ and a standard deviation $\sigma = 18.42$. So, the person had a tendency of throwing almost 4 cm too far.

Now, we would like to answer the following question: "Did the person learn anything while performing this task?".

This can be done in many ways, but we shall compare three different ones:

(1) Comparing the averages in the beginning and at the end.

Defining "beginning" and "end" is kind of arbitrary, but let's give it a try by taking the first 10 attempts and the last 10.

throws	μ	σ
1..10	8.9	31.8793
91..100	-2.95	10.53435

There is an obvious improvement: at the end, the pebble landed closer to the line *and* the person threw less spread out (the standard deviation was much smaller).

Of course, this could be "just a coincidence". Statisticians like to stick a probability to their results: what are the odds that this difference is due to pure chance?

In this case, an unpaired one-tailed Student's t-test is appropriate. More information, e.g., en.wikipedia.org/wiki/Student%27s_t-test.
This can be performed online, for example, using
www.graphpad.com/quickcalcs/ttest2.
This produces $t = 1.0961$ with 18 degrees of freedom, and standard error of difference = 10.811 and a two-tailed p value of 0.2875. Since our hypothesis was that the second average would be less than the first, we have to use the one-tailed p (0.14375). This means that there is still a 14% possibility that the "improvement" was due to pure luck.

Remark: We could have compared also the first and the last 20 attempts, or 30 or 50, etc. and of course that would give different t and hence p values, so... the exact value of this p should always be taken with a serious grain of salt.

(2) The Kendall tau test

Comparing averages is not wrong, but you throw away a lot of possibly interesting data. A test that makes a more complete use of the available information is the Kendall τ test (see p. 84).

If there is a learning process going on, we would expect that if we compare $|y_2|$ with $|y_{28}|$ or $|y_{13}|$ with $|y_{74}|$, etc., in most cases the latter is smaller than the first.   That is what the τ test does: consider all the couples $(|y_i|, |y_j|)$ with $i > j$, add +1 if $|y_i| > |y_j|$ and -1 if $|y_i| < |y_j|$. Then divide the result by the total number of comparisons ($n \cdot (n - 1)/2$). A τ of 1 would mean that the results worsen every time, -1 would be 'perfect', and 0 would mean no learning effect is detected.
So, in this situation, τ can be written as

$$\tau = \frac{2}{n \cdot (n-1)} \sum_{i>j} \mathrm{sgn}\big(|y_i| - |y_j|\big)$$

This calculation can be done online, for example at
www.gigacalculator.com/calculators/correlation-coefficient-calculator.php
In the left column, the attempt numbers (1–100) must be entered, and in the right column, the $|y_i|$ ($i = 1$–100). This calculator makes a small correction if some of the values are the same, but that doesn't make a big difference here.

We obtain $\tau \approx -0.2006$, which confirms our hypothesis that in more cases the second $|y|$ is smaller than the first.

Now, again, this can be a pure coincidence. What are the odds that it isn't? In order to find that, we should shuffle the y values in all possible ways and see in how many cases τ is "better" (meaning here, more negative) than -0.2006.
Gigacalculator uses an approximation and gives us a z score of -2.964, corresponding to a one-tailed p value of 0.00151799. This tells us that there is only a 0.15% chance that this result is due to sheer luck. So, we may be pretty confident that our test person did learn something!

(3) Regression analysis

Another way of formulating a hypothesis would be as follows: with every attempt, the person should improve his result (on average of course) with a certain percentage.

That means that the y values should fit in a descending exponential pattern:

$$y = b \cdot a^x + c$$

with, in our case $c = 0$ (the test person tries to achieve perfection: $y = 0$ after an infinite number of attempts) and $0 < a < 1$, and x is the attempt number. Fig. 4.180 shows the best fit.

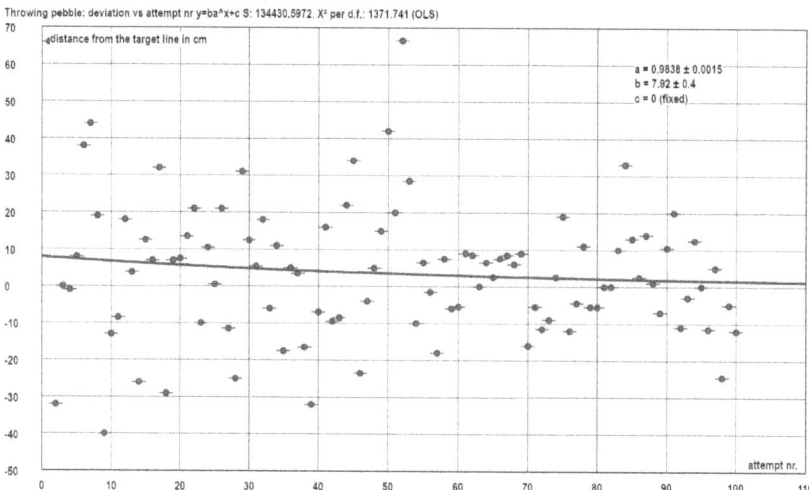

Fig. 4.180. The measurements and the best fitting exponential curve.

Remark: The entry of "measurement errors" σ for both x and y values is required if we do this with FittingKVdm. For the x values, they were set to 1, why? Since all the attempts were done consecutively, x corresponds more or less to the time in units of the time needed for each attempt. To assume that the time between two consecutive attempts didn't vary more than 1, such unit seems reasonable. Anyway, if they are set all the same, all the measurements get the same weight, and the value of σ_x and σ_y influences only the estimation of the uncertainties in the fitted parameters a and b. For the y values, an uncertainty of 0.5 cm was entered (measuring the exact distance from the center of gravity of the pebble to the line more precisely was not realistic in our experiment).

By the way, most other software programs will crash when processing these data, because of the negative y values. Unfortunately, the use of MDLS is not possible here since the data are above and below the horizontal asymptote.

So, with OLS, we got
$a = 0.9838 \pm 0.0015$
$b = 7.92 \pm 0.40$
This means that with every attempt, there is an improvement of about 1.62% ($= (1 - 0.9838) \cdot 100$).
According to this model, the average throwing deviation in the beginning was $y(1) \approx 7.7961$ cm, and at the end: $y(100) \approx 1.5459$ cm. This is an improvement of about $(1 - 1.5459/7.7961) \cdot 100 \approx 80\%$.
A high χ_y^2 value of 1371.74 indicates of course that there is a lot of randomness in the data.

Remark: Maybe you would think that just a simple linear regression trend line might do just as well, but no, that's wrong: extrapolating such a line would predict an endless worsening after the moment of "perfection"!

So, we can conclude that all three methods indicate that a learning effect occurred during the experiment.
The comparison of averages is the simplest but it wastes a lot of the available information. Kendall's tau is good in the sense that it gives a simple probability for the following question: "Did the person learn anything or not?". But the regression provided a good estimation for the following question: "How much did the person learn (at least in this time span)?".

FittingKVdm example data file: "Throwing a pebble.dta1".

4.4.9. Introversion and extraversion

The most popular character trait among psychologists seems to be the degree of extraversion of a person.
Collins dictionary explains as follows: "Someone who is *extravert* enjoys interacting with other people" and for the opposite, "An *introvert* is a person who enjoys solitary activities and calm environments, preferring to interact with individuals or in small groups".
If you ask me, this is a rather vague definition, like many definitions in psychology. I like to share ideas and knowledge and to discuss with many people, but I also like to sit quietly and read and do research on my own, and in some periods in my life, I prefer the first, and in others, the second. Does that make me an extrovert or an introvert?

Psychologists try to predict this extraversion using questionnaires [e.g., Eysenck 1965]. A public dataset with such questions and answers, by Yam Peleg, can be found at
www.kaggle.com/datasets/yamqwe/introversionextraversion-scales.

The test contained 91 questions and they were answered by 7184 people. The questions were presented one at a time in a random order. Each question was a statement for which the participant could select a response in a Likert scale: "1 = Disagree, 2 = Slightly disagree, 3 = Neutral, 4 = Slightly agree, 5 = Agree". After the main question sequence, the participant was asked, "Do you identify as either an introvert or extravert?" and the answer could be: 1 = "Yes, introvert"; 2 = "Yes, extravert"; 3 = "No".
My first remarks are as follows:
(1) It would seem to make more sense to me if we would replace the first five values with -2, -1, 0, 1, and 2, and the answers to the last question with 1, -1, and 0, since 0 looks more like "neutral/don't know". You might think it makes no difference for the regression, but no, it makes things easier if you want a model function that goes through the "neutral" position (origin).
(2) Letting the user choose whether he/she "identifies as introvert or extravert" must cause a lot of "noise" in this research! First, the user has to know the meaning of the words very well, and second, he has to know himself enough to choose the answer.
(3) Translating a Likert scale into numbers is a subjective act. Maybe the mental difference between "neutral" and "slightly agree" is smaller than between "slightly agree" and "agree", who knows? A "Cantril

ladder" might be more realistic since the participant gives the numbers (see p. 347).

If the answers are somehow correlated with this character trait, the answers to the questions can be expected to be correlated too since positive answers to some questions and negative answers to other questions are supposed to indicate extraversion.

I selected two statements, Q4: "I would hate living with room mates", and Q49: "I really like dancing", because I was curious to find out how they were related.

The scatterplot doesn't reveal much, except that all 25 combinations of answers occur many times. Just like in the simple example with the four points (see p. 101), a linear MDLS regression (forced through the origin or not) has two minima, more or less coinciding with the diagonals, but the line with the negative slope corresponds to a much deeper minimum of S.

So, $y = ax$ ($x =$ answer to Q4 and $y =$ answer to Q49) with $a = -1.0241 \pm 0.0062$ (Fig. 4.181).

OLS produces, as usual, a much flatter slope: $a = -0.1696 \pm 0.0035$ (Fig. 4.182).

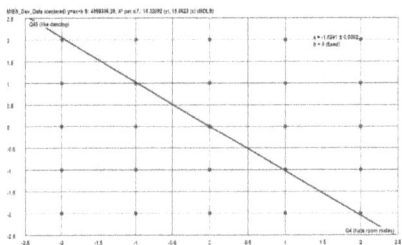

Fig. 4.181. Best linear fit through the origin (MDLS).   **Fig. 4.182.** Best linear fit through the origin (OLS).

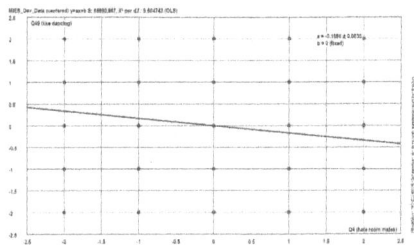

So, what causes this diagonal MDLS fit? It becomes (a bit) clearer if we look at the number of occurrences of all 25 possible combinations of answers (the number of dots on top of each other in the scatterplot):

Q4: -2	Q49: -2	-1	0	1	2
-2	311	181	184	280	460
-1	337	213	159	336	334
0	274	168	187	199	199
1	392	215	167	282	268
2	850	254	233	303	399

Fig. 4.183. The number of occurrences of each combination of answers.

The most occurring combinations were "I hate room mates" and "I don't like dancing" (850 times), and the opposite answers (460 times), so that explains the negative slope. But the pattern is not very clear: the combination of "I hate room mates" and "I like dancing" is also quite frequent (399 times).

The Pearson r value of -0.1587 and Kendall $\tau = -0.1038$ also indicate that the correlation and the ordering are very weak.

Are those numbers just accidental then? Well, if you put them in a χ^2 test to compare them with the average of the numbers (286.42), you get $\chi^2 = 1696.75$, which is very high for 24 degrees of freedom ($25-1$ because the average is calculated from the 25 values). So, the probability that such high deviations from the average occur by chance is astronomically low. But the pattern in this human behavior might be too complex to summarize in a simple regression...

So, is the answer to the questions related to the extraversion?

If we use the answer to Q4 as x, and the extraversion rating $(-1, 0, 1)$ as y, we get Kendall $\tau \approx -0.1549$, Pearson $r \approx -0.2758$, which indicates a small negative correlation.

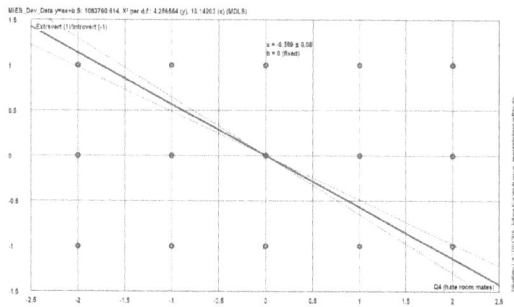

Fig. 4.184. Disliking roommates (Q4) and extraversion, best fitting line through the neutral point (MDLS).

The simplest applicable model here is $y = ax$. There are too few different x values and the data are too noisy to try any more complicated model.

MDLS produces $a = -0.569 \pm 0.080$ (Fig. 4.184).

OLS, as usual, produces a flatter slope: $a = -0.1657 \pm 0.0043$.
So, the best we can conclude is that there is a small tendency that people who don't like roommates are introvert, which makes sense to me.

For Q49, we get a small positive correlation – not very surprising:
Kendall $\tau \approx 0.1543$,
Pearson $r \approx 0.2753$.
MDLS produces
$a = 0.553 \pm 0.080$
(Fig. 4.185);
OLS produces
$a = 0.1440 \pm 0.0039$.

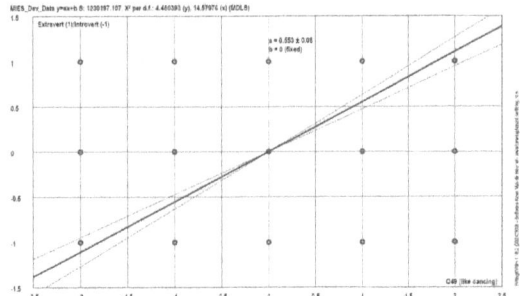

Fig. 4.185. Liking to dance (Q49) and extraversion, best fitting line through the neutral point (MDLS).

This doesn't have much predictive value: from all the 1656 people who said they liked dancing, only 417 (25.18%) said they felt "extravert".
It's even worse; 715 of the dancers (43.18%) said they felt "introvert"!
Only by combining the predictions from many questions (multilinear modeling), you might improve the probability to predict the extraversion correctly... for this dataset at least, without a guarantee it will work for new data. The more questions you include, the bigger the risk for over-fitting too.

References:
• Eysenck, H. J. and Eysenck, S. G. B. (Nov. 1965), "The Eysenck Personality Inventory", *British Journal of Educational Studies*, 14 (1): 140, doi:10.2307/3119050.

• A chi-squared (χ^2) test can be found online, for example at
https://homepage.divms.uiowa.edu/~mbognar/applets/chisq.html

• If you like to participate in similar tests, have a look at
https://openpsychometrics.org/

FittingKVdm example data files:
"Extraversion Q4-Q49.dta1", "Extrav-Q4.dta1", "Extrav-Q49.dta1".

4.5. Economy, etc.

4.5.1. How much do we spend on food?

Another thought experiment: if x is the total of a person's annual expenses (or income), and y the budget spent on food, what function might describe the relationship $y = f(x)$ between the two? Similar to the "writer"s vocabulary" case study (see p. 354) is that y can never be more than x, and that poor people spend a proportionately larger share of their money on food. A beggar will probably spend almost 100% of his money on food. This suggests that the same logarithmic function might be applicable.

We can test this reasoning on data from people from one particular region, or we can also compare countries in this way. You can find the necessary information for the second at
https://ourworldindata.org/grapher/food-expenditure-share-gdp.
The data and best fit are shown in Fig. 4.186.

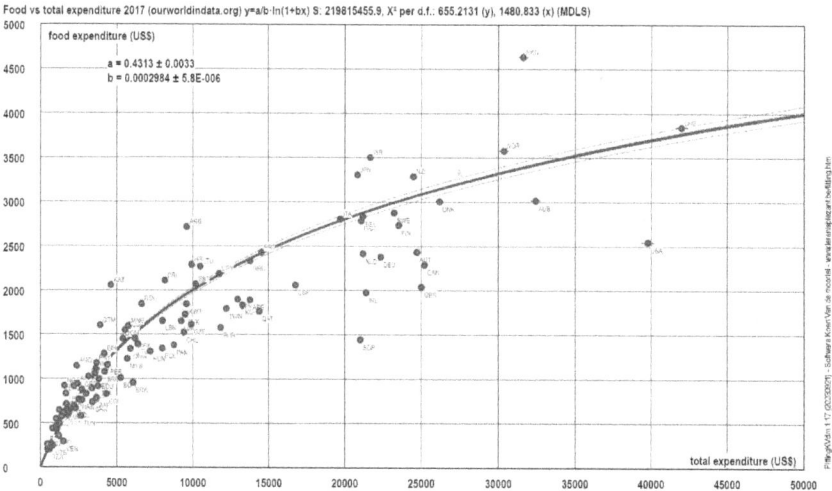

Fig. 4.186. Food budget vs. total budget for different countries, and the best fitting shifted logarithmic function.

The data do not lie neatly on the best-fitting curve, and there are at least two sensible reasons for this:

- Food prices and incomes do not have the same relationship everywhere (but that doesn't mean that every kind of food is more expensive in richer than in poorer countries; just think of the fact that the Netherlands can export onions profitably to Africa!).

- Some countries have a more gastronomic culture than others.
The latter is very clearly visible in the graph (data from 2017): the worst "junk food" eaters are far below the global curve: the US, Canada, Ireland, Great Britain, etc. As expected, the Netherlands is also below the slightly more Burgundian Belgium. Israel and Japan have a reputation for sophisticated cooking, and apparently they seem to cook well in Hong Kong too (haven't been there – should check it out!). Singapore looks very bad, but they have started to catch up to improve their food quality, or so I have heard.

FittingKVdm example data file: "Food vs total expenditure 2017.dta1".

4.5.2. Wine ratings

When we want to drink a good bottle of wine, we expect to pay the price for it, and vice versa. Now, how could the price be related to the quality? For that, we need to define "quality" first! There are a number of variables that can be measured objectively: chemical substances that should or shouldn't be there, like tannins, nitrates, sulfites, etc., the alcohol percentage, the acidity, etc. But usually people care most about the taste, and how do you rate that objectively? "De gustibus et coloribus non disputandum est", they say. Everybody has a different taste. You may even have noticed that you liked the same wine better or not, depending on what you ate with it, the temperature, the age, the time of the day, etc.
But fortunately, there are some lucky bastards who do wine tasting for a living (my god, what a job!). They have a fine nose and they are used to comparing tastes, so we hope we can rely on their judgments that are published on sites like, for example,
www.bordoverview.com/?year=2018&bank=both.
This site also shows the average price when the bottles came on the shelves, so that's interesting for model makers like us!

There is something strange about the ratings these sommeliers give: some have a scale 0–100, but less than 80 is rarely given; others have a scale 0–20, and then less than 10 is not often given. You can't just compare different scales... We'll just pick one for our test.
Our variables are x = the price of a 2018 Bordeaux wine (in €) when it came out and y = the rating by Lisa Perrotti-Brown for robertparker.com (on a scale of 0–100). A rating of "95–97" was entered as "96 ± 1"; "95–97+" was converted to "96.1 ± 1".

Could the relationship between x and y be linear? Definitely not! The rating can never be more than 100, so $y = 100$ has to be the horizontal asymptote, and we expect expensive wines to be rated near that limit. The function should be ascending, but should it go through the origin? You might say that a very cheap or "free" wine can still have some quality, so it shouldn't go through the origin, but on the other hand... a free wine (meaning: market value zero, not "a present") would never be put in a store. If it were impossible to sell, it would mean that the quality was really bad.
If we make the first choice, a homographic relationship like
$$y = 100 - \frac{b}{x + c}$$
is a good candidate and it fits reasonably, although with a lot of "noise"; and the uncertainty on the parameters is very big (Fig. 4.187). And a wine of 0€ would still get a rating of 83? That looks very suspicious. Some ratings of cheap wines would have been welcome to stabilize the regression, but still, we should consider the option "through" the origin too! The simplest model would then be
$$y = \frac{100x}{x + b}$$

The fit seems more realistic now on the cheaper side of the spectrum (Fig. 4.188). But the parameter estimation would certainly be better if some cheaper wines were tasted too.

What's the *use* of such a model? Well, suppose you want to choose a wine with a good quality for the price, you can scale the graph according to your budget, (and in FittingKVdm, switch the labels on) and look which wines are far above the curve. For example, in Fig. 4.189, we see that "Les Gravières" and "Fontenil" seem to be excellent, and "Poujeaux" and "Sociando-Mallet" seem to be disappointing for the price, according to Lisa Perrotti-Brown. Strangely enough, the latter are a few of my favorite wines, so apparently Lisa doesn't have the same taste as me... You could do this exercise with ratings from other tasters and find out whose advice you like to follow the most.

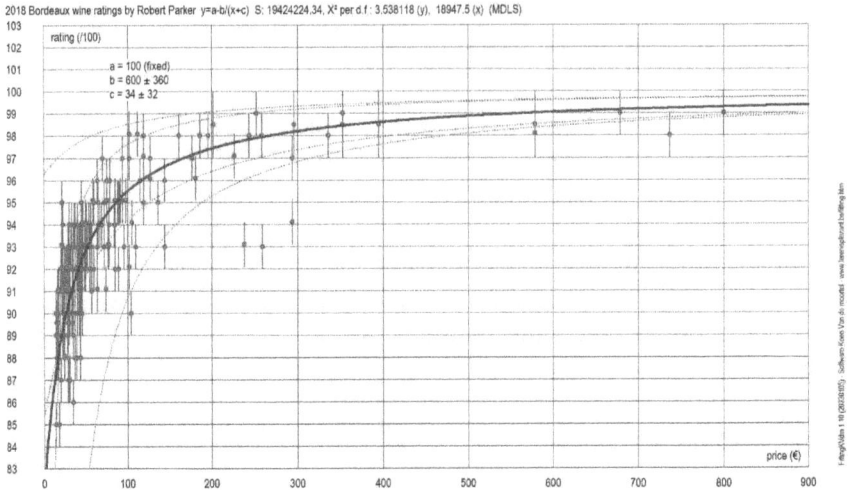

Fig. 4.187. Bordeaux wine ratings by Lisa Perrotti-Brown for Robert Parker. Model NOT through (0,0).

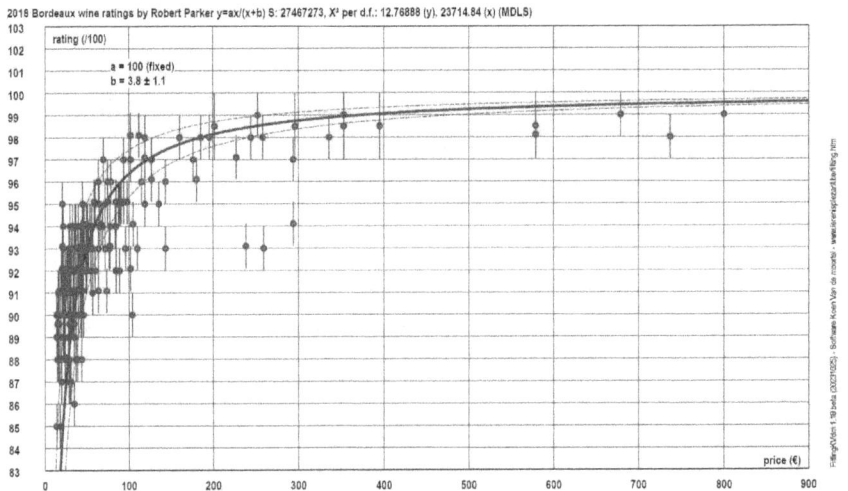

Fig. 4.188. Bordeaux wine ratings by Lisa Perrotti-Brown for Robert Parker. Model function through (0,0).

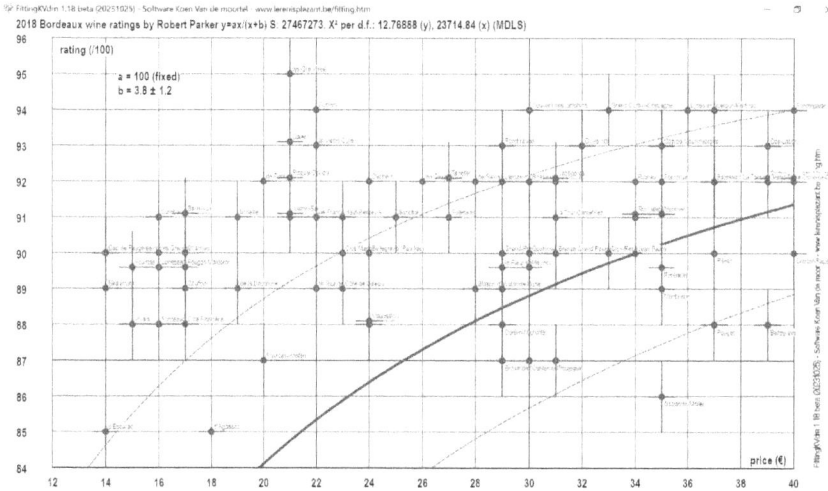

Fig. 4.189. Ratings for wines in the modest price class.

Do the other guys have a very different taste? Apparently they do! We did the comparison with Tim Atkin. On a scatterplot of the judgements of both sommeliers, we don't really see a line but rather a cloud. There is *some* correlation, but Kendall's τ is only 0.356.

The best fitting *linear* model is shown in Fig. 4.190. I know, the extrapolation would not make much sense here, but I don't know any better way to compare the two.

The MDLS method was used of course, because there is absolutely no preference for choosing Tim as x and Lisa as y or vice versa; in both cases, we should get the same line (mirrored). Using OLS, this is NOT the case.

Remark: We might ask ourselves if it could be better to use average ratings from *consumer forums* on the internet? The problem there is that people who are not used to drink very special wines might give 4/5 to a 10€ wine, while experienced drinkers might give it 2/5. And it's also not sure wether people give a good rating if the wine is simply good, or good for its price. Besides that, many mistakes can be made here since some wines have very similar names, and everybody can enter ratings that have not been double checked.

FittingKVdm example data files:
"Wine ratings vs price Bordeaux 2018 RP.dta1", "Wine rating comparison.dta1".

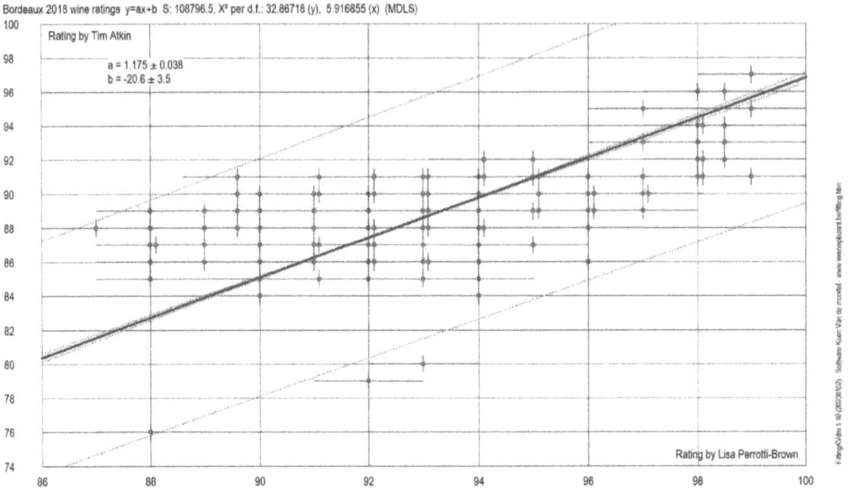

Fig. 4.190. Wine ratings of two tasters compared.

4.5.3. Income distributions

How are the incomes of all the households in a country distributed? Can it be described by a "normal" Gauss curve? Most probably not since the number of people with a low and modest income is much higher than that with higher incomes, so the distribution can be expected to be highly asymmetrical.

The Argentine econometrist Camilo Dagum (1925–2005) proposed a function that "would be flexible enough to describe this distribution"; see his article "Wealth distribution models: analysis and applications", *Statistica*, 66(3), 235–268 (2006). It can be found online at https://doi.org/10.6092/issn.1973-2201/1243.
For the formula of the Dagum distribution, see p. 41.

I tested this claim with income data I found from Belgium and the USA. They can be found at
https://statbel.fgov.be/nl/themas/huishoudens/fiscale-inkomens/plus and www.census.gov/data/tables/time-series/demo/income-poverty/cps-hinc/hinc-06.html.

The variables here are x = income class, and y = number of people in that class. Unfortunately, for the US, a large group (income > 200000$) had to be omitted since they were all thrown in the same bin and that would disrupt the graph.

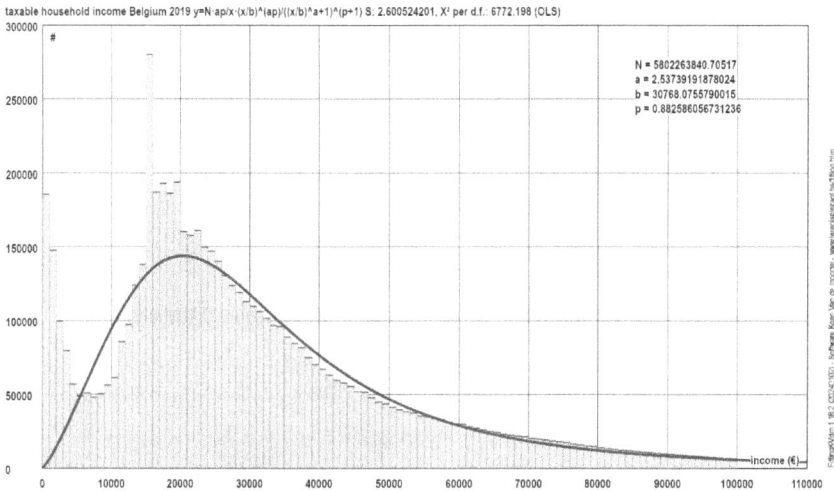

Fig. 4.191. Household incomes in Belgium 2019, and the best fitting Dagum distribution.

I'm not really impressed with the goodness-of-fit of the Belgian data (Fig. 4.191). They look more like a mixture of different skewed peaks. That is understandable since there are a lot of employed people earning a minimum wage, a group of unemployed people who receive about 800...1300€ monthly, and then the leftmost group might consist of the many self-employed people who often struggle to survive and might even have negative incomes (but those are hidden from the statistics). Retired people also often have very low incomes.

The American data (Fig. 4.192) fit better, but on the left side is also something wrong.

FittingKVdm example data files: "Income distribution Belgium 2019.dta1", "Income distribution of households USA 2020.dta1".

Household income USA 2020 y=N·ap/x·(x/b)^(ap)/((x/b)^a+1)^(p+1) S: 0.2069388928, X² per d.f.: 35926.89 (OLS)

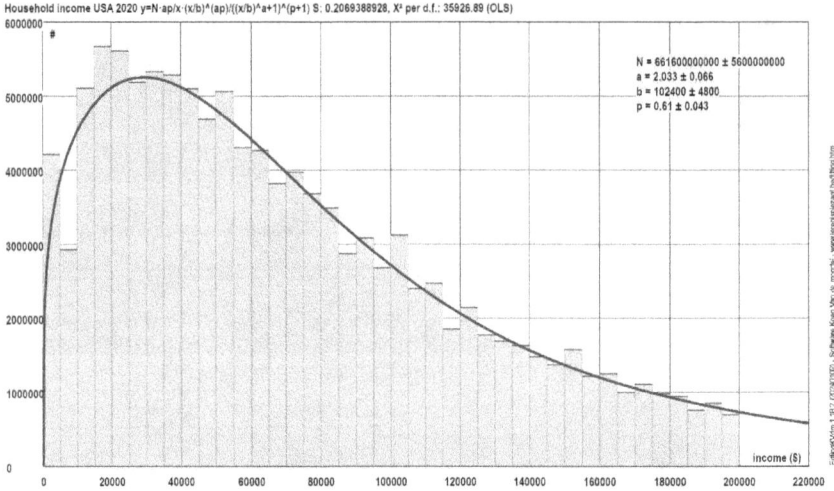

Fig. 4.192. Household incomes in the USA 2020, and the best fitting Dagum distribution.

4.5.4. Mobile phone usage evolution

A typical example of a phenomenon in society that came "out of the blue" and rose like exponentially, but then started to "saturate", is the usage of mobile phones. You could say it started around WW II with the walkie-talkies, but the real network connected mobile phones started as an oddity in 1980.

You can study the evolution at the website of the International Telecommunication Union (ITU) World Telecommunication/ICT Indicators Database: https://data.worldbank.org/indicator/IT.CEL.SETS.
To provide you with an example, I took the data for the entire world from there. So, our variables are x = time (calendar year, so ± 0.5) and y = mobile cellular subscriptions per 100 people. The uncertainty on the y data was unknown, and set to the same value (± 0.001).

A logistic model (with base value 0) seems very plausible:

$$y = \frac{a}{1 + e^{-k(x-c)}} \qquad k > 0$$

You might think the saturation value (*a*) would be 100%, but apparently it's higher! And some countries, like Libya, Antigua and Barbuda, and the United Arab Emirates, have about two cell phones per person! One for their normal and one for their "secret" affairs? We can only guess!

The logistic function seems to describe the evolution quite well (Fig. 4.193), and *if* we can believe it, we are now (2024) close to the saturation, about 1.08 phone subscriptions per person, and the period of maximal growth was around 2007–2008 (c ≈ 2007.56). The high chi-squared values suggest that the error estimations were too small or that the model is not perfect (of course, since many factors influence the numbers, like income evolutions in all the countries).

Any extrapolation should, again, be taken with a grain of salt. In the poorest countries, there is still a lot of growth possible (e.g., Mozambique: 42/100, Djibouti, 46/100 in 2022), and that will pull up the average saturation value. And maybe in a hundred years from now we won't need them anymore if we all have chips implanted?

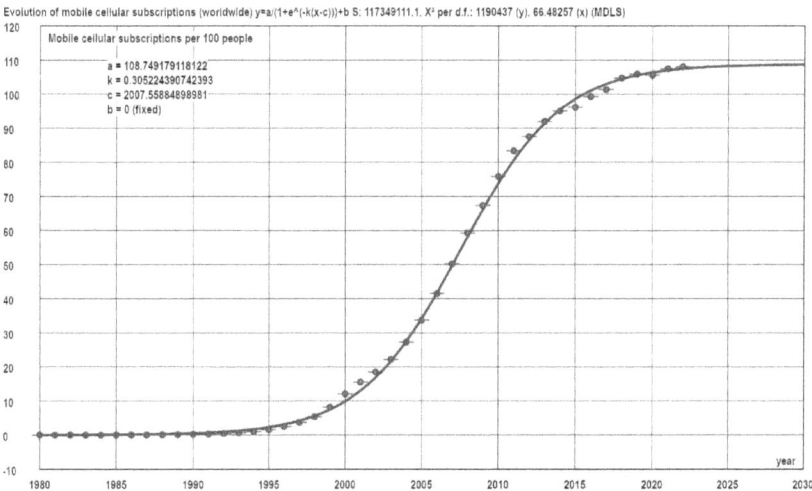

Fig. 4.193. Mobile phone evolution and the best fitting logistic function.

FittingKVdm example data file: "Mobile phone usage.dta1".

4.5.5. Noble prizes and chocolate

In 2020, a certain Aloys Leo Prinz from the University of Münster (D), published an article: "Chocolate consumption and Noble laureates", *Social Sciences & Humanities Open*, 2(1). You can read it at www.sciencedirect.com/science/article/pii/S2590291120300711#bib22. The data are included, so you can see for yourself. I have put them in a file for you: *x* = chocolate consumption (kg/year/capita); *y* = number of Nobel laureates (up to the year 2018) per 10 million inhabitants, for 27 countries.
And yes!!! Chocolate makes you more intelligent, it appears clearly!

There is a positive correlation indeed: Kendall $\tau \approx 0.450$ and Pearson $r \approx$ 0.685. But, I quote from the article, "It remains unclear whether the correlation is spurious or an indication for hidden variables".

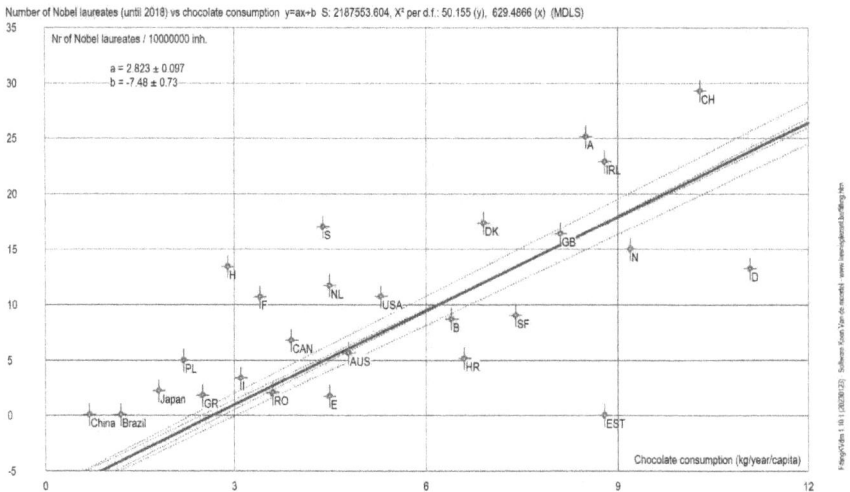

Fig. 4.194. Noble prizes vs. chocolate consumption and best fitting line.

This is one of those tricky cases... The most probable explanation is of course: rich countries have more means for scientific research, and chocolate is consumed more in rich countries, so... But then why is Japan ranked so low? They do have the highest average IQ in the world, and yet... Obviously from the scatterplot, and as it is well known, Switzerland has the *best* chocolate in the world! Although the taste of the Belgian chocolate is also legendary, apparently it is not so effective for the brain!

We can laugh about this, but a scientist always has to stay open. I'm not sure how reputed this website is as a scientific source, but I quote, "Research suggests that certain compounds found in chocolate, such as flavonoids, may help to boost cognitive function, reduce inflammation in the brain, and even enhance mood.". See www.markys.com/blog/the-science-behind-chocolate-how-it-affects-our-brain-and-mood. A lot of food for thought, I would say, pun intended!

FittingKVdm example data file: "Noble prizes and chocolate.dta1".

Fig. 4.195. The only "artificial intelligence" you will find in this book: "Alfred Nobel eating chocolate".
By starryai.com, April 2024.

You have now learned something about the absolute basics of AI: regression analysis. Cheers!

Index

www.ingramcontent.com/pod-product-compliance
Lightning Source LLC
Chambersburg PA
CBHW072256210326
41458CB00074B/1797